I0489266

Methods, Quality Assurance, and Data for Assessing Atmospheric Deposition of Pesticides in the Central Valley of California

By Celia Zamora, Michael S. Majewski, and William T. Foreman

Scientific Investigations Report 2013-5023

U.S. Department of the Interior
U.S. Geological Survey

U.S. Department of the Interior
KEN SALAZAR, Secretary

U.S. Geological Survey
Suzette M. Kimball, Acting Director

U.S. Geological Survey, Reston, Virginia: 2013

For more information on the USGS—the Federal source for science about the Earth, its natural and living resources, natural hazards, and the environment, visit http://www.usgs.gov or call 1–888–ASK–USGS.

For an overview of USGS information products, including maps, imagery, and publications, visit http://www.usgs.gov/pubprod

To order this and other USGS information products, visit http://store.usgs.gov

Suggested citation:
Zamora, Celia, Majewski, M.S., and Foreman, W.T., 2013, Methods, quality assurance, and data for assessing atmospheric deposition of pesticides in the Central Valley of California: U.S. Geological Survey Scientific Investigations Report 2013-5023, 180 p.

Acknowledgments

The California Department of Pesticide Regulation and the California Regional Water Quality Control Board provided funding for this study.

Connie Clapton, Henry Miyashita, and Patricia Orlando (USGS California Water Science Center) helped collect and process samples for this study. Serena Skaates, James Madsen, Max Stroppel, and Frank Wiebe (USGS National Water Quality Laboratory) helped with methods development and sample analysis. Jo Ann Gronberg and Donna Knifong assisted with figures and compilation of pesticide application data.

Contents

Contents—Continued

Contents—Continued

Figures

Tables

Tables—Continued

Conversion Factors

SI to Inch/Pound

Multiply	By	To obtain
Length		
centimeter (cm)	0.3937	inch (in.)
millimeter (mm)	0.03937	inch (in.)
meter (m)	3.281	foot (ft)
kilometer (km)	0.6214	mile (mi)
Area		
square meter (m^2)	10.76	square foot (ft^2)
square kilometer (km^2)	0.3861	square mile (mi^2)
Volume		
liter (L)	0.2642	gallon (gal)
liter (L)	61.02	cubic inch (in^3)
Mass		
gram (g)	0.03527	ounce, avoirdupois (oz)
kilogram (kg)	2.205	pound avoirdupois (lb)
Loading Rate		
micrograms per square meter per day [($\mu g/m^2$)/day]	0.00324	pounds per acre per year [(lb/acre)/yr]

Temperature in degrees Celsius (°C) may be converted to degrees Fahrenheit (°F) as follows:

$$°F=(1.8\times°C)+32$$

Altitude, as used in this report, refers to distance above the vertical datum.

Vertical coordinate information is referenced to North American Vertical Datum of 1988 (NAVD 88).

Horizontal coordinate information is referenced to North American Datum of 1983 (NAD 83).

*Transmissivity: The standard unit for transmissivity is cubic foot per day per square foot times foot of aquifer thickness [(ft^3/d)/ft^2]ft. In this report, the mathematically reduced form, foot squared per day (ft^2/d), is used for convenience.

Specific conductance is given in microsiemens per centimeter at 25 degrees Celsius (μS/cm at 25°C).

Concentrations of chemical constituents in water are given either in milligrams per liter (mg/L) or micrograms per liter (μg/L).

Abbreviations

B-H	biased-high
B-L	biased-low
CAWSC	California Water Science Center
CDPR	California Department of Pesticide Regulation
D-U	value deleted by NWQL because compound could not be determined due to matrix interference
E	estimated reported laboratory value
(E)	estimated compound concentration due to recognized laboratory performance issues
EPA	Environmental Protection Agency
GC-MS	gas chromatography-mass spectrometry
H	Henry's Law constant
$\log K_{oc}$	soil-organic carbon partition coefficient
LRB	laboratory reagent blank
LRL	laboratory reporting limit
LRS	laboratory reagent spike
LT-MDL	long-term method detection limit
MDL	method detection limit
MID	Modesto Irrigation District
NEMI	National Environmental Methods Index
NWQL	National Water Quality Laboratory
OP	organophosphate
PBW	pesticide-grade blank water
PL	sub-cooled liquid vapor-phase pressure
QA	quality assurance
QC	quality control
QMS	Quality management system
RPD	relative percent difference
SJR	San Joaquin River
SL	water solubility
SPE	solid phase-extraction
ss-IRL	sample-specific interim reporting limit
USGS	U.S. Geological Survey

Methods, Quality Assurance, and Data for Assessing Atmospheric Deposition of Pesticides in the Central Valley of California

By Celia Zamora, Michael S. Majewski, and William T. Foreman

Abstract

The U.S. Geological Survey monitored atmospheric deposition of pesticides in the Central Valley of California during two studies in 2001 and 2002–04. The 2001 study sampled wet deposition (rain) and storm-drain runoff in the Modesto, California, area during the orchard dormant-spray season to examine the contribution of pesticide concentrations to storm runoff from rainfall. In the 2002–04 study, the number and extent of collection sites in the Central Valley were increased to determine the areal distribution of organophosphate insecticides and other pesticides, and also five more sample types were collected. These were dry deposition, bulk deposition, and three sample types collected from a soil box: aqueous phase in runoff, suspended sediment in runoff, and surficial-soil samples. This report provides concentration data and describes methods and quality assurance of sample collection and laboratory analysis for pesticide compounds in all samples collected from 16 sites. Each sample was analyzed for 41 currently used pesticides and 23 pesticide degradates, including oxygen analogs (oxons) of 9 organophosphate insecticides. Analytical results are presented by sample type and study period.

The median concentrations of both chlorpyrifos and diazinon sampled at four urban (0.067 micrograms per liter [µg/L] and 0.515 µg/L, respectively) and four agricultural sites (0.079 µg/L and 0.583 µg/L, respectively) during a January 2001 storm event in and around Modesto, Calif., were nearly identical, indicating that the overall atmospheric burden in the region appeared to be fairly similar during the sampling event. Comparisons of median concentrations in the rainfall to those in the McHenry storm-drain runoff showed that, for some compounds, rainfall contributed a substantial percentage of the concentration in the runoff; for other compounds, the concentrations in rainfall were much greater than in the runoff. For example, diazinon concentrations in rainfall were about 70 percent of the diazinon concentration in the runoff, whereas the chlorpyrifos concentration in the rain was 1.8 times greater than in the runoff. The more

water-soluble pesticides—carbaryl, metolachlor, napropamide, and simazine—followed the same pattern as diazinon and had lower concentrations in rain compared to runoff. Similar to chlorpyrifos, compounds with low water solubilities and higher soil-organic carbon partition coefficients, including dacthal, pendimethalin, and trifluralin, were found to have higher concentrations in rain than in runoff water and were presumed to partition to the suspended sediments and organic matter on the ground.

During the 2002–04 study period, the herbicide dacthal had the highest detection frequencies for all sample types collected from the Central Valley sites (67–100 percent). The most frequently detected compounds in the wet-deposition samples were dacthal, diazinon, chlorpyrifos, and simazine (greater than 90 percent). The median wet-deposition amounts for these compounds were 0.044 micrograms per square meter per day ($\mu g/m^2/day$), 0.209 $\mu g/m^2/day$, 0.079 $\mu g/m^2/day$, and 0.172 $\mu g/m^2/day$, respectively. For the dry-deposition samples, detection frequencies were greater than 73 percent for the compounds dacthal, metolachor, and chlorpyrifos, and median deposition amounts were an order of magnitude less than for wet deposition. The differences between wet deposition and dry deposition appeared to be closely related to the Henry's Law (H) constant of each compound, although the mass deposited by dry deposition takes place over a much longer time frame.

Pesticides detected in rainfall usually were detected in the aqueous phase of the soil-box runoff water, and the runoff concentrations were generally similar to those in the rainfall. For compounds detected in the aqueous phase and suspended-sediment samples of soil-box runoff, concentrations of pesticides in the aqueous phase generally were detected in low concentrations and had few corresponding detections in the suspended-sediment samples. Dacthal, diazinon, chlorpyrifos, and simazine were the most frequently detected pesticides (greater than 83 percent) in the aqueous-phase samples, with median concentrations of 0.010 µg/L, 0.045 µg/L, 0.016 µg/L, and 0.077 µg/L, respectively. Simazine was the most frequently detected compound in the suspended-sediment samples (69 percent), with a median concentration of 0.232 µg/L.

Results for compounds detected in the surficial-soil samples collected throughout the study period showed that there was an increase in concentration for some compounds, indicating atmospheric deposition of these compounds onto the soil-box surface. In the San Joaquin Valley, the compounds chlorpyrifos, dacthal, and iprodione were detected at higher concentrations (between 1.4 and 2 times greater) than were found in the background samples collected from the San Joaquin Valley soil-box sites. In the Sacramento Valley, the compounds chlorpyrifos, dacthal, iprodione, parathion-methyl, and its oxygen analog, paraoxon-methyl, were detected in samples collected during the study period in low concentrations, but were not detected in the background concentration of the Sacramento Valley soil mix.

Introduction

California is one of the world's leading agricultural areas, and many thousands of metric tons of pesticides are used each year on many different crops (Majewski and Baston, 2002). Pesticides have been recognized as potential air pollutants since 1946 (Daines, 1952), and a wide variety have been detected in California air (Baker and others, 1996; Majewski and Baston, 2002; Zamora and others, 2003; Majewski and others, 2005). In most cases, pesticides in agriculture are applied by spraying an aqueous suspension. As a result, a portion of the sprayed compound does not reach the target area, but is transported by wind beyond the application site as direct drift. The droplets transported by the drift are either deposited on soil or plants close to the treated area or are transported in the atmosphere over longer distances, depending on their size. A loss of pesticide after application caused by volatilization or wind erosion of soil to the atmosphere is called "indirect drift." These direct and indirect sources to drift are the main input paths of pesticides in the atmosphere (Epple and others, 2002). Post-application volatilization from treated surfaces is often a major dissipation pathway for many pesticides (Glotfelty, 1978; Cliath and others, 1980; Glotfelty and others, 1990; Risebrough, 1990; Majewski, 1991; Majewski and others, 1993; Majewski and Capel, 1995; Seiber and Woodrow, 1995, Wania and Mackay, 1996; Majewski and Baston, 2002).

Atmospheric transport and subsequent deposition of pesticides can affect the quality of streams and other surface waters adversely. Residues of pesticides in surface waters of the Central Valley have been evaluated in many previous studies (Kuivala and Foe, 1995, Domagalski, 1997a, b; Domalgalski and others, 1997; Panshin and others, 1998; Kratzer, 1998; Kratzer, 1999; Kratzer and others, 2002). Under section 303(d) of the 1972 Clean Water Act, states are required to develop lists of impaired waters that do not meet the water-quality standards set by states. The 303(d) list shows that several streams in the Central Valley are impaired because of pesticides (U.S. Environmental Protection Agency, *http://www.epa.gov/region09/water/tmdl/303d-pdf/*

ca-06-303d-list-final-06-28-07-combined.pdf, accessed August 2012). The most frequent impairments have been attributed to the organophosphate (OP) insecticides diazinon and chlorpyrifos. Most of the agricultural applications of diazinon and chlorpyrifos take place from December through February. They are applied to dormant orchards of several stone fruits and nuts in the San Joaquin Basin, primarily almonds (Panshin and others, 1998). These insecticides are applied during extended dry periods during the dormant-spray period, and then rainfall events after spraying cause most of the unintentional transport of pesticides from fields to streams with rainfall-induced runoff. Atmospheric transport and subsequent deposition of pesticides are most likely to affect stream water quality when rain and direct surface runoff are major sources of streamflow.

Study Areas

The Central Valley is a large flat valley that dominates the central portion of California and is one of California's most productive agricultural regions. It is about 400 miles long, averages 50 miles in width, and is composed of four hydrographic subregions or drainage basins named for the major natural surface-water feature in each subregion (fig. 1). Sacramento Valley, the northernmost third of the Central Valley, has an area of about 4,400 square miles (mi^2) and is drained by its namesake, the Sacramento River. San Joaquin Valley, the southern two-thirds of the Central Valley, has two subregions: the San Joaquin Basin and, at the southern end, a basin of interior drainage called the Tulare Basin after a Pleistocene lake that occupied most of the area. The fourth hydrographic subregion is the delta, a low lying area that drains directly to the Sacramento–San Joaquin Delta rather than to either river. The lower part of the delta subregion consists of wetlands interspersed with hundreds of miles of channels and numerous islands (Bertoldi and others, 1991). In this report, the study area only includes the San Joaquin Valley and Sacramento Valley, and results are summarized by the respective geographic region.

Sacramento Valley

The Sacramento Valley is geographically contiguous with the San Joaquin Valley to the south, but is defined by its distinct drainage basin. The generally flat valley floor covers about 5,000 mi^2, and its elevation decreases from about 300 ft. at its northern end to near sea level in the delta. The major land uses in the watershed are forestry, agriculture, urban, and mining. Agriculture is the dominant land use on the valley floor, followed by urban development. Orchards—principally, walnut, almond, prune, and peach—are found along the river channels where they take advantage of well-drained soils. The hot summer and temperate winter climate, combined with the availability of water for irrigation during the normally dry summer months, allows for a variety of crops to be grown.

Figure 1. The Central Valley of California and its four hydrographic subregions.

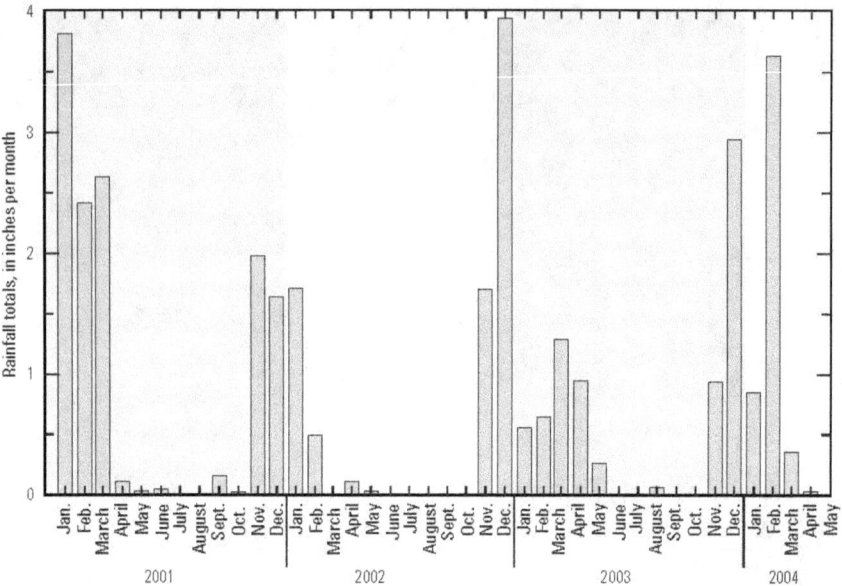

Figure 2. Monthly rainfall totals from the rainfall gage at Oroville Dam (site 15) in California, 2001–04.

The major crops are rice, fruits and nuts, tomatoes, sugar beets, corn, alfalfa, and wheat. Dairy products also are important agricultural commodities. The soils of the Sacramento Valley are mostly clay with very slow to slow infiltration rates (Soil Conservation Service, 1993). Because of the widespread presence of clay soils and associated slow infiltration rates, and the availability of sufficient irrigation water from the Sacramento River, rice farming is possible. Rice production includes the seasonal creation of temporary wetlands.

The Sacramento Valley has an arid-to-semiarid climate characterized by hot summers and mild winters, and average precipitation ranges from 14 to 25 in. Most rain falls in the watershed from November through April, and the largest amount, on average, falls during the month of January. Mean annual rainfall on the valley floor tends to increase with latitude and elevation, ranging from 17.9 in. at Sacramento to 28.8 in. at Oroville (*http://cdo.ncdc.noaa.gov/cgi-bin/ climatenormals/climatenormals.pl,* accessed May 27, 2008). Figure 2 depicts the monthly rainfall totals for the study period from the Oroville Dam rainfall gage (site 15).

San Joaquin Valley

The San Joaquin Valley floor is about 3,750 mi², and the altitude ranges from about 1,000 ft in the south to near sea level in the north at the Sacramento–San Joaquin Delta (fig. 1).

The predominant land use on the valley floor is agriculture. The distribution of crops in the valley generally reflects the distribution of soil texture and chemistry, a long growing season, and supply of water for irrigation (Kratzer and others, 2011). Orchards and vineyards primarily are grown on the well-drained alluvial-fan soils of the eastside. Cotton, a salt tolerant crop, is the principal crop grown on the basin deposits at the southern end of the basin. Row crops, such as beans, are primarily grown on the well-drained alluvial fans of the westside. Land along the eastside of San Joaquin River is primarily used for corn, alfalfa, pasture, and dairies (Kratzer and others, 2011). The San Joaquin Valley has an arid-to-semiarid climate that is characterized by hot summers and mild winters. Annual precipitation on the valley floor ranges from 7 inches (in.) in the south to 15 in. in the north (National Center for Atmospheric Research, 2003). The eastern slopes of the Coast Ranges and the valley are in the rain shadow of the Coast Ranges. The major source of water entering the basin is wet deposition, or rainfall and snow, on the western slope of the Sierra Nevada. The overall long-term average annual rainfall for the basin is 28 in. (13 inches in the valley portion only), which mostly falls from November through March (National Center for Atmospheric Research, 2003). Figure 3 depicts the monthly rainfall totals for the study period taken from a City of Modesto rainfall gage in downtown Modesto, California (site 4).

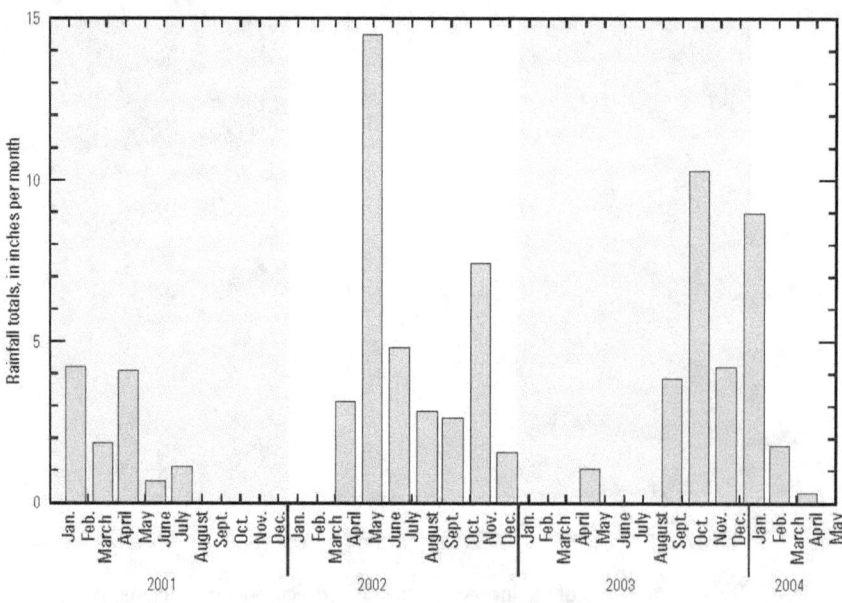

Figure 3. Monthly rainfall totals from the City of Modesto rainfall gage in downtown Modesto, California (site 4), 2001–04.

Study Objective and Design

The data and methods presented in this report are from two separate studies that took place over a 3.5-year sampling period. For the 2001 study, the objective was to determine if agricultural pesticides applied to crops during the orchard dormant-spray season (January through February 2001) were drifting into urban areas and being deposited, resulting in pesticides introduced to urban storm runoff. For the 2002–04 study, the objective was to determine the areal distribution of OP insecticides and other pesticides through wet and dry atmospheric deposition and to further investigate contribution of pesticides in rainfall to pesticides in runoff from soil surfaces. Figure 4 depicts the study area, land use, and sampling sites for the 2001 and 2002–04 sampling periods in the Central Valley. The sampling media, location setting, and sample type collected during the respective study periods are presented in table 1.

2001 Study Design

Wet deposition (rainfall) from a single storm event in January 2001 was sampled, and concentrations in rainfall were measured at four urban and four agricultural sites (fig. 4, sites 1–8) to examine the overall atmospheric burden in the Modesto, Calif., region. At the same time, storm-runoff samples were collected from a Modesto storm drain (site 9) to examine the contribution of pesticide concentrations to storm runoff from rainfall. The storm drain and two of the urban sites (sites 3 and 4) were located in a small, urban watershed (McHenry storm drain watershed). The major land use within a 4-km radius of each of the rainfall sampling site is given in table 2. The sites were distributed spatially to be representative of the predominant land uses in the study area. The urban sampling sites were located on rooftops of buildings to reduce the possibility of ground-level contamination and to secure them from public access. Each of the urban sites had varying degrees of agricultural activities within a 4-km radius (table 2). The agricultural sites were predominately surrounded by orchards or other land use (row crops, pasture, and native vegetation), and only one agricultural sampling location (site 5) was located on a rooftop.

2002–04 Study Design

For the 2002–04 study, the collection sites were extended throughout the Central Valley to determine the areal distribution of OP insecticides and other pesticides. Six sites were located in the San Joaquin Valley: two on the west side of the San Joaquin River (SJR) in an area of row crops and orchards (fig. 4, sites 10 and 11), one in a rural residential area (site 12), and another at a small rural airport in an agricultural area (site 13). Two of the sites sampled in 2001 (sites 4 and 5) were retained for the 2002–04 study period. The locations of sites 4 and 5 represented an urban and agricultural setting,

respectively. In addition, two additional sites were sampled in the adjacent northern Sacramento Valley hydrologic basin: an urban site located in Gridley, Calif. (site 16), and a down-wind site in the Sierra-Nevada foothills (site 15). Table 3 presents the major land use within a 4-km radius of each of the 2002–04 sampling sites for both Sacramento and San Joaquin Valley study areas. The sample collection period for the 2002–04 study began in January 2002 and ended in August 2004.

The samples collected during this study period included wet-deposition, dry-deposition, and bulk-deposition samples. In addition, a soil box was added to the 2002–04 study to compare the pesticide concentrations in rainfall to that in runoff (aqueous phase and suspended sediment) from the soil, the concentrations in aqueous phase with those on the surface soil mobilized by the rainfall runoff (suspended sediment), and the concentrations of dry deposition collected on the surficial soil.

Purpose and Scope

The purpose of this report is to describe sampling and analytical methods used to measure the concentrations of pesticide compounds in wet deposition, dry deposition, and bulk-deposition samples collected from 16 sites in the Central Valley of California. The sampling and analytical methods also are presented for samples collected from soil boxes at three sites. The pesticide concentration data for each of these sample types are presented for the 2001 study and the 2002–04 study. The two studies were part of a larger U.S. Geological Survey (USGS) project designed to identify all sources of pesticides in the San Joaquin and Sacramento River watersheds (Majewski and others, 2005; Zamora and others, 2003).

Methods of Monitoring

Description of methods includes information about field collection of the different sample types, sample analyses, and quality-control procedures in the laboratory and the field.

Field Sampling Methods

Seven types of samples were collected over the combined 3.5-year sampling period of the two studies (January 25, 2001, to August 7, 2004). The seven sample types collected during the two studies are defined and presented separately.

Cleaning Procedures for Sampling Media

The cleaning procedure for the funnels, collection bottles, stainless steel buckets, and the soil box sampling spoon was as follows: sampling equipment was rinsed with American Society for Testing and Materials (ASTM) II quality deionized water (source: California Water Science Center Laboratory,

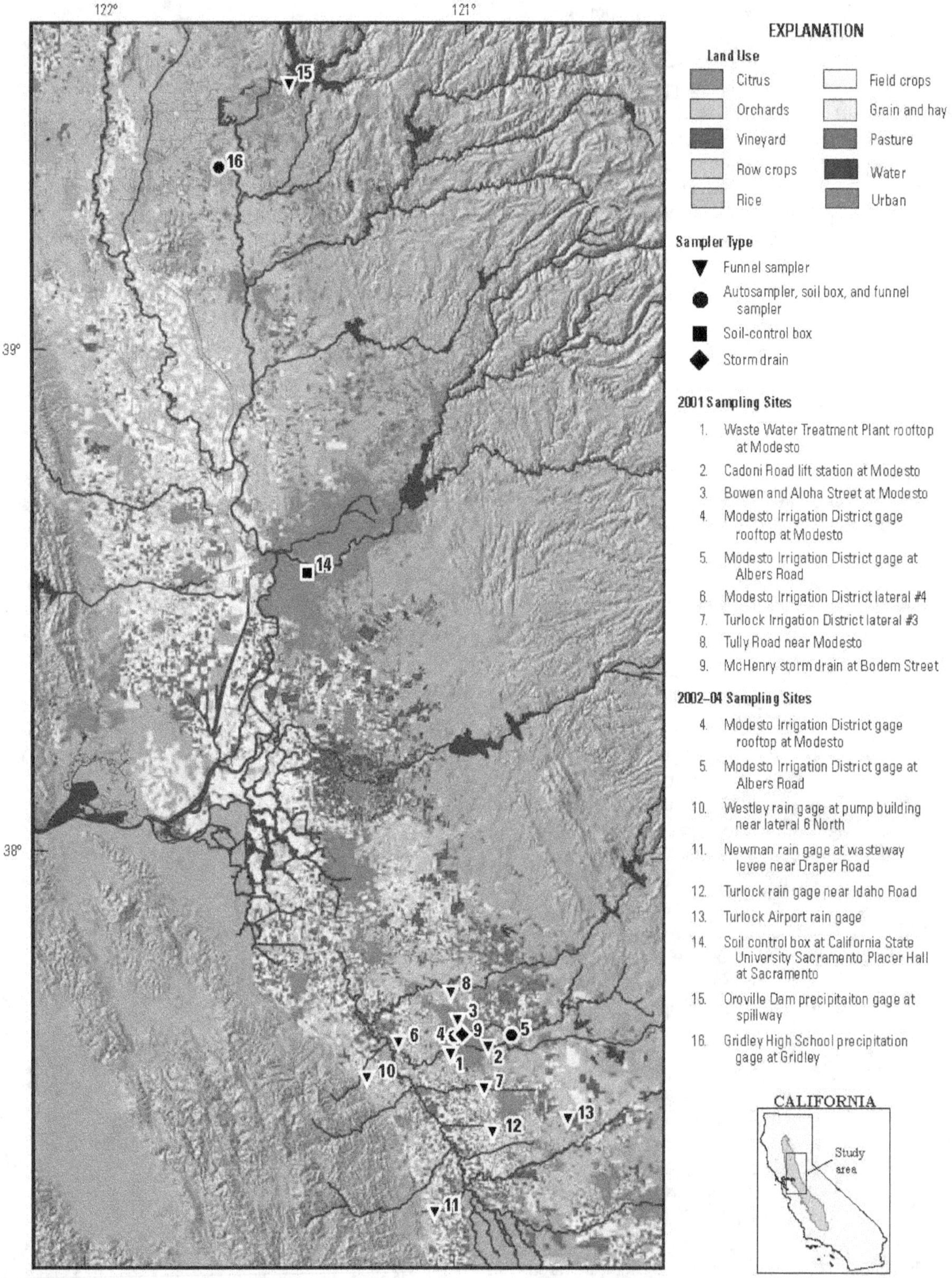

Figure 4.   Study area, land use, and sampling sites for the 2001 and 2002–04 sampling periods in the Central Valley of California.

Table 1. Site number, site name, study period, sampling equipment, location setting, and sample type collected during the 2001 and 2002–04 sampling periods in the Central Valley of California.

[Abbreviations: CSUS, California State University, Sacramento]

Site number	Site name	Site identification number	Study period	Sampling equipment	Sampler location setting	Sample type
1	Waste Water Treatment Plant rooftop at Modesto	373637121004601	2001	Teflon-lined funnel	Urban	Wet deposition.
2	Cadoni Road lift station at Modesto	373725120543701	2001	Teflon-lined funnel	Urban	Wet deposition.
3	Bowen and Aloha Street at Modesto	374028120594301	2001	Teflon-lined funnel	Urban	Wet deposition.
4	Modesto Irrigation District gage rooftop at Modesto	373834121000601	2001, 2002–04	Teflon-lined funnel and automatic wet-dry sampler and soil box[1]	Urban	Wet, dry, and bulk deposition[2], soil runoff (aqueous phase and suspended sediment), and surficial soil.
5	Modesto Irrigation District gage rooftop at Albers Road	373841120504801	2001, 2002–04	Teflon-lined funnel and automatic wet-dry sampler and soil box[1]	Agricultural	Wet, dry, and bulk deposition[2], soil runoff (aqueous phase and suspended sediment), and surficial soil.
6	Modesto Irrigation District lateral #4	373750121092601	2001	Teflon-lined funnel	Agricultural	Wet deposition.
7	Turlock Irrigation District lateral #3	373228120551201	2001	Teflon-lined funnel	Agricultural	Wet deposition.
8	Tully Road near Modesto	374351121004701	2001	Teflon-lined funnel	Agricultural	Wet deposition.
9	McHenry storm drain at Bodem Street	373847120590801	2001	ISCO sampler[3]	Urban	Storm-drain runoff.
10	Westley rain gage at pump building near lateral 6 North	373335121143001	2002–04	Teflon-lined funnel	Agricultural	Wet, dry, and bulk deposition[2].
11	Newman rain gage at wasteway levee near Draper Road	371735121031201	2002–04	Teflon-lined funnel	Agricultural	Wet, dry, and bulk deposition[2].
12	Turlock rain gage near Idaho Road	372713120534901	2002–04	Teflon-lined funnel	Agricultural	Wet, dry, and bulk deposition[2].
13	Turlock Airport rain gage	372857120414001	2002–04	Teflon-lined funnel	Agricultural	Wet, dry, and bulk deposition[2].
14	Soil control box at CSUS Placer Hall at Sacramento	383343121252501	2002–04	Soil-box control	Urban	Surficial soil.
15	Oroville Dam precipitation gage at spillway	393234121292701	2002–04	Teflon-lined funnel	Agricultural	Wet, dry, and bulk deposition[2].
16	Gridley High School precipitation gage at Gridley	392205121410201	2002–04	Teflon-lined funnel and automatic wet-dry sampler and soil box	Urban	Wet, dry, and bulk deposition[2], soil runoff (aqueous phase and suspended sediment), and surficial soil.

[1] Soil boxes added to the existing 2001 site during the 2002–04 study period.

[2] Bulk deposition includes bulk wet and bulk dry-deposition samples as defined in text.

[3] Storm-drain samples collected by using an ISCO automated sampler (Teledyne ISCO Company, Lincoln, Nebr.) that was programmed to collect hourly samples in 1-liter glass collection bottles.

Table 2.   Predominant land-use classification within a 4-kilometer radius surrounding each of the 2001 rainfall sampling sites in the area of Modesto, California.

[Urban land-use classification includes residential, commercial, industrial, and other areas. Agricultural land-use classification includes orchards, vineyards, and other areas. Other land use includes corn and other row crops, pasture, and native vegetation]

Site number	Site name	Percentage urban	Percentage orchard	Percentage vineyards	Percentage other
	Urban sites				
1	Waste Water Treatment Plant rooftop at Modesto	56.0	23.5	3.9	16.8
2	Cadoni Road lift station at Modesto	38.8	38.4	2.1	22.1
3	Bowen and Aloha Street at Modesto	88.3	4.9	1.6	5.3
4	Modesto Irrigation District rooftop at Modesto	89.7	3.4	2.4	4.4
	Agricultural sites				
5	Modesto Irrigation District gage rooftop at Albers Road	6.3	45.8	10.3	43.9
6	Modesto Irrigation District lateral #4	0.5	11.9	0.5	87.1
7	Turlock Irrigation District lateral #3	10.9	49.2	10.1	29.8
8	Tully Road near Modesto	12.3	61.7	3.8	22.3

Table 3.   Predominant land-use classification within a 4-kilometer radius surrounding each of the 2002–04 rainfall sampling sites in the Central Valley of California.

[Urban land-use classification includes residential, commercial, industrial, and other areas. Agricultural land-use classification includes orchards, vineyards, corn/alfalfa/vegetables, and other areas. Other land use includes corn and other row crops, pasture, and native vegetation]

Site number	Site name	Percentage urban	Percentage orchard	Percentage vineyards	Percentage corn/alfalfa/ vegetables	Percentage other
4	Modesto Irrigation District gage rooftop at Modesto	91.4	2.06	2.25	1.02	3.30
5	Modesto Irrigation District gage rooftop at Albers Road	8.27	40.9	12.8	10.3	27.7
10	Westley rain gage at pump building near lateral 6 North	3.83	31.6	0.44	27.9	36.2
11	Newman rain gage at wasteway levee near Draper Road	8.61	36.0	None	27.4	28.0
12	Turlock rain gage near Idaho Road	5.90	15.2	0.89	58.4	19.7
13	Turlock Airport rain gage	3.1	52.2	2.50	13.1	29.1
15	Oroville Dam precipitation gage at spillway	11.9	0.64	0.01	None	87.4
16	Gridley High School precipitation gage at Gridley	20.9	54.9	0.01	2.39	21.8

CAWSC Laboratory) followed by a thorough wash with a dilute Liquinox solution and a sterilized sponge; the wash solution was then rinsed by using deionized water, followed by a methanol rinse, and then a final rinse of pesticide-grade blank water (PBW) purchased and quality assured through the National Water Quality Laboratory (NWQL).

Storm-Drain Runoff

Urban storm runoff samples were collected from a city storm drain in Modesto on an hourly basis beginning on January 25, 2001, at 6 p.m. and ending on January 26, 2001, at 2 a.m. by using a ISCO (Teledyne ISCO Company, Lincoln Nebr.) automated sampler outfitted with a pressure transducer that measured water depth. Discharge was determined from the collected water-depth data by using a depth-discharge relation provided by the City of Modesto. The storm sampling at this location was designed to determine the presence and concentrations of pesticides in samples collected from the urban storm-water runoff and to allow for comparison of these results with results from the wet-deposition samples collected from agricultural sites during the same storm event.

Wet Deposition

For this study, a wet-deposition sample is defined as having fewer than 5 days of dry-deposition exposure—that is, periods during the composite sample-collection period when it was not raining. For example, a sample collected at the Newman rain gage (site 11) has a sample composite time of 15 days (February 4–19, 2004) and is considered a wet-deposition sample (rainfall) because there were five rain events during that period (approximately every 2.5 days). Wet-deposition (rain) samples were collected by using either a Teflon-coated funnel (funnel sampler) or an automated wet-dry sampler (autosampler). The funnel sampler collected rainfall by using a funnel-bottle assembly consisting of a 32-centimeter (cm) diameter funnel placed in the opening of a cleaned, 4-L amber glass bottle (collection bottle). The funnel-bottle assembly was supported by an appropriate length of plastic irrigation pipe attached to a small wooden table (fig. 5). Chicken wire covered each funnel to intercept large debris and to keep birds and other wildlife from entering the collection area, but one or more small insects (flies, bees, and spiders) were present in many samples. During the 2001 study period, the wet deposition was collected from a funnel sampler at sampling sites 1–8. During the 2002–04 study period, the funnel sampler collected wet-deposition, dry-deposition, and bulk-deposition samples at sampling sites 4, 5, 10–13, 15, and 16. (fig. 4 and table 1).

Clean funnels and collection bottles were placed at each sampling site for no more than 24 hours before a rain event. The samples were collected at the end of the rain event. The samples were immediately stored on ice and transported to the California Water Science Center Laboratory in Sacramento, Calif., for processing. There, the samples were

Figure 5. Funnel sampler at the Newman rain gage site at wasteway levee near Draper Road (site 11) in the Central Valley of California. Photograph by Michael Majewski, U.S. Geological Survey.

weighed in the collection bottle to the nearest 0.1 gram and poured into 1-liter (L) glass amber bottles, packed in ice, and shipped to the USGS NWQL in Denver, Colo., within 24 hours of field collection.

The autosamplers used 27-cm diameter stainless-steel buckets to collect the wet-deposition samples during the 2001 study period and wet deposition, dry deposition, and bulk deposition during the 2002–04 study period (fig. 6). The sampler was controlled by a moisture sensor. During dry periods, a moveable lid covered the "wet" collection bucket, and the "dry" collection bucket remained open to collect dry deposition from the atmosphere. At the onset of rainfall, the moisture sensor activated a motor that moved the lid from the "wet" collection bucket to the "dry" collection bucket, which enabled the collection of rain and kept the collected dry deposition free of moisture. Wet-deposition samples were collected from the autosampler at the end of a rain event and poured into a clean collection bottle and processed by using the same procedures as the funnel-collected samples.

Dry Deposition

Dry-deposition samples were collected by using a funnel sampler and the autosampler during the 2002–04 study period. The funnels were exposed to the atmosphere for periods of 3 weeks to approximately 3 months. At collection, the wire screen was removed, and approximately 50 milliliters (mL) of PBW from a Teflon squeeze bottle was used to rinse deposited material off the sides of the funnel into the collection bottle, followed by a rinse composed of approximately 50 mL of an equal part mixture of hexane and ethyl acetate (the primary extraction solvent mixture used for dry-deposition samples collected prior to November 2003) to transfer any remaining particles into the sample-collection bottle.

Figure 6.   Autosampler at Gridley, California (site 16). Photograph by Michael Majewski, U.S. Geological Survey.

The method for sample extraction and subsequent extract processing of the collected dry deposition, and bulk dry-deposition samples (see next section) was modified during the study in an effort to improve laboratory method performance (see details under NWQL research method LS8054 in appendix 1). These modifications necessitated a change in the solvents used for field-rinsing of the collection funnel or autosampler bucket. Samples collected prior to November 2003 contained at least one or more organic solvents from point of collection through any storage period prior to sample preparation. Samples with begin dates of November 2003 or later were collected by rinsing the funnel with less than 50 mL of PBW (no organic solvent rinse) or with PBW and a small volume (less than or equal to 17 mL) of a 25-percent ethyl acetate in hexane solution. All rinses were collected in the collection bottle on which the funnel rested and drained. The same rinse procedure was applied to the stainless-steel buckets of the autosampler, with each collected sample poured into a clean 4-L glass amber collection bottle. All collected samples were immediately stored on ice and shipped to NWQL within 24 hours. For samples with begin dates of November 2003 or later, a 50-mL aliquot of dichloromethane was added to each sample immediately upon sample receipt at the NWQL as the primary extraction solvent and as a preservative.

Bulk Deposition (Bulk Wet Deposition and Bulk Dry Deposition)

Bulk deposition samples are defined as samples that contained both wet and dry atmospheric deposition. Two types of bulk-deposition samples are presented and discussed separately in this report. The first type, a bulk wet-deposition sample, was intended to be a dry-deposition sample only; however, for these sample types, one of the following was

true: (1) there were several small storm events during a composite sample time, which resulted in periods of wet deposition along with periods of dry deposition that were greater than 5 days; (2) there was a large storm event at the end of a long dry period that compromised an intended dry-deposition sample; or (3) the autosampler failed to cover the "wet" or "dry" collection bucket, and the autosampler collected both wet and dry deposition over an extended period. Because of the larger volume of water collected (median volume was 1.04 L for 38 bulk wet samples), these samples were analyzed by using two laboratory analytical methods for pesticides and degradates in filtered water: methods O-1126-95 (Zaugg and others, 1995) and O-2002-01 (Sandstrom and others, 2001) with pesticide isolation by C-18 solid phase-extraction (SPE) and analysis by gas chromatography-mass spectrometry (GC-MS).

The second type of bulk-deposition sample, referred to as a bulk dry-deposition sample is defined as a sample that was intended to be only a dry-deposition sample; however, there was an unpredicted rain event and some wet deposition was collected also. Because of the small volume of wet deposition collected (median volume was 0.541 L for 26 samples), the samples were prepared and analyzed by using the dry-deposition analytical method (NWQL research method LS8054, appendix 1).

Deposition to Soil Box

Each soil box was approximately 1-meter (m) square and divided in half. The inside of each section was lined with aluminum sheeting and filled with a composite mixture of soil. The composite soil mixture was composed of soil collected near site 10 (mix 1), site 11 (mix 2), and site 9 (mix 3). The collected soil from each of the sites was composited and placed in the soil boxes located in the San Joaquin Valley (fig. 4, sites 4 and 5). The soil in the Sacramento Valley soil box (mix 4) was a well-mixed sample collected from one location in an agriculture field adjacent to Gridley High School (site 16).

Soil-Box Runoff (Aqueous Phase and Suspended Sediment)

Each soil box was set up in a north-south orientation and inclined slightly in the southerly direction to direct and facilitate collection of surface runoff. The runoff samples were collected from only one half of the box into a clean collection bottle secured to one of the sampler legs (fig. 7). Sample bottles containing soil-box runoff were collected at the end of a rain event that had sufficient intensity and volume to produce surficial runoff. Collected samples were immediately stored on ice. Runoff samples were processed at the CAWSC Laboratory by first weighing the volume of sample collected, then, after shaking, filtering the sample through one or more 142-millimeter (mm) diameter 0.7-micron glass-fiber filters (GFFs)

Figure 7. Soil box at the Modesto Irrigation District gage at Albers Road (site 5) in the Central Valley of California. Photograph by Michael Majewski, U.S. Geological Survey.

as described by Sandstrom (1995). Prior to use, the GFFs were precleaned by baking at 400 degrees Celcius (°C) for 3 hours, and then they were weighed to the nearest 0.1 gram. The number of GFFs required for filtration ranged from 1 (for 17 samples) to 38, depending on volume filtered and the amount and particle size of the suspended sediment. Following sample filtration, GFFs were folded in half (particles inward), wrapped in aluminum foil, placed in a resealable plastic storage bag, and stored at –15°C both prior to shipping to the NWQL on ice and after receipt at the NWQL, until they were prepared for analysis. The two sample types, the aqueous phase (filtered sample volume) and suspended sediment (filters), were shipped to NWQL within 24 hours of field sample collection. Laboratory sample preparation and analysis details are provided in appendix 1 (NWQL research method LS7503).

Surficial Soil

Surficial-soil samples were taken from one side of the soil box three times during the study and analyzed for the same suite of pesticides as the rain samples. No runoff samples were collected from this side of the box. Subsamples of the soil were carefully scooped with a solvent-rinsed spoon from the top layer (approximately 0.5–1 cm) of soil. These subsamples were collected across the surface of the soil and composited as one sample into a 500-mL wide mouth glass jar with Teflon-lined lid and stored on ice after collection. The sample configuration used to sample the soil surface was photographed

Figure 8. Surficial soil box sampling array. Photograph by Michael Majewski, U.S. Geological Survey.

and recorded in the field notes so that the same locations on the soil surface would not be resampled during subsequent sampling events (fig. 8). The soil in the side of the box where the runoff was collected was left undisturbed (not sampled) for the duration of the study. Collected soil samples were stored in a freezer (–15°C) at the CAWSC Laboratory and shipped to NWQL at the end of the study period (August 2004). Laboratory sample preparation and analysis details are provided in appendix 1 (NWQL research method LS5503).

A control soil box was set up outdoors on a building rooftop (site 14) to examine the transformation or degradation of compounds that were not exposed to sunlight or atmospheric deposition. This soil box contained a composite mixture of the soils collected in the San Joaquin Valley (mix 1, mix 2, and mix 3), and one side was covered with a piece of plywood. The covered side had an 0.5 inch gap to allow for soil-air exchange and indirect particle deposition (particle deposition from the side). Three times during the study, the cover was removed, and surficial samples were collected for laboratory analysis.

Laboratory Analytical Methods

All samples types were analyzed at the USGS NWQL in Denver, Colo. Laboratory analytical methods are presented by study period and sample type (table 4). Some methods used in the report can be found in the National Environmental Methods Index (NEMI) at *http://www.nemi.gov/*.

2001 Study Period

The wet-deposition and storm-drain runoff samples were analyzed for pesticides in water by C-18 SPE and GC-MS according to laboratory analytical method O-1126-95 (Zaugg and others, 1995).

Table 4.   Laboratory analytical method by sample type and study period.

[Abbreviations: NWQL, National Water Quality Laboratory]

Sample type	Laboratory analytical method	Study period	Citation
Storm-drain runoff	O-1126-95	2001	Zaug and others, 1995; Lindley and others, 1996.
Wet deposition, bulk wet deposition	O-1126-95	2001	Zaug and others, 1995; Lindley and others, 1996.
Wet deposition, bulk wet deposition	O-1126-95 and O-2002-O	2002–04	Zaug and others, 1995; Lindley and others, 1996; Sandstrom and others, 2001.
Wet deposition, bulk wet deposition	O-1126-95 and O-2002-O and O-1126-02	2002–04	Zaug and others, 1995; Lindley and others, 1996; Sandstrom and others, 2001; Madsen and others, 2003.
Dry deposition, bulk dry deposition	NWQL research analytical method LS8054	2002–04	Described in this report, appendix 1.
Soil-box runoff (aqueous phase)	O-1126-95 and O-2002-O and O-1126-02	2002–04	Zaug and others, 1995; Lindley and others, 1996; Sandstrom and others, 2001; Madsen and others, 2003.
Soil-box runoff (suspended sediment)	NWQL research analytical method LS7503	2002–04	Described in this report, appendix 1.
Soil-box surficial soil	NWQL research analytical method LS5503	2002–04	Described in this report, appendix 1.

2002–04 Study Period

The wet-deposition, bulk wet-deposition, and filtered soil-box runoff (aqueous phase) sample types collected between January 25, 2002, and April 29, 2004, were analyzed for pesticides in water by using analytical method O-1126-95 (Zaugg and others, 1995) and for moderate-use pesticides and selected degradates in water by using analytical method O-2002-01 (Sandstrom and others, 2001) with C-18 SPE and GC-MS. These sample types were also analyzed for concentrations of acetochlor in water by automated SPE and GC-MS (Lindley and others, 1996). Samples collected between November 7, 2002, and April 29, 2004, were analyzed for fipronil and fipronil degradates in water by using GC-MS, according to analytical method O-1126-02 (Madsen and others, 2003).

The dry-deposition, bulk dry-deposition, surficial-soil, and the suspended-sediment sample types were analyzed for the same compounds as the wet-deposition samples collected during the 2002–4 study period by using NWQL analytical research methods. The dry and bulk-dry sample types were analyzed by using NWQL research analytical method LS8054. The surficial-soil and suspended-sediment sample types were analyzed by using NWQL research analytical methods LS7503 and LS5503, respectively. Each of these analytical methods are described by sample type in appendix 1.

Reporting Levels

Two types of reporting levels were used in this study: a laboratory reporting limit (LRL) for approved analytical methods and a sample-specific interim reporting level (ss-IRL) for research analytical methods. The NWQL uses an extension of the Environmental Protection Agency (EPA) method detection limit (MDL) procedure to determine the LRL of compounds with approved analytical methods (U.S. Environmental Protection Agency, 1992). Updates to the NWQL's MDLs and LRLs based on a statistical Levene's-test analysis are made only if a new long-term method detection level (LT-MDL) for a compound differs from the preceding year's LT-MDL or if a compound exhibits poor recovery (Childress and others, 1999). As a result, the LRLs for a number of compounds differed during the study period (table 5).

The ss-IRL was applied to the suspended-sediment samples and surficial-soil samples collected from the soil box, and sample results are presented as µg/L and µg/Kg, respectively (table 5). The ss-IRL for the suspended-sediment sample type was calculated by dividing the mass of compound detected in the suspended-sediment sample by 0.63 L, which was the median volume of soil-box runoff from samples collected during the study period (sample number of 30). The dry and bulk deposition data are presented with respect to limited method performance quality-control data obtained for the applied research methods. For some compounds, this performance data indicated likely greater quantitative uncertainty and, thus, are reported as estimated concentrations, and no ss-IRL was applicable.

For all mass-spectrometry analysis, concentrations of a compound less than the laboratory reporting level or lowest calibration standard (whichever is higher) are reported as estimated or having a higher degree of uncertainty (coded with an E preceding laboratory value). In these cases, the presence of the pesticide has been verified, but the concentration is estimated because it falls below the range of the calibration standards. Some compounds (table 5) are always reported as estimated (E) regardless of the concentration because of recognized performance issues (Childress and others, 1999; Sandstrom and others, 2001). In the remainder of this report, estimated concentrations are treated as normal detected concentrations for all sample types for the purpose of interpreting the concentration data.

Table 5. Pesticides analyzed and the laboratory reporting limits for samples collected in the Central Valley of California during the 2001 and 2002–04 study periods.

[Abbreviations: (E), estimated compound because of poor performance by applied method (Childress and others, 1999; Sandstrom and others, 2001); D, Degradate; F, Fungicide; H, Herbicide; I, Insecticide; LRL, Laboratory Reporting Limit; NA, not applicable; ss-IRL, sample specific-Interim Reporting Limit; *, current LRL limit; μg/L, microgram per liter; μg/Kg, micrograms per kilogram; –, compound not analyzed for given sample type]

Compound	Chemical Abstract Service Registry Number	Type of compound	Sample types		
			Storm-drain runoff, wet deposition, bulk wet deposition, and soil-box runoff (aqueous phase)	Soil-box runoff (suspended sediment)	Surficial soil (soil box)
			LRL, μg/L	ss-IRL, μg/L	ss-IRL, μg/kg
Acetochlor	34256-82-1	H	0.006*, 0.1	0.08	1
Alachlor	15972-60-8	H	0.004*, 0.005	0.08	1
2-Chloro-2,6-diethylacetanilide	6967-29-9	D	0.005	0.08	1
2,6-Diethylaniline (E)	579-66-8	D	0.006	2.38	30
alpha-Hexachlorocyclohexane (HCH)	319-84-6	I	0.005	–	–
2-Amino-N-isopropylbenzamide	30391-89-0	D	0.005	–	–
Atrazine	1912-24-9	H	0.007*, 0.008, 0.01, 0.031, 0.2	0.08	1
Deethylatrazine, CIAT	6190-65-4	D	0.005. 0.006*	0.16	2
Azinphos-methyl (E)	86-50-0	I	0.05	0.4	5
Azinphos-methyl-oxon (E)	NA	D	0.02*, 0.03	0.24	30
Benfluralin	1861-40-1	H	0.01	0.08	1
Bifenthrin	82657-04-3	I	0.005	–	–
Butylate	2008-41-5	H	0.002	–	–
Carbaryl (E)	63-25-2	I	0.041*, 0.045	0.16	2
1-Naphthol (E)	90-15-3	D	0.09	0.79	10
1,4-Naphthoquinone	130-15-4	D	0.05	–	–
Carbofuran	1563-66-2	H	0.02	–	–
4-Chloro-2-methylphenol (E)	1570-64-5	D	0.006*, 0.007, 0.008, 0.009	0.79	10
Chlorpyrifos	2921-88-2	I	0.005*, 0.006, 0.011	0.08	1
Chlorpyrifos, oxon (E)	5598-15-2	D	0.06*, 0.03, 0.01	2.38	30
Cyanazine	21725-46-2	H	0.018	–	–
Cycloate	1134-23-2	H	0.005	–	–
Cyfluthrin (E)	68359-37-5	I	0.008*, 0.07	1.59	20
Cyhalothrin-lambda	91465-08-6	I	0.009	–	–
Cypermethrin (E)	52315-07-8	I	0.009*, 0.015	1.59	20
Dacthal	1861-32-1	H	0.003	0.08	1
p,p'-Dichlorodiphenyldichloroethylene (DDE)	72-55-9	D	0.003*, 0.004, 0.007, 0.021, 0.03	–	–
Diazinon	333-41-5	I	0.005*, 0.006, 0.015	0.08	1
Diazoxon (E)	962-58-3	D	0.006	0.4	5
2,5-Dichloroaniline	95-82-9	D	0.03	–	–
4,4-Dichlorobenzophenone	90-98-2	D	0.003	–	–
Dichlorvos (E)	62-73-7	I	0.01*, 0.03	2.38	30
Dicrotophos (E)	141-66-2	I	0.08	0.24	3
Dieldrin	60-57-1	I	0.005*, 0.009, 0.01	0.16	2
Dimethoate	60-51-5	I	0.006*, 0.007, 0.01, 0.015, 0.01	0.16	2
(E)-Dimethomorph	110488-70-5	F	0.02	–	–
(Z)-Dimethomorph	110488-70-5	F	0.05	–	–
Disulfoton	298-04-4	I	0.02	–	–
Disulfoton sulfone	2497-06-5	D	0.02	–	–
Disulfoton sulfoxide	2497-07-6	D	0.002	–	–
Endosulfan-alpha	959-98-8	I	0.005	–	–
Endosulfan-beta	33213-65-9	I	0.01	–	–
Endosulfan ether	3369-52-6	D	0.004	–	–
Endosulfan sulfate	1031-07-8	D	0.006	–	–
EPTC	759-94-4	H	0.002	–	–
Ethalfluralin	55283-68-6	H	0.009	–	–
Ethion	563-12-2	I	0.004*, 0.006	0.16	2

Table 5.   Pesticides analyzed and the laboratory reporting limits for samples collected in the Central Valley of California during the 2001 and 2002–04 study periods.—Continued

[Abbreviations: (E), estimated compound because of poor performance by applied method (Childress and others, 1999; Sandstrom and others, 2001); D, Degradate; F, Fungicide; H, Herbicide; I, Insecticide; LRL, Laboratory Reporting Limit; NA, not applicable; ss-IRL, sample specific-Interim Reporting Limit; *, current LRL limit; μg/L, microgram per liter; μg/Kg, micrograms per kilogram; –, compound not analyzed for given sample type]

Compound	Chemical Abstract Service Registry Number	Type of compound	Sample types		
			Storm-drain runoff, wet deposition, bulk wet deposition, and soil-box runoff (aqueous phase)	Soil-box runoff (suspended sediment)	Surficial soil (soil box)
			LRL, μg/L	ss-IRL, μg/L	ss-IRL, μg/kg
Ethion monoxon	17356-42-2	D	0.03*, 0.01	0.16	2
Ethoprop	13194-48-4	I	0.005	–	–
O-Ethyl-O-methyl-S-propylphosphorothioate	76960-87-7	D	0.008	–	–
Fenamiphos (E)	22224-92-6	N	0.03	2.38	30
Fenamiphos sulfone (E)	31972-44-8	D	0.008*, 0.01, 0.012, 0.031, 0.312	0.79	10
Fenamiphos sulfoxide (E)	31972-43-7	D	0.03*, 0.12	0.79	10
Fenthion	55-38-9	I	0.02	–	–
Fenthion sulfoxide	3761-41-9	D	0.008	–	–
Fipronil (E)	120068-37-3	I	0.02	0.08	1
Fipronil sulfide	120067-83-6	D	0.013	0.08	1
Fipronil sulfone	120068-36-2	D	0.024	0.08	1
Friponil, desulfinyl (Desulfinylfipronil)	NA	D	0.012	0.08	1
Friponil amide, desulfinyl (Desulfinylfipronil amide) (E)	NA	D	0.029	0.08	1
Flumetralin	62924-70-3	Plant growth regulator	0.004	–	–
Fonofos	944-22-9	I	0.003	0.08	1
Fonofos, oxon	NA	D	0.002*, 0.007, 0.011	0.4	5
Hexazinone	51235-04-2	H	0.013	0.08	1
Iprodione (E)	36734-19-7	F	1*, 0.035	0.79	10
3,5-Dichloroaniline	626-43-7	D	0.005	–	–
Isofenphos	25311-71-1	I	0.003	0.16	2
Lindane	58-89-9	I	0.004	–	–
Linuron	330-55-2	H	0.035	–	–
3,4-Dichloroaniline (E)	95-76-1	D	*0.004, 0.006, 0.007, 0.009, 0.011	5.95	75
Malathion	121-75-5	I	0.027	0.16	2
Malaoxon (E)	1634-78-2	D	0.008*, 0.01, 0.015, 0.025, 0.125, 0.312	0.4	5
Metalaxyl	57837-19-1	F	0.005*, 0.007, 0.125	0.6	1
Methidathion	950-37-8	I	0.006*, 0.007, 0.008, 0.021	0.16	2
Metolachlor	51218-45-2	H	0.013	0.08	1
2-Ethyl-6-methylaniline (E)	24549-06-2	D	0.001*, 0.004	2.38	30
Metribuzin	21087-64-9	H	0.006*, 0.125, 0.312	0.32	4
Molinate	2212-67-1	H	0.002	–	–
Myclobutanil (E)	88671-89-0	F	0.008*, 0.011, 0.015 0.025, 0.031	0.08	1
Napropamide	15299-99-7	H	0.007*, 0.01	–	–
Oxyfluorfen	42874-03-3	H	0.007*, 0.011	–	–
Parathion	56-38-2	I	0.01	–	–
Paraoxon	311-45-5	D	0.008	–	–
Parathion-methyl	298-00-0	I	0.006*, 0.015, 0.01	0.16	2
Paraoxon-methyl (E)	950-35-6	D	0.03	0.4	5
Pebulate	1114-71-2	H	0.004	–	–
Pendimethalin	40487-42-1	H	0.022*, 0.021, 0.031	0.11	1
cis-Permethrin	54774-45-7	I	0.006*, 0.007, 0.021, 0.01, 0.07, 0.08, 0.2	0.4	5
trans-Permethrin	NA	I	–	0.4	5
cis-Propiconazole	NA	F	0.008	–	–

Table 5. Pesticides analyzed and the laboratory reporting limits for samples collected in the Central Valley of California during the 2001 and 2002–04 study periods.—Continued

[Abbreviations: (E), estimated compound because of poor performance by applied method (Childress and others, 1999; Sandstrom and others, 2001); D, Degradate; F, Fungicide; H, Herbicide; I, Insecticide; LRL, Laboratory Reporting Limit; NA, not applicable; ss-IRL, sample specific-Interim Reporting Limit; *, current LRL limit; μg/L, microgram per liter; μg/Kg, micrograms per kilogram; –, compound not analyzed for given sample type]

Compound	Chemical Abstract Service Registry Number	Type of compound	Sample types		
			Storm-drain runoff, wet deposition, bulk wet deposition, and soil-box runoff (aqueous phase)	Soil-box runoff (suspended sediment)	Surficial soil (soil box)
			LRL, μg/L	ss-IRL, μg/L	ss-IRL, μg/kg
trans-Propiconazole	NA	F	0.01	–	–
Phorate (E)	298-02-2	I	0.011	0.4	5
Phorate, oxon (E)	2600-69-3	D	0.1*, 0.12	0.56	7
Phosmet	732-11-6	I	0.008*, 0.01, 0.015, 0.025, 0.2	2.38	30
Phosmet oxon	3735-33-9	D	0.06	0.79	10
Phostebupirim	96182-53-5	I	0.005	–	–
Profenofos	41198-08-7	I	0.006	–	–
Prometon (E)	1610-18-0	H	0.01	0.22	2
Prometryn	7287-19-6	H	0.005*, 0.007	0.16	2
Pronamide (Propyzamide)	23950-58-5	H	0.004*, 0.005, 0.006, 0.01	0.16	2
Propachlor	1918-16-7	H	0.01	–	–
Propanil	709-98-8	H	0.011	–	–
Propargite	2312-35-8	I	0.02	–	–
2-(4-*tert*-Butylphenoxy)-cyclohexanol	1942-71-8	D	0.01*, 0.02, 0.03	–	–
Propetamphos	31218-83-4	I	0.004	–	–
trans-Propiconazole	NA	F	0.01	–	–
Simazine	122-34-9	H	0.005*, 0.006, 0.007, 0.01, 0.021, 0.04	0.16	2
Sulfotepp	3689-24-5	I	0.003	–	–
Sulprofos	35400-43-2	I	0.02	–	–
Tebupirimphos, oxon	NA	D	0.006	–	–
Tebuthiuron (E)	34014-18-1	H	0.02	0.24	3
Tefluthrin	79538-32-2	I	0.008	–	–
Temephos	3383-96-8	I	0.3	–	–
Terbacil	5902-51-2	H	0.034	–	–
Terbufos	13071-79-9	I	0.02	0.24	1
Terbufos-oxon-sulfone (E)	56070-15-6	D	0.07*, 0.12	0.4	5
Terbuthylazine	5915-41-3	H	0.01	0.72	1
Thiobencarb	28249-77-6	H	0.005	–	–
Triallate	2303-17-5	H	0.002	–	–
Tribuphos	78-48-8	Plant growth regulator	0.004	–	–
3-(Trifluoromethyl)aniline	98-16-8	D	0.01	–	–
Trifluralin	1582-09-8	H	0.009*, 0.01	0.08	1

Laboratory and Field Quality Control

Data obtained from quality-control (QC) samples are used to estimate the variability and bias from sample collection, processing, and laboratory analysis. Variability is defined as random error in independent measurements that can result with repeated application of the process under specified conditions. Bias refers to a systematic error manifested as a consistent positive or negative deviation from the known or true value. The results of the laboratory and field quality-control measurements are presented separately.

Laboratory Quality-Control Methods

The NWQL uses a quality management system (QMS) that provides a consistent system of protocols and procedures for carrying out required quality assurance (QA) and QC in compliance with the standards set by the National Environmental Laboratory Accreditation Conference (Maloney, 2005). As part of the QMS system, three types of laboratory quality-control measurements are prepared and analyzed with each group of environmental samples to monitor method performance. The three types are defined as follows:

- Laboratory Reagent Spike (LRS). A LRS is a sample where known concentrations of the compounds of interest are added, most commonly, into a synthetic matrix (usually blank water or clean sand). An evaluation of the LRS yields information about the method performance. The LRS analytical results are compared to acceptable criteria for each method to assess potential bias of environmental sample results. The calculation for spike recovery is as follows:

$$\left[\frac{\left(C_{spike} - C_{environmental} \right)}{C_{expected}} \right] \times 100 \text{ percent} = \text{percent recovery} \quad (1)$$

where

C_{spike} is the concentration of analyte found in spiked sample, in $\mu g/L$,

$C_{environmental}$ is the concentration of analyte in environmental "background" sample, in $\mu g/L$, and

$C_{expected}$ is the concentration of analyte expected in spike sample.

- Laboratory Reagent Blank (LRB). A LRB is prepared by using an appropriate matrix (typically blank water or clean sand) and is processed in the same way as the accompanying environmental sample. It is used to monitor random contamination introduced from laboratory processing.

- Laboratory Surrogates. A surrogate is a compound that is expected to perform similarly to the compound being analyzed in a laboratory method. The surrogate is not normally found in the environment and, therefore, can be used to monitor the recovery efficiency of analytical processes (U.S. Geological Survey, 1998). Two surrogate compounds, α-HCH-d6 and diazinon-d10, were added to all sample types submitted in this study.

The LRS and LRB data for the wet deposition, bulk wet deposition, and aqueous phase of the soil runoff samples are not presented because these methods are established, and the NWQL performs the data assessment and assesses the potential bias of an environmental sample result (*http://bqs.usgs.gov/ltmdl/*, accessed February 22, 2011). The minimum, maximum, and median percentage recoveries for laboratory surrogates prepared and analyzed with all the environmental sample types collected in during the entire study period are presented in table 6. The percentage recoveries for the LRSs,

Table 6. Minimum, maximum, and median percentage recoveries for laboratory surrogates prepared and analyzed with all environmental-sample types submitted to the National Water Quality Laboratory (NWQL).

[All compounds were analyzed by gas chromatography-mass spectrometry (GC-MS), at the U.S. Geological Survey NWQL]

Sample type	*alpha*-HCH-d6	Diazinon-d10
Storm run off		
Minimum percentage recovery	76.8	97.4
Maximum percentage recovery	118	153
Median percentage recovery	91.9	105.5
Wet-deposition samples		
Minimum percentage recovery	58.4	77.8
Maximum percentage recovery	112	127
Median percentage recovery	91	102
Dry-deposition samples		
Minimum percentage recovery	42	60
Maximum percentage recovery	99	111
Median percentage recovery	80	89
Bulk-depositions samples		
Minimum percentage recovery	7.1	8.7
Maximum percentage recovery	112	120
Median percentage recovery	77.4	85.6
Surficial-soil samples		
Minimum percentage recovery	81.4	117
Maximum percentage recovery	102	167
Median percentage recovery	97.2	140
Suspended sediment		
Minimum percentage recovery	11.1	16.9
Maximum percentage recovery	66.5	100.9
Median percentage recovery	47.9	73.2
Soil run off		
Minimum percentage recovery	75.7	86.7
Maximum percentage recovery	103	109
Median percentage recovery	87.2	98.6

and the detections found in the LRBs, are discussed in the following sections by sample type for the dry-deposition, bulk dry-deposition, suspended-sediment, and surficial-soil sample types. These sample types are considered research methods and, therefore, required an evaluation of the laboratory QA and QC data. Median percentage recoveries greater than 50 percent for the LRSs were considered acceptable for the study.

Dry-Deposition and Bulk Dry-Deposition Samples

Nine LRSs and seven LRBs were analyzed with the dry-deposition and the bulk dry-deposition environmental samples. Table 7 describes the minimums, medians, maximums, and standard deviations of percentage recoveries for LRSs. Because the analytical techniques used to analyze these sample types were being developed and refined during the study, LRS recoveries for many compounds were lower and more variable than LRS recoveries for similar compounds in other media that were obtained by using approved analytical methods (table 4). As a result, only 29 of the 64 compounds analyzed had median percentage recoveries greater than or equal to 50 percent. For compounds with median percent of recoveries less than 50 percent, the concentrations in the environmental samples are considered usable but qualified as biased-low (B-L) on the basis of low and variable recovery (table 7). For compounds that are biased low, it is possible that small concentrations present in a sample could escape detection and be reported as less than the reporting level (appendix 5 and 6).

Two compounds (azinphos-methyl and phosmet) were detected in the LRBs. These detections were likely a result of laboratory analysis carryover. A LRB was run at the end of each set of environmental samples (seven sets total). For each set of environmental samples, concentrations in the associated LRBs were approximately an order of magnitude less than the concentrations observed in environmental samples with detections of the same compounds (table 8).

Soil-Box Suspended Sediment

Four LRSs and two LRBs were analyzed along with the environmental samples for the suspended-sediment samples. The minimums, medians, maximums, and standard deviations of percentage recoveries for the LRSs are presented in table 9. Median percentage recoveries greater than or equal to 50 percent were found for 44 of the 63 compounds analyzed. The remaining 19 compounds had percentage recoveries of less than 50 percent and are qualified as B-L. Malaoxon had spike recoveries of zero percent and was qualified as B-L (appendix 8); however, this compound was detected in an environmental sample (site 16, December 18–23, 2003) despite the low bias. No compounds were detected in the suspended-sediment LRBs.

Soil-Box Surficial Soil

Duplicate environmental aliquots were extracted and analyzed from the soil-box mix 1, 2, 3, and 4 samples to examine laboratory method performance. Table 10 presents the concentration in micrograms per kilogram (μg/Kg) for the environmental aliquot, the duplicate environmental aliquot, and relative percentage difference (RPD) of duplicates. The RPD was calculated as follows:

$$RPD = \left(\frac{|\text{environmental aliquot} - \text{duplicate environmental aliquot}|}{\text{mean of environmental aliqout and duplicate environmental aliquot}} \right) \times 100 \text{ percent} \qquad (2)$$

Overall, the RPD for compounds with detections in both samples was less than 25 percent. For compounds with a RPD greater than 25 percent, the greater variability was presumed to be a result of the inherent heterogeneity of the soil matrix, an indication of the variability of the compound in the soil matrix, or both.

The quality-control samples analyzed along with the environmental surficial soil-box samples included a laboratory matrix spike (LMS), an LRS, and an LRB. The LMS was performed on an aliquot of soil-box mix 2, and the LRS and LRB were performed on aliquots of

Table 7.   Minimum, maximum, median, and standard deviation of percentage recoveries for laboratory reagent spikes extracted along with the dry-deposition and bulk dry-deposition environmental samples.

[Abbreviations: B-L, biased low based on low and variable recoveries; na, none applied; (E), estimated compound because of known performance issues by applied method (Childress and others, 1999; Sandstrom and others, 2001); ±, plus or minus]

Compound (common name)	Minimum percentage recovery	Median percentage recovery	Maximum percentage recovery	Standard deviation	Detection qualifier
1-Naphthol (E)	0	5.83	27.4	±10.1	B-L
2,6-Diethylanaline[1] (E)	0	31.7	86.5	±23.6	B-L
2-[(2-ethyl-6-methylphenyl)-amino]-1-propanol[1]	0	32.6	45.5	±19.45	B-L
2-chloro-2,6-diethylacetanilide	32.7	83.9	92.3	±19.2	na
2-ethyl-6 methylaniline[1]	0	36.3	86.3	±24.5	B-L
3,4-Dichloroaniline (E)	5.7	47.5	80.9	±25.3	B-L
Acetochlor	38.1	70.3	106	±18.3	na
Alachlor	37.6	68.2	88.6	±14.7	na
Atrazine	36.7	85.4	91.9	±17.4	na
Azinphos-methyl (E)	0	39.3	69.9	±23.7	B-L
Azinphos-methyl-oxon (E)	0	7.3	61.2	±23.0	B-L
Benfluralin[1]	14.4	27.3	56.6	±13.2	B-L
Carbaryl (E)	41.8	53.0	83.1	±15.3	na
Chlorpyrifos	42.3	71.4	95.3	±16.0	na
Chlorpyrifos, oxon (E)	0	38.8	64.3	±24.7	B-L
Cyfluthrin (E)	20.0	80.8	91.2	±25.7	na
Cypermethrin (E)	30.0	76.3	82.8	±20.9	na
Dacthal	49.4	76.4	109	±16.5	na
Deethylatrazine, CIAT	19.8	45.3	84.0	±20.2	B-L
Diazinon	41.8	68.8	97.1	±18.0	na
Diazoxon (E)	2.3	40.2	83.6	±24.6	B-L
Dichlorvos	0	36.4	82.7	±32.6	B-L
Dicrotophos[1] (E)	0	0	70.0	±24.5	B-L
Dieldrin	44.8	87.6	157	±30.9	na
Dimethoate	4.9	20.2	74.0	±22.3	B-L
Ethion	30.2	66.5	149	±33.2	na
Ethion monoxon	6.9	56.5	97.1	±28.3	na
Fenamiphos sulfone[1] (E)	2.0	26.3	67.0	±21.8	B-L
Fenamiphos sulfoxide (E)	0	52.9	140	±49.5	na
Fenamiphos[1] (E)	0	10.8	69.8	±24.2	B-L
Fipronil[1] (E)	20.3	34.2	75.0	±17.7	B-L
Fipronil sulfide	37.1	71.6	79.8	±17.4	na
Fipronil sulfone[1]	0	34.5	70.1	±21.5	B-L
Desulfinylfipronil	38.1	74.1	94.3	±19.4	na
Desulfinylfipronil amide[1]	8.2	48.2	93.4	±30.4	B-L
Fonofos	36.0	63.8	76.6	±15.2	na
Fonofos oxon[1]	0	42.2	73.3	±27.9	B-L
Hexazinone	16.6	22.6	68.7	±22.5	B-L
Iprodione (E)	0	31.7	87.6	±23.1	B-L
Isofenphos	39.2	60.2	90.8	±18.2	na
Malathion	34.9	66.8	94.7	±21.4	na
Malaoxon (E)	0	28.5	72.7	±24.4	B-L
Methidathion	29.8	67.8	123	±24.7	na
Metolachlor	2.0	70.8	95.2	±25.4	na
Metribuzin	10.9	22.9	75.4	±22.7	B-L
Myclobutanil	9.0	66.9	118	±31.3	na

Table 7. Minimum, maximum, median, and standard deviation of percentage recoveries for laboratory reagent spikes extracted along with the dry-deposition and bulk dry-deposition environmental samples.—Continued

[Abbreviations: B-L, biased low based on low and variable recoveries; na, none applied; (E), estimated compound because of known performance issues by applied method (Childress and others, 1999; Sandstrom and others, 2001); ±, plus or minus]

Compound (common name)	Minimum percentage recovery	Median percentage recovery	Maximum percentage recovery	Standard deviation	Detection qualifier
Parathion-methyl	25.3	45.6	69.6	±14.9	B-L
Paraoxon-methyl (E)	0	24.1	63.3	±23	B-L
Pendimethilan	12.1	33.9	77.1	±19.4	B-L
cis-permethrin	39.5	90.1	110	±23.6	na
trans-permethrin	32.7	75.6	95.6	±22.8	na
Phorate oxon[1] (E)	0	15.8	62.2	±19.3	B-L
Phorate[1]	0	22.4	72.7	±19.8	B-L
Phosmet (E)	0	8.4	40.0	±13.1	B-L
Phosmet oxon (E)	0	0	29.0	±10.3	B-L
Prometon (E)	7.0	71.0	91.7	±25.6	na
Prometryn	2.0	51.6	89.4	±22.6	na
Pronamide	32.2	67.0	80.4	±15.4	na
Simazine	31.9	69.0	96.9	±19.6	na
Tebuthiuron (E)	0	27.2	93.0	±26.9	B-L
Terbufos-oxon-sulfone (E)	0	26.9	63.9	±22.7	B-L
Terbufos[1]	0	49.8	73.2	±23.7	B-L
Terbuthylazine	42.0	82.2	97.8	±16.3	na
Trifluralin	15.5	29.9	74.0	±17.7	B-L

[1]Compound was not detected in environmental samples and is not presented in appendix 5 and 6 for the dry and bulk dry deposition sample types, respectively, as there are no reporting limits for these sample types; however, it is possible that small concentrations of these compounds may be present in a sample and were not detected based on the low and variable laboratory reagent spike recovery.

Table 8.   Total number and detection mass for the laboratory reagent blanks extracted along with the dry deposition and bulk-dry deposition environmental samples.

[Abbreviations: (E), estimated compound because of known performance issues by applied method (Childress and others, 1999; Sandstrom and others, 2001); E, estimated; <, actual value is known to be less than value shown; µg/ sample, microgram per sample; –, no detection for given compound]

Compound	Number of blanks	Number of detections	Detection in laboratory blank (µg/sample)	Detection in environmental sample (µg/sample)
1-Naphthol (E)	7	0	–	–
2,6-Diethylanaline	7	0	–	–
2-[(2-ethyl-6-methylphenyl)-amino]-1-propanol	7	0	–	–
2-chloro-2,6-diethylacetanilide	7	0	–	–
2-ethyl-6 methylaniline	7	0	–	–
3,4-Dichloroaniline	7	0	–	–
Acetochlor	7	0	–	–
Alachlor	7	0	–	–
Atrazine	7	0	–	–
Azinphos-methyl (E)	7	2	E0.008, E0.006	<0.06
Azinphos-methyl-oxon (E)	7	0	–	–
Benfluralin	7	0	–	–
Carbaryl (E)	7	0	–	–
Chlorpyrifos	7	0	–	–
Chlorpyrifos oxon (E)	7	0	–	–
Cyfluthrin (E)	7	0	–	–
Cypermethrin (E)	7	0	–	–
Dacthal	7	0	–	–
Deethylatrazine, CIAT	7	0	–	–
Diazinon	7	0	–	–
Diazoxon (E)	7	0	–	–
Dichlorvos (E)	7	0	–	–
Dicrotophos (E)	7	0	–	–
Dieldrin	7	0	–	–
Dimethoate	7	0	–	–
Ethion	7	0	–	–
Ethion monoxon	7	0	–	–
Fenamiphos (E)	7	0	–	–
Fenamiphos sulfone (E)	7	0	–	–
Fenamiphos sulfoxide (E)	7	0	–	–
Fipronil (E)	7	0	–	–
Fipronil sulfide	7	0	–	–
Fipronil sulfone	7	0	–	–
Desulfinylfipronil	7	0	–	–
Desulfinylfipronil amide	7	0	–	–
Fonofos	7	0	–	–
Fonofos oxon	7	0	–	–
Hexazinone	7	0	–	–
Iprodione (E)	7	0	–	–
Isofenphos	7	0	–	–
Malathion	7	0	–	–
Malaoxon (E)	7	0	–	–
Methidathion	7	0	–	–
Metolachlor	7	0	–	–
Metribuzin	7	0	–	–
Myclobutanil (E)	7	0	–	–
Parathion-methyl	7	0	–	–
Paraoxon-methyl (E)	7	0	–	–
Pendimethilan	7	0	–	–
cis-permethrin	7	0	–	–
trans-permethrin	7	0	–	–

Table 8.   Total number and detection mass for the laboratory reagent blanks extracted along with the dry deposition and bulk-dry deposition environmental samples.—Continued

[Abbreviations: (E), estimated compound because of known performance issues by applied method (Childress and others, 1999; Sandstrom and others, 2001); E, estimated; <, actual value is known to be less than value shown; μg/ sample, microgram per sample; –, no detection for given compound]

Compound	Number of blanks	Number of detections	Detection in laboratory blank (μg/sample)	Detection in environmental sample (μg/sample)
Phorate (E)	7	0	–	–
Phorate oxon (E)	7	0	–	–
Phosmet (E)	7	1	E0.005	E0.05
Phosmet oxon (E)	7	0	–	–
Prometon (E)	7	0	–	–
Prometryn	7	0	–	–
Pronamide	7	0	–	–
Simazine	7	0	–	–
Tebuthiuron (E)	7	0	–	–
Terbufos	7	0	–	–
Terbufos-oxon-sulfone (E)	7	0	–	–
Terbuthylazine	7	0	–	–
Trifluralin	7	0	–	–

Table 9. Minimum, maximum, median, and standard deviation of percentage recoveries for laboratory-reagent spikes in the soil-box suspended-sediment samples.

[Abbreviations: B-L, biased low; na, none applied; (E), estimated compound because of known performance issues by applied method (Childress and others, 1999; Sandstrom and others, 2001); ±, plus or minus]

Compound	Minimum percentage recovery	Median percentage recovery	Maximum percentage recovery	Standard deviation	Environmental detection qualifier code
1-Naphthol (E)	3.9	4.0	4.2	±0.2	B-L
2,6-diethylaniline (E)	23.2	52.5	56.9	±15.5	na
2-chloro-2,6-diethylacetanilide	56.6	63.5	70.3	±9.60	na
2-ethyl-6-methylaniline (E)	20.3	36.7	53.2	±23.3	B-L
3,4-dichloroaniline (E)	23.8	34.8	45.8	±15.5	B-L
4-chloro-2-methylphenol (E)	13.0	29.8	46.7	±23.8	B-L
Acetochlor	52.5	71.1	82.7	±14.8	na
Alachlor	45.3	61.6	70.6	±12.4	na
Atrazine	46.2	63.8	71.1	±11.3	na
Azinphos-methyl (E)	40.7	50.2	52.9	±5.5	na
Azinphos-methyl-oxon (E)	22.1	23.6	25.0	±2.0	B-L
Benfluralin	52.8	60.8	65.7	±5.50	na
Carbaryl (E)	0	2.8	19.6	±9.3	B-L
Chlorpyrifos	46.0	59.5	63.2	±7.9	na
Chlorpyrifos oxon (E)	43.6	45.8	48.1	±3.2	B-L
Cyfluthrin (E)	0	22.3	44.6	±31.5	B-L
Cypermethrin (E)	0	28.7	57.5	±40.6	B-L
Dacthal	56.5	73.5	85.2	±13.3	na
Deethylatrazine, CIAT	53.6	82.3	101.1	±23.5	na
Diazinon	46.4	60.5	70.1	±10.3	na
Diazoxon (E)	49.7	53.4	57.0	±5.2	na
Dichlorvos (E)	36.3	36.9	37.4	±0.8	B-L
Dicrotophos (E)	53.3	61.3	69.3	±11.3	na
Dieldrin	59.6	79.0	92.5	±17.3	na
Dimethoate	40.2	48.5	56.9	±11.8	B-L
Ethion	55.0	58.3	61.7	±4.7	na
Ethion monoxon	52.1	52.1	52.1	±0.01	na
Fenamiphos (E)	26.4	33.2	40.0	±9.6	B-L
Fenamiphos sulfone (E)	67.5	70.1	72.7	±3.7	na
Fenamiphos sulfoxide (E)	101.9	103.0	104.1	±1.6	na
Fipronil (E)	49.1	64.6	73.7	±12.6	na
Fipronil sulfide	51.4	75.3	90.9	±19.0	na
Fipronil sulfone	35.8	60.7	83.3	±25.0	na
Desulfinylfipronil	44.0	61.0	68.6	±10.8	na
Desulfinylfipronil amide (E)	63.7	86.5	98.4	±16.5	na
Fonofos	40.4	56.8	64.1	±11.1	na
Fonofos oxon	53.2	58.3	63.5	±7.3	na
Hexazinone	62.3	69.3	76.3	±9.9	na
Iprodione (E)	7.1	8.0	8.8	±1.2	B-L
Isofenphos	48	53.8	59.6	±8.2	na
Malathion	0	6.7	19.2	±9.7	B-L
Malaoxon (E)	0	0	0	0	B-L
Metalaxyl	49.8	57.7	65.5	±11.1	na
Methidathion	51.4	53.1	54.8	±2.5	na
Metolachlor	45.7	63.4	72.4	±12.0	na
Metribuzin	50.5	69.5	87.9	±19.8	na
Myclobutanil (E)	60.1	64.4	68.7	±6.1	na
Parathion-methyl	49.3	63.0	76.4	±13.6	na
Paraoxon-methyl (E)	49.0	50.3	51.6	±1.8	na
Pendimethalin	49.4	72.4	93.3	±23.1	na
cis-permethrin	62.5	82.4	95.6	±13.9	na

Table 9. Minimum, maximum, median, and standard deviation of percentage recoveries for laboratory-reagent spikes in the soil-box suspended-sediment samples.—Continued

[Abbreviations: B-L, biased low; na, none applied; (E), estimated compound because of known performance issues by applied method (Childress and others, 1999; Sandstrom and others, 2001); ±, plus or minus]

Compound	Minimum percentage recovery	Median percentage recovery	Maximum percentage recovery	Standard deviation	Environmental detection qualifier code
trans-permethrin	55.5	71.3	87.2	±22.4	na
Phorate (E)	32.6	45.7	60.0	±11.2	B-L
Phorate oxon (E)	27.1	37.2	47.4	±14.4	B-L
Prometon (E)	54.0	70.2	80.3	±12.9	na
Prometryn	47.0	52.4	57.7	±7.6	na
Pronamide	46.3	58.3	70.8	±11.9	na
Simazine	52.2	75.9	89.7	±18.3	na
Tebuthiuron (E)	45.8	69.4	90.2	±24.5	na
Terbufos	19.4	36.8	38.7	±9.1	B-L
Terbufos-oxon-sulfone (E)	32.8	33.4	34.1	±0.9	B-L
Terbuthylazine	59.3	66.4	73.5	±10.0	na
Trifluralin	51.4	63.9	76.5	±11.1	na

Table 10. Concentration in micrograms per kilogram (µg/Kg) of duplicate environmental aliquots of soil-box mix 1, mix 2, mix 3, and mix 4, and the relative percentage difference between each of the duplicate sets.

[Mix 1 collected near site 10, Westley rain gage at pump house, San Joaquin Valley; mix 2 collected from near site 11, Newman rain gage at wasteway levee, San Joaquin Valley; mix 3 collected from near site 5, Modest Irrigation District at Albers Road, San Joaquin Valley; mix 4 collected near site 16, Gridley High School precipitation gage, Sacramento Valley. Abbreviations: E, estimated reported laboratory value; (E), estimated compound because of poor performance by applied method (Childress and others, 1999; Sandstrom and others, 2001); RPD, relative percentage difference; <, less than laboratory reporting value; µg/kg, micrograms per kilogram; –, not calculated because compound was not detected in environmental and duplicate sample; NA, not applicable to calculate]

Compound	Mix 1			Mix 2		
	Environmental sample (µg/kg)	Duplicate environmental sample (µg/kg)	RPD of duplicates (in percent)	Environmental sample (µg/kg)	Duplicate environmental sample (µg/kg)	RPD of duplicates (in percent)
1-Naphthol (E)	<10	<10	–	<10	<10	–
2,6-Diethylaniline (E)	<30	<30	–	<30	<30	–
2-Chloro-2,6-diethylacetanilide	<1	<1	–	<1	<1	–
2-Ethyl-6-methylaniline (E)	<30	<30	–	<2	<2	–
3,4-Dichloroaniline (E)	5.4	3.6	38.2	<30	<30	–
4-Chloro-2-methylphenol (E)	<10	<10	–	6.10	4.08	39.4
Acetochlor	<1	<1	–	<10	<10	–
Alachlor	<1	<1	–	<1	<1	–
Atrazine	<1	<1	–	<1	<1	–
Azinphos-methyl (E)	<5	<5	–	<5	<5	–
Azinphos-methyl-oxon (E)	<30	<30	–	<30	<30	–
Benfluralin	<1	<1	–	<1	<1	–
Carbaryl (E)	1.5	1.6	9.84	2.10	1.94	8.40
Chlorpyrifos	<1	<1	–	E 0.68	E 0.63	7.90
Chlorpyrifos oxon (E)	<30	<30	–	<30	<30	–
Cyfluthrin (E)	<20	<20	–	<20	<20	–
Cypermethrin (E)	<20	<20	–	0.59	0.53	9.81
Dacthal	E 0.56	E 0.54	4.91	<1	<1	–
Deethylatrazine , CIAT	<2	<2	–	<5	<5	–
Diazinon	<1	<1	–	3.70	3.38	7.95
Diazoxon (E)	<5	<5	–	<2.8	0	–
Dichlorvos (E)	<30	<30	–	<6.1	<6.5	–
Dicrotophos (E)	<3	<3	–	<30	<30	–
Dieldrin	3.1	E 1.91	47.2	<10	<10	–
Dimethoate	E 18	E 15.2	16.3	<10	<10	–
Ethion	<2	<2	–	<1	<1	–
Ethion monoxon	<3.1	<3.3	–	<1	<1	–
Fenamiphos (E)	<30	<30	–	1.20	1.11	7.79
Fenamiphos sulfone (E)	<10	<10	–	0.62	0.56	9.80
Fenamiphos sulfoxide (E)	<10	<10	–	<1	<1	–
Fipronil (E)	<1	<1	–	<1	<1	–
Fipronil sulfide	<1	<1	–	<5	<5	–
Fipronil sulfone	1.2	1.92	50.2	<1	<1	–
Desulfinylfipronil	E 0.61	E 0.59	2.17	<30	<30	–
Desulfinylfipronil amide (E)	<1	<1	–	<3	<3	–
Fonofos	<1	<1	–	E 8.9	E 7.51	16.49
Fonofos oxon	<5	<5	–	<2	<2	–
Hexazinone	<1	<1	–	25.0	24.4	2.83
Iprodione (E)	5.4	6.44	17.9	<5	<5	–
Isofenphos	<2	<2	–	<1	<1	–
Malathion	18	17.8	2.22	<2	<2	–
Malaoxon (E)	<5	<5	–	130	128	0.00
Metalaxyl	<1	<1	–	E 23	E 24.1	3.81
Methidathion	<2	<2	–	E 14	E 13.7	3.58
Metolachlor	58	68.9	18.0	<2	<2	–
Metribuzin	E 15	E 15.5	4.62	<5	<5	–

Table 10. Concentration in micrograms per kilogram (µg/Kg) of duplicate environmental aliquots of soil-box mix 1, mix 2, mix 3, and mix 4, and the relative percentage difference between each of the duplicate sets.—Continued

[Mix 1 collected near site 10, Westley rain gage at pump house, San Joaquin Valley; mix 2 collected from near site 11, Newman rain gage at wasteway levee, San Joaquin Valley; mix 3 collected from near site 5, Modest Irrigation District at Albers Road, San Joaquin Valley; mix 4 collected near site 16, Gridley High School precipitation gage, Sacramento Valley. Abbreviations: E, estimated reported laboratory value; (E), estimated compound because of poor performance by applied method (Childress and others, 1999; Sandstrom and others, 2001); RPD, relative percentage difference; <, less than laboratory reporting value; µg/kg, micrograms per kilogram; –, not calculated because compound was not detected in environmental and duplicate sample; NA, not applicable to calculate]

Compound	Mix 1			Mix 2		
	Environmental sample (µg/kg)	Duplicate environmental sample (µg/kg)	RPD of duplicates (in percent)	Environmental sample (µg/kg)	Duplicate environmental sample (µg/kg)	RPD of duplicates (in percent)
Myclobutanil (E)	18	12.7	32.3	48.0	57	18.0
Parathion-methyl	<2	<2	–	<7	<7	–
Paraoxon-methyl	<5	<5	–	<5	<5	–
Pendimethalin	58	47.1	20.4	<30	<30	–
cis-Permethrin	E 11	E 5.8	64.5	<20	<20	–
trans-Permethrin	6.1	E 4.04	40.0	135	140	3.98
Phorate (E)	<5	<5	–	<42	<35	–
Phorate oxon (E)	<7	<7	–	1.40	1.49	3.41
Phosmet	<30	<30	–	<2	<2	–
Phosmet oxon	<29	<34	–	<5	<5	–
Prometon (E)	1.2	1.2	2.53	<5	<5	–
Prometryn	<2	<2	–	<2	<2	–
Pronamide	<2	<2	–	70.0	71.5	2.12
Simazine	99	83.5	16.8	<3	<3	–
Tebuthiuron (E)	<3	<3	–	<5	<5	–
Terbufos	<3	<3	–	<1	<1	–
Terbufos-oxon-sulfone (E)	<5	<5	–	31.0	33.4	7.78
Terbuthylazine	<1	<1	–	101	100	1.17
Trifluralin	17	19.3	10.4	42.2	45.0	NA
alpha-HCH-d6	93.0	86.9	6.86	<1	<1	–
Diazinon-d10	135	127	5.76	26.0	29	12.1
Sample amount, sched 5503, solid, grams	40.9	45.1	NA	<3	<3	–

Table 10.   Concentration in micrograms per kilogram (µg/Kg) of duplicate environmental aliquots of soil-box mix 1, mix 2, mix 3, and mix 4, and the relative percentage difference between each of the duplicate sets.—Continued

[Mix 1 collected near site 10, Westley rain gage at pump house, San Joaquin Valley; mix 2 collected from near site 11, Newman rain gage at wasteway levee, San Joaquin Valley; mix 3 collected from near site 5, Modest Irrigation District at Albers Road, San Joaquin Valley; mix 4 collected near site 16, Gridley High School precipitation gage, Sacramento Valley. Abbreviations: E, estimated reported laboratory value; (E), estimated compound because of poor performance by applied method (Childress and others, 1999; Sandstrom and others, 2001); RPD, relative percentage difference; <, less than laboratory reporting value; µg/kg, micrograms per kilogram; –, not calculated because compound was not detected in environmental and duplicate sample; NA, not applicable to calculate]

Compound	Mix 3			Mix 4		
	Environmental sample (µg/kg)	Duplicate environmental sample (µg/kg)	RPD of duplicates (in percent)	Environmental sample (µg/kg)	Duplicate environmental sample (µg/kg)	RPD of duplicates (in percent)
1-Naphthol (E)	<10	<10	–	<10	<10	–
2,6-Diethylaniline (E)	<30	<30	–	<30	<30	–
2-Chloro-2,6-diethylacetanilide	<1	<1	–	<1	<1	–
2-Ethyl-6-methylaniline (E)	<2	<2	–	<2	<2	–
3,4-Dichloroaniline (E)	<30	<30	–	<30	<30	–
4-Chloro-2-methylphenol (E)	4	4.07	2.99	0.9	1.09	19.9
Acetochlor	<10	<10	–	<10	<10	–
Alachlor	<1	<1	–	<1	<1	–
Atrazine	<1.2	<1	–	<1.1	<1.2	–
Azinphos-methyl (E)	<5	<5	–	<5	<5	–
Azinphos-methyl-oxon (E)	<30	<30	–	<30	<30	–
Benfluralin	<1	<1	–	<1	<1	–
Carbaryl (E)	1.6	1.23	27.4	1.9	2.6	30.6
Chlorpyrifos	E 0.86	<1	–	<1	<1	–
Chlorpyrifos oxon (E)	<30	<30		<30	<30	–
Cyfluthrin (E)	<20	<20	–	<20	<20	–
Cypermethrin (E)	0.79	0.62	23.9	<1	<1	–
Dacthal	<1	<1	–	<1	<1	–
Deethylatrazine , CIAT	<5	<5	–	<5	<5	–
Diazinon	3.1	3.02	2.29	E 0.43	E 0.60	30.0
Diazoxon (E)	<2	<2	–	<2	<2	–
Dichlorvos (E)	<4.4	<3.8	–	<2	<2	–
Dicrotophos (E)	<30	<30	–	<30	<30	–
Dieldrin	<10	<10	–	<10	<10	–
Dimethoate	<10	<10	–	<10	<10	–
Ethion	<1	<1	–	<1	<1	–
Ethion monoxon	<1	<1	–	<1	<1	–
Fenamiphos (E)	1.6	1.31	21.8	<1	<1	–
Fenamiphos sulfone (E)	0.83	<1	200	<1	<1	–
Fenamiphos sulfoxide (E)	<1	<1	–	<1	<1	–
Fipronil (E)	<1	<1	–	<1	<1	–
Fipronil sulfide	<5	<5	–	<5	<5	–
Fipronil sulfone	<1	<1	–	<1	<1	–
Desulfinylfipronil	<30	<30	–	<30	<30	–
Desulfinylfipronil amide (E)	<3	<3	–	<3	<3	–
Fonofos	E 7.9	E 6.2	23.3	<10	<10	–
Fonofos oxon	<2	<2	–	<2	<2	–
Hexazinone	27	23.2	13.7	<2	<2	–
Iprodione (E)	<5	<5	–	<5	<5	–
Isofenphos	<1	<1	–	<1	<1	–
Malathion	<2	<2	–	<2	<2	–
Malaoxon (E)	110	95.8	12.0	<1	<1	–
Metalaxyl	E 25	E 17.9	31.9	<4	<4	–
Methidathion	E 13	E 13.4	6.95	<1	<1	–
Metolachlor	<2	<2	–	<2	<2	–
Metribuzin	<5	<5	–	<5	<5	–

Table 10. Concentration in micrograms per kilogram (µg/Kg) of duplicate environmental aliquots of soil-box mix 1, mix 2, mix 3, and mix 4, and the relative percentage difference between each of the duplicate sets.—Continued

[Mix 1 collected near site 10, Westley rain gage at pump house, San Joaquin Valley; mix 2 collected from near site 11, Newman rain gage at wasteway levee, San Joaquin Valley; mix 3 collected from near site 5, Modest Irrigation District at Albers Road, San Joaquin Valley; mix 4 collected near site 16, Gridley High School precipitation gage, Sacramento Valley. Abbreviations: E, estimated reported laboratory value; (E), estimated compound because of poor performance by applied method (Childress and others, 1999; Sandstrom and others, 2001); RPD, relative percentage difference; <, less than laboratory reporting value; µg/kg, micrograms per kilogram; –, not calculated because compound was not detected in environmental and duplicate sample; NA, not applicable to calculate]

Compound	Mix 3			Mix 4		
	Environmental sample (µg/kg)	Duplicate environmental sample (µg/kg)	RPD of duplicates (in percent)	Environmental sample (µg/kg)	Duplicate environmental sample (µg/kg)	RPD of duplicates (in percent)
Myclobutanil (E)	47	51.9	9.27	<1	<1	–
Parathion-methyl	<7	<7	–	<7	<7	–
Paraoxon-methyl	<5	<5	–	<5	<5	–
Pendimethalin	<30	<30	–	<30	<30	–
cis-Permethrin	<20	<20	–	<20	<20	–
trans-Permethrin	136	127	6.74	155	119	26.2
Phorate (E)	<33	< 30	–	<10	<10	–
Phorate oxon (E)	1.6	1.21	25.3	2.1	2.48	14.7
Phosmet	<2	<2	–	<2	<2	–
Phosmet oxon	<5	<5	–	<5	<5	–
Prometon (E)	<5	<5	–	<5	<5	–
Prometryn	<2	<2	–	<2	<2	–
Pronamide	130	74.9	52.3	10	11.1	7.48
Simazine	<3	<3	–	<3	<3	–
Tebuthiuron (E)	<5	<5	–	<5	<5	–
Terbufos	<1	<1	–	<1	<1	–
Terbufos-oxon-sulfone (E)	25	24.8	2.39	0.11	0.08	38.6
Terbuthylazine	98.0	92.9	5.29	81.4	86.5	6.05
Trifluralin	34.6	40.7	NA	28.3	30.1	NA
alpha-HCH-d6	<1	<1	–	<1	<1	–
Diazinon-d10	E 23	E 23.9	4.71	<2	<2	–
Sample amount, sched 5503, solid, grams	<3	<3	–	<3	<3	–

laboratory-grade Ottawa Sand (clean dry sand). The average ambient concentration from the duplicate analysis performed on the soil-box mix 2 matrix was compared to the matrix-spike concentration that was recovered (table 11).

Substantially lower recoveries for some compounds in the soil-box mix 2 LMS compared to the LRS prepared by using clean, dry sand appeared largely influenced by the presence of ambient or "background" concentrations in the unspiked soil-box mix 2 sample (unspiked mix 2) that were subtracted from the determined LMS concentration in the LMS recovery calculation. Spike recoveries for a number of compounds from the soil-box mix 2 LMS and the LRS exhibited reasonable agreement, which is defined as a relative percent difference (RPD) in recoveries that is less than or equal to 25 percent (table 11). For some compounds, however, the spike recoveries from the soil-box mix 2 LMS were substantially greater or less than from the LRS. As a result of the bias, variability, or both, as indicated by the recoveries in both the LMS and LRS matrices, the compounds are designated with the following qualifier codes in table 11 and appendix 9:

- Acceptable Data—Data for environmental samples were considered acceptable and were not qualified for compounds with LMS and LRS recoveries between 50–150 percent. Likewise, environmental data were considered acceptable and were not qualified for those compounds that had (1) LMS recoveries that were coded U-R, which is defined as unable to calculate a recovery value because the concentration in the unspiked mix 2 sample was equivalent to or greater than the fortification concentration, or (2) unspiked mix 2 sample concentrations that were at least 25 percent of the LMS fortification concentration (for example, dieldrin), but that had LRS recoveries greater than 50 percent.

- Biased Low—Environmental-sample data were qualified as possibly being biased low (B-L) if the LRS or LMS recovery was less than 50 percent and the low LMS recovery was not a result of having an unspiked mix 2 sample concentration that was greater than 25 percent of the LMS fortification concentration. For compounds coded B-L, it was assumed that the analytical method probably would not adequately recover the compound as a result of matrix interference, adsorptive losses on the dry sand, or other procedural problems. For compounds that were qualified as biased low, it is possible that small, true concentrations present in a sample near the detection level were not detected because of low bias method recovery and, thus, were reported as less than the reporting level. Those compounds exhibiting low-biased method performance are included in appendix 9 nevertheless. For example, dichlorvos was not detected in the environmental samples; its recoveries in the LMS and LRS were 45.1 and 24.6 percent, respectively.

- Biased High—Environmental sample data were qualified as possibly biased high (B-H) for those compounds with an LMS recovery greater than 150 percent. Cyfluthrin is an example for which the determined concentration from the LMS (190 percent recovery) was nearly twice the fortification concentration and substantially greater than from the LRS (82 percent).

- Unable to Determine—D-U, for unable to determine, was assigned to those compounds for which the analytical method did not determine the LMS or sample concentration successfully because of matrix interference or other method limitations. Phosmet oxon is an example of a compound that was not detected (0 percent recovery) in the LRS and for which there was interference in the LMS, and a recovery could not be determined. Results for phosmet oxon are not presented in appendix 9.

No compounds were detected in the laboratory reagent blank sample. This could be a consequence of the substantially low recoveries for some compounds in the spiked clean, dry sand (the LRS). For the majority of compounds, the LRS recoveries were substantially lower than those from the soil-box mix 2 LMS (table 11). These low LRS recoveries are believed to be due to strong adsorption to active sites on the dry sand on the basis of separate method testing (unpublished results). This was especially the case for chlorpyrifos oxon, diazoxon, iprodione, phorate, phorate oxon, and prometon. Although there could be active sites on the soil box mix 2 matrix, their numbers could be substantially less than in the dry sand as a result of residual moisture in this environmental-sample matrix. A small amount of moisture can dramatically alter the adsorptive capacity for organic compounds because water strongly binds to active sorptive sites on the mineral surface, thereby deactivating those sites from strong adsorption of the organic compounds (Chiou, 2002). In addition, the recoveries for the two surrogates, α-HCH-d6 and diazion-d10, were lower in the clean, dry sand compared to the soil-mix 2 matrix, indicating overall lower recoveries in the dry sand because of more procedural losses relative to the soil-mix 2 matrix (table 11). The method applied to this study was subsequently modified to improve recovery performance by prewetting the reagent sand matrix with a small amount of PBW prior to spiking. This change resulted in acceptable recoveries for more method compounds (Mast and others, 2006).

Field Quality-Control Methods

The field quality-control methods for the project included the collection and processing of equipment blanks, field blanks, and split replicates. There was no QC performed on surficial-soil samples collected from the soil boxes. It was impractical to collect field blanks or apply field spikes to soils because of the inherent heterogeneity of the soil matrix used in the soil boxes and the lack of appropriate "blank" matrix

Table 11. Ambient concentrations, spike concentrations, calculated recoveries for the laboratory matrix spike and laboratory reagent spike, and the relative percentage difference between recoveries of the soil-box surficial-soil samples.

[Laboratory reagent spikes were extracted along with the environmental surficial soil samples. Abbreviations: B-H, biased high; B-L, biased low; D-U, compound deleted due to matrix interference; na, none applied; (E), estimated compound due to poor performance by applied method (Childress and others, 1999; Sandstrom and others, 2001); LMS, Laboratory Matrix Spike; LRS, Laboratory Reagent Spike; RPD, relative percent difference; U-R, unable to determine recovery because the concentration in the unspiked sample is comparable to or higher than the fortification concentration; μg/kg, microgram per kilogram; –, compound not detected]

Compound (common name)	Average ambient concentration of soil mix 2 and duplicate, (μg/kg)	LMS concentration recovered, (μg/kg)	LMS, percentage recovery	LRS, clean dry sand, percentage recovery	RPD between LRS and LMS recoveries, less than 25 percent	Environmental detection qualifer code
1-Naphthol (E)	–	1.02	19.4	16.7	14.9	B-L
2,6-Diethylaniline (E)	–	0.41	7.9	7.6	3.5	B-L
2-Chloro-2,6-diethylacetanilide	–	7.05	134.0	80.4	–	na
2-Ethyl-6-methylaniline (E)	–	0.35	6.7	0	–	B-L
3,4-Dichloroaniline (E)	5.08	5.21	U-R	12.6	–	na
4-Chloro-2-methylphenol (E)	–	3.58	68.1	0	–	B-L
Acetochlor	–	7.19	136.7	92.3	–	na
Alachlor	–	6.08	115.6	81.8	–	na
Atrazine	–	5.08	96.6	102.5	6.0	na
Azinphos-methyl (E)	–	9.23	175.5	54.4	–	B-H
Azinphos-methyl-oxon (E)	–	0.00	0	63.4	–	B-L
Benfluralin	–	7.44	141.4	78.3	–	na
Carbaryl (E)	2.03	10.1	151.9	113.4	–	B-H
Chlorpyrifos	0.66	6.00	101.1	82.2	20.6	na
Chlorpyrifos oxon (E)	–	5.08	96.6	0	–	B-L
Cyfluthrin (E)	–	9.98	189.7	82.4	–	B-H
Cypermethrin (E)	–	8.78	166.9	100.4	–	B-H
Dacthal	0.56	3.72	59.5	85.3	–	na
Deethylatrazine (CIAT)	–	7.59	144.3	89.6	–	na
Diazinon	–	5.85	111.2	74	–	na
Diazoxon (E)	–	5.50	104.6	0	–	B-L
Dichlorvos (E)	–	2.37	45.1	24.6	–	B-L
Dicrotophos (E)	–	0.00	0.0	77.2	–	B-L
Dieldrin	3.52	6.08	48.6	92.8	–	na
Dimethoate	27.4	23.6	U-R	95.7	–	na
Ethion	–	6.49	123.4	101.3	19.7	na
Ethion monoxon	–	10.1	192.0	98.2	–	B-H
Fenamiphos (E)	–	5.19	98.7	21	–	B-L
Fenamiphos sulfone (E)	–	5.49	104.4	99.6	4.7	na
Fenamiphos sulfoxide (E)	–	2.89	54.9	79.3	–	na
Fipronil	–	10.2	193.9	102.8	–	B-H
Fipronil sulfide	–	9.99	189.9	118.3	–	B-H
Fipronil sulfone	1.16	4.38	60.5	116.2	–	na
Desulfinylfipronil	0.59	6.70	115.6	89.2	–	na
Desulfinylfipronil amide (E)	–	9.98	189.7	104.6	–	B-H
Fonofos	–	4.02	76.4	61.7	21.3	na
Fonofos oxon	–	7.11	135.2	69.5	–	na
Hexazinone	–	4.82	91.6	103.3	12.0	na
Iprodione	8.19	12.5	69.2	2	–	B-L
Isofenphos	–	10.1	192.0	106	–	B-H
Malathion	24.8	28.7	68.4	111.1	–	na
Malaoxon (E)	–	9.22	175.3	55.3	–	B-H

Table 11.   Ambient concentrations, spike concentrations, calculated recoveries for the laboratory matrix spike and laboratory reagent spike, and the relative percentage difference between recoveries of the soil-box surficial-soil samples.—Continued

[Laboratory reagent spikes were extracted along with the environmental surficial soil samples. Abbreviations: B-H, biased high; B-L, biased low; D-U, compound deleted due to matrix interference; na, none applied; (E), estimated compound due to poor performance by applied method (Childress and others, 1999; Sandstrom and others, 2001); LMS, Laboratory Matrix Spike; LRS, Laboratory Reagent Spike; RPD, relative percent difference; U-R, unable to determine recovery because the concentration in the unspiked sample is comparable to or higher than the fortification concentration; μg/kg, microgram per kilogram; –, compound not detected]

Compound (common name)	Average ambient concentration of soil mix 2 and duplicate, (μg/kg)	LMS concentration recovered, (μg/kg)	LMS, percentage recovery	LRS, clean dry sand, percentage recovery	RPD between LRS and LMS recoveries, less than 25 percent	Environmental detection qualifer code
Metalaxyl	–	5.05	96.0	83.6	13.8	na
Methidathion	–	8.02	152.5	101	–	B-H
Metolachlor	128	140	U-R	100.3	–	na
Metribuzin	23.7	25.8	49.4	121.9	–	na
Myclobutanil (E)	14.0	19.6	102.7	93.2	9.7	na
Parathion-methyl	–	7.80	148.3	101	–	na
Paraoxon-methyl	–	5.83	110.8	112.7	1.7	na
Pendimethalin	52.3	45.4	U-R	91.2	–	na
cis-Permethrin	–	9.48	180.2	117.5	–	B-H
trans-Permethrin	–	9.51	180.8	105.3	–	B-H
Phorate (E)	–	4.96	94.3	16.9	–	B-L
Phorate oxon (E)	–	4.68	89.0	12.8	–	B-L
Phosmet	–	5.35	101.7	0	–	B-L
Phosmet oxon	–	36.0	D-U	0	–	D-U
Prometon (E)	1.47	6.35	93.3	47	–	na
Prometryn	–	4.51	85.7	67.9	23.2	na
Pronamide	–	6.48	123.2	72.5	–	na
Simazine	70.8	D-U	U-R	79.5	–	na
Tebuthiuron (E)	–	7.87	149.6	75.4	–	na
Terbufos	–	6.62	125.8	0	–	B-L
Terbufos-oxon-sulfone (E)	–	12.3	233.8	50.7	–	B-H
Terbuthylazine	–	5.03	95.6	73.1	–	na
Trifluralin	32.15	26.1	U-R	63.4	–	na
alpha-HCH-d6	100.6	97.8	97.8	55.2	–	na
Diazinon-d10	137.3	116.0	116.0	67.3	–	na
Sample amount, grams	43.6	38.0	38.0	50	Not applicable	na

material. Instead, two types of QC samples were collected and used for comparison with environmental samples, which are discussed separately. All QC samples collected and reported in this study were analyzed at the NWQL and are presented by study period.

2001 Study Period

One equipment blank and two field blanks were processed during the 2001 study period with no detections on the ISCO (Teledyne ISCO Company, Lincoln, Nebraska) automated sampler and collection bottles. The equipment blank was processed prior to field deployment, and two field blanks were collected during storm-drain runoff sampling (appendix 2).

2002–04 Study Period

Equipment and Field Blanks

Equipment blanks were run on the sample collection equipment prior to deployment in the field to confirm the suitability of equipment to provide samples within the data-quality objectives of the project. Field blanks were used to evaluate bias resulting from contamination of environmental samples by the compounds of interest during sample collection, processing, shipping, or analysis. There were no compounds detected in the 24 equipment blanks and 19 field blanks processed and collected during the 2002–04 study period (table 12).

NWQL PBW was used as the blank solution to assess the degree of potential contamination introduced during field processing and handling, as well as laboratory handling of samples and equipment. The field-blank procedure for the funnel sampler was to place the funnel on top of the sample-collection bottle in the funnel-bottle assembly, following the standard cleaning procedure, and then to pour approximately 1-L of blank solution into the funnel and collection bottle. The collected blank solution was then poured into a 1-L amber glass bottle and sent to NWQL with the environmental samples. Field equipment blanks on the autosampler collection pots were done in the same manner, following the standard cleaning procedure, except that the PBW was poured directly into the collection pot, swirled around, and then poured into a 1-L amber glass bottle.

The equipment blanks were processed in the same manner but at the CAWSC Laboratory. Each site had a dedicated funnel and 4-L sample collection bottle. The equipment blanks were completed prior to the initial deployment at a site and periodically throughout the study period to reduce the possibility of contamination of the environmental samples by the equipment as the funnel and collection bottle were reused throughout the study period. Individual equipment blanks were processed and submitted for the funnel and the 4-L sample-collection bottle (two equipment blanks total per site).

Table 12. Environmental and field quality-control sample types and totals collected during the 2001 and 2002–04 study periods in the Central Valley of California.

[Abbreviations: –, no field quality control collected]

2001 Study period				
Environmental sample type	Environmental sample total	Blanks		Split replicates
		Field	Equipment	
Storm-drain samples	10	2	1	0
Wet deposition	8	0	0	0
2002–04 Study period				
Environmental sample type	Environmental sample total	Blanks		Split replicates
		Field	Equipment	
Wet deposition	184	12	16	19
Bulk wet deposition	38	–[1]	–[1]	–[1]
Dry deposition	37	5	6	0
Bulk dry deposition	26	–[2]	–[2]	–[2]
Soil-box run off samples	30	2	2	1
Suspended sediment	30	0	0	0
Surficial-soil samples	8	0	0	0

[1]No field quality control collected for bulk wet-deposition sample because it was collected on the intended sample type: wet deposition.

[2]No field quality control collected for bulk dry-deposition sample because it was collected on the intended sample type: dry deposition.

Split Replicates

Replicate samples were used to identify and quantify the variability inherent to sampling and analysis. Replicate samples were split from the environmental samples at various sites to characterize the reproducibility of the complete sample processing and analytical process. The sample split from the environmental sample was considered the QC sample or the split replicate. Both samples (environmental and split replicate) were defined as the replicate set. RPD was used to describe the variability within the replicate set. In at least one sample of the replicate sets, 28 compounds were detected (table 13). With the exception of metalaxyl and dichlorovos, the median RPD ranged from 0 to 17.7 percent. The higher median percentage values for metalaxyl (72.8 percent) and dichlorovos (66.7 percent) were a result of a non-detection in the replicate set. If one value in a replicate pair was reported as a non-detection and the other value was greater than the laboratory reporting limit, then the non-detection value was set to one-half of the LT-MDL, and an RPD was calculated.

QA for Soil-Box Samples

There were no QC samples collected from the soil boxes. It was impractical to collect field blanks or do field spikes because of the heterogeneity of the soil used for the soil boxes. Samples of native soil were analyzed to determine background levels of compounds of interest in the native soil. Aliquots of 30–40 grams from each mix (mix 1, mix 2, mix 3, and mix 4) were analyzed prior to exposure and compared

Table 13. Minimum, maximum, and median relative percentage difference values for compounds detected in at least one sample of the replicate set.

[If one value in a sample pair was reported as a non-detection and the other value was greater than the laboratory reporting limit, then the non-detection value was set equal to one-half of the laboratory reporting limit and the relative percent difference (RPD) calculated.]

Compound detected in split replicate sets	Total number of replicate sets with detections	Minimum RPD, in percent	Maximum RPD, in percent	Median RPD, in percent
1-Naphthol	12	0.0	85.7	0.0
3,4-Dichloroaniline	7	0.0	120.0	6.9
4-Chloro-2-methylphenol	6	0.0	63.6	0.0
Azinphos-methyl	5	0.0	13.3	8.4
Carbaryl	14	0.0	127.0	10.8
Chlorpyrifos	18	0.0	104.0	2.6
Chlorpyrifos oxon	9	0.0	40.0	0.0
Dacthal	20	0.0	35.3	2.4
Diazinon	20	0.0	52.6	5.2
Diazoxon	12	0.0	40.0	0.0
Dichlorvos	3	0.0	75.0	66.7
Hexazinone	6	0.0	28.9	5.4
Iprodione	12	0.0	125.3	13.2
Malathion	10	0.0	37.8	7.7
Malaoxon	6	0.0	57.1	6.1
Metalaxyl	2	0.0	145.7	72.8
Methidathion	9	0.0	131.8	6.3
Metolachlor	12	0.0	150.0	5.0
Myclobutanil	12	0.0	113.7	10.2
Parathion-methyl	5	0.0	5.0	3.4
Paraoxon-methyl	3	0.0	0.0	0.0
Pendimethalin	16	0.0	152.0	9.8
Phosmet	5	10.5	81.0	13.3
Phosmet oxon	2	0.0	0.0	0.0
Prometryn	6	0.0	61.1	0.0
Pronamide	5	0.0	22.2	0.0
Simazine	15	0.0	148.2	8.7
Trifluralin	12	0.0	107.7	17.7

to environmental samples collected during the study period. In addition, a separate soil box was assembled to serve as a source of "control samples" (fig. 4, site 14). The control soil box was set up outdoors on a building rooftop to examine the transformation or degradation of compounds with no exposure to sunlight or atmospheric deposition. This soil box contained a composite mixture of the soils collected in the San Joaquin Valley (mix 1, mix 2, mix 3), and one side was covered with a piece of plywood. The covered side had a 0.5 inch gap to allow air flow, but minimize atmospheric deposition (both wet and dry). Three times during the study, the cover was removed, and surficial samples were collected for laboratory analysis. The results of these analyses are presented in appendix 9 with the environmental data for comparison.

Concentrations of detections in the control were nearly identical to the background concentrations, indicating there was no photolysis or hydrolysis over the 2-year period. The concentrations of some compounds, however, decreased over time, possibly as a result of volatilization or biodegradation. Simazine is an example of a compound for which the initial concentration collected from the control was the greatest, and concentrations decreased in subsequent samples. The half-life of simazine in aerobic soil conditions is approximately 90 days (*http://sitem.herts.ac.uk/aeru/footprint/en/index.htm*, accessed April 5, 2011).

Results of Monitoring

2001 Study Period (Rainfall and Storm-Drain Runoff)

A single storm event on January 25–26, 2001, was sampled, with rainfall collected at four urban and four agricultural sites in the Modesto, Calif., region (fig. 4, sites 1–8), while simultaneously hourly storm-drain runoff samples were collected from a Modesto storm drain (site 9). The average rainfall at the urban and agricultural sites was about 0.52 inches. Appendix 2 presents the measured concentrations of compounds with at least one detection during the storm sampling event for the urban sites, agricultural sites, and urban storm drain. Detected pesticide concentrations in rainfall at the urban sites were fairly uniform, with the exception of site 1, which was located immediately downwind of a recent orchard pesticide application of diazinon that probably influenced the results. Registered agricultural diazinon application data from the California Department of Pesticide Regulation (CDPR) indicated approximately 100–200 kilograms of active ingredient was applied within 3 kilometers (km) of site 1 (Majewski and others, 2005; *http://pubs.usgs.gov/of/2005/1307/*, accessed April 19, 2011). Detected pesticide concentrations at the agricultural sites were variable, but correlated with proximity to orchards and recent pesticide applications.

Chlorpyrifos, dacthal, diazinon, pendimethalin, and simazine were detected in all rainfall samples collected from urban and agricultural sites. Median concentrations of chlorpyrifos and diazinon in agricultural and urban rainfall were similar, indicating that the overall atmospheric level of these pesticides in the region was fairly evenly distributed during the sampling event (fig. 9). The median concentrations of chlorpyrifos, diazinon, and simazine collected from the agricultural sites (0.079 µg/L, 0.583 µg/L, and 0.051 µg/L, respectively) were slightly greater than the median concentrations collected from the urban sites (0.067 µg/L, 0.515 µg/L, and 0.031 µg/L, respectively; table 14). Metolachlor and napropamide had higher detection frequencies in samples collected from the

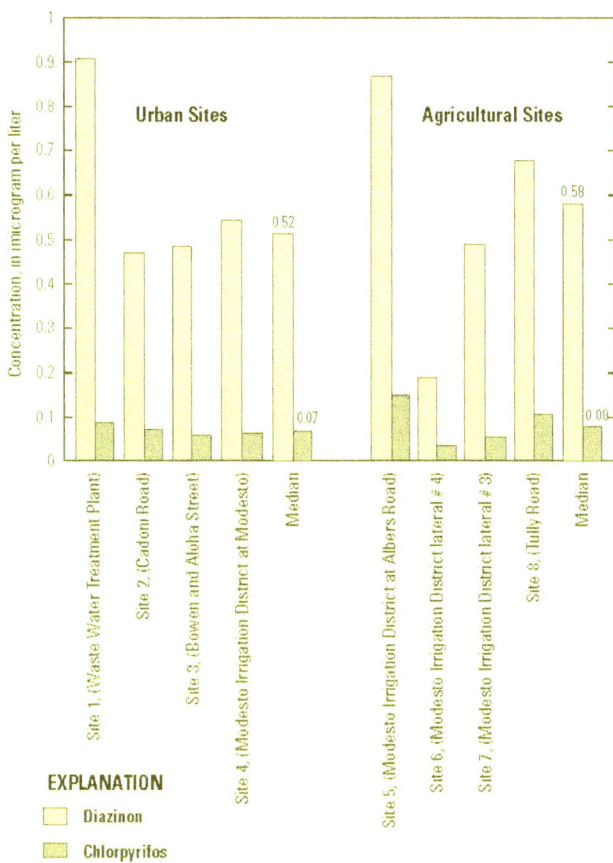

Figure 9. Concentrations of diazinon and chlorpyrifos in rainfall measured at the four urban and four agricultural sites during the January 25–26, 2001, storm event in the area of Modesto, California.

Table 14. Median concentrations of compounds detected in samples collected from the urban and agricultural rainfall collection sites in Modesto, California, during the January 25–26, 2001, sampling event.

[Abbreviations: E, estimated reported laboratory value; μg/L, micrograms per liter; <, less than]

Compound	Median concentrations for storm sampling event, μg/L	
	Urban sites	Agricultural sites
Carbaryl	<0.1	<0.09
Chlorpyrifos	0.067	0.079
Dacthal	0.01	0.009
Diazinon	0.515	0.583
Metolachlor	E0.004	<0.013
Pendimethalin	0.122	0.170
Napropamide	0.016	[1]0.032
Simazine	0.031	0.051
Trifluralin	[1]0.036	0.008

[1]Concentration represents detections in 50 percent of samples.

urban sites than those collected from the agricultural sites (100 percent compared to 50 percent, appendix 2). Pesticides applied in agricultural areas could have drifted into the urban environment, but it is unknown how many, if any, urban applications took place during this study that contributed to the observed urban concentrations.

Many of the same pesticides detected in rainfall samples collected in the McHenry urban watershed (sites 3 and 4) were detected in the storm-drain runoff samples from the urban watershed also (appendix 2). The median concentrations of these compounds detected in both rainfall and storm drain runoff samples are presented in table 15. Comparison of median concentrations in the rainfall to those in the McHenry storm drain runoff indicates that rainfall contributed to a substantial percentage of the concentration found in the runoff for some compounds, whereas for other compounds, the concentrations in rainfall were much greater than in the runoff. The median diazinon concentration in rain was 69 percent of the median in runoff, whereas the median chlorpyrifos concentration in rain was 194 percent of the median in runoff. Other water-soluble pesticides—carbaryl, metolachlor, napropamide, and simazine—were similar to diazinon, with rainfall containing less than 100 percent of the median concentration in runoff. Malathion was detected in the storm-drain runoff water, but not in any rainfall sample. The results are explained by the physical and chemical properties of these pesticides. Table 15 includes the water solubility (S_L), subcooled liquid vapor-phase pressure (P_L), and soil organic-carbon partition coefficient (log K_{oc}) for the compounds detected (greater than 50 percent detection frequency) in both the rainfall and storm drain runoff samples (Worthing and Walker, 1987; Wauchope and others, 1992; Mackay and others, 1997; Suntio and others, 1998; U.S. Department of Agriculture, 2005; U.S. Environmental Protection Agency, 2005). Diazinon is much more water soluble than chlorpyrifos, whereas chlorpyrifos has a higher organic-carbon partition coefficient. Since the runoff samples were filtered, and the filters were not analyzed, it is likely that most of the chlorpyrifos had been sorbed onto the suspended particulate matter and was removed by filtration, whereas diazinon stayed in solution (fig. 10).

2002–04 Study Period

The wet-deposition, bulk wet-deposition, dry-deposition, and bulk-dry-deposition samples collected during the 2002–04 study period were analyzed and presented as flux in micrograms per square meter per day ($\mu g/m^2/day$) in appendix 3 through 6, respectively. These sample types are presented as a flux for this study period to further examine the areal distribution of atmospheric loading rates (mass/area/time) of pesticides to the Central Valley, Calif. Compounds with non-detections were not calculated and are not presented.

Table 15. Comparison of median concentrations and the physical and chemical properties of detected pesticides in rainfall (sites 3 and 4) and storm-drain runoff (site 9) during the January 25–26, 2001, sampling event in Modesto, California.

[Abbreviations: E, estimated reported laboratory value; Pa, Pascals; mg/L, milligrams per liter; μg/L, micrograms per liter]

Compound	Median rain concentration (sites 3 and 4), (μg/L)	Median storm-drain runoff concentration (site 9), (μg/L)	Water solubility, S_L (mg/L)	[9]Subcooled liquid vapor pressure, P_L (Pa)	Soil-organic carbon partition coefficient, log K_{OC}
Carbaryl	[1]<0.1	E0.18	[2]120	0.003	[8]2.36
Chlorypyrifos	0.06	0.031	[3]0.73	0.002	[2]3.78
Dacthal	0.011	0.009	[6]0.5	0.007	[6]3.75
Diazinon	0.515	0.742	[2]60	0.008	[8]2.76
Metolachlor	E0.004	E0.005	[4]530	0.002	[2]2.30
Napropamide	0.011	0.019	[5]74	0.002	[5]3.74
Pendimethalin	0.122	0.061	[6]0.275	0.008	[6]4.13
Simazine	0.026	0.053	[4]5	0.001	[2]2.11
Trifluralin	[1]0.025	0.006	[7]0.05–40	0.010	[8]4.14

[1]Concentration represents detections in 50 percent of samples.

[2]Wauchope and others, 1992 (referenced from compilation in Mackay and others, 1997).

[3]Bowman and Sans, 1983 (referenced from compilation in Mackay and others, 1997).

[4]Worthing and Walker, 1987 (referenced from compilation in Mackay and others, 1997).

[5]U.S. Environmental Protection Agency, 2005.

[6]U.S. Department of Agriculture, 2005.

[7]According to Mackay and others, 1997, "the reported values for this quantity vary considerably."

[8]Howard and others, 1991 (referenced from compilation in Mackay and others, 1997).

[9]Calculated using δ water solubility values of 56 joules per mole Kelvin from Suntio and others, 1988.

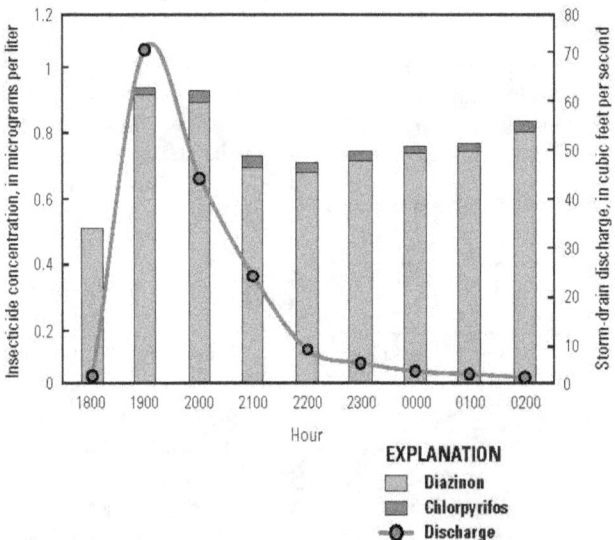

Figure 10. Measured discharge and concentration of insecticides diazinon and chlorpyrifos detected in the hourly storm-drain runoff samples collected during the January 25–26, 2001, storm event in Modesto, California.

Wet-Deposition Sample Type

A total of 184 wet-deposition samples were analyzed for 41 currently used pesticides and 23 pesticide degradates, including oxygen analogs (oxons) of 9 organophosphate insecticides. Fifty-three compounds were detected, and detection frequencies ranged from of 0.5 percent (representing one detection) for four compounds to 97.8 percent for dacthal. The most frequently detected compounds in both the San Joaquin and Sacramento Valleys were dacthal, diazinon, chlorpyrifos, simazine, and pendimethalin (table 16). The median wet-deposition flux for these compounds was greater in samples collected from the San Joaquin Valley (between 1.2 and 4.0 times more), than samples collected from the Sacramento Valley (fig. 11 and table 17).

In the San Joaquin Valley, the compounds with the greatest wet-deposition rates were chlorypyrifos (10.1 µg/m²/day, site 5), iprodione (77.6 µg/m²/day, site 10), simazine (48.9 µg/m²/day, site 13), and trifluralin (29.0 µg/m²/day, site 11; appendix 3). The greatest wet-deposition rates in the Sacramento Valley were for diazinon, at 50.6 µg/m²/day and 45.6 µg/m²/day, which were collected from the funnel and autosampler, respectively, at urban site 16 following the February 7–14, 2003, storm event (appendix 3). Diazinon and dacthal were detected in every sample collected from this site (February 7, 2003, to February 20, 2004), with median deposition amounts of 0.225 and 0.033 µg/m²/day, respectively. At the Sierran-Nevada foothills site 15, located downwind from the urban site 16, dacthal also was detected in every sample and had a slightly greater median deposition amount at 0.044 µg/m²/day.

Wet-deposition samples collected in the San Joaquin Valley following the January 9–14, 2003, storm event resulted in similar rates of diazinon wet-deposition at urban site 4, located in downtown Modesto, and the agricultural site 5, located in rural Modesto adjacent to agricultural land (appendix 3). At the downtown Modesto site, 12.0 µg/m²/day and 12.7 µg/m²/day were collected from the funnel and autosampler, respectively. At the agricultural site, 14.4 µg/m²/day and 11.7 µg/m²/day were collected from the funnel and autosampler, respectively. These samples were collected during the orchard dormant-spraying season and indicate that atmospheric levels and deposition in the region could have been fairly constant for the compound throughout the sampling event.

Measured concentrations of chlorpyrifos, dacthal, and diazinon collected from the autosampler and the funnel sampler during the same sampling events were compared to examine differences in concentrations between the samplers (fig. 12). Measured concentrations collected from the funnel were expected to be slightly greater than from the autosampler because the funnel was exposed to the atmosphere between rain events for a sampling period equal to or less than 5 days, whereas the autosampler lid closed following a rain event, and

therefore, the sample was not exposed to potential dry-deposition. For the San Joaquin Valley sites, concentrations for the samplers at site 5 were similar (fig. 12B); however, at site 4, concentrations of chlorpyrifos and dacthal collected from the funnel sampler were greater than from the autosampler (fig. 12A). Higher concentrations were also collected from the funnel sampler than the autosampler in the Sacramento Valley from site 16 for chlorpyrifos and dacthal (fig. 12C). The land use surrounding site 16 is predominately orchards, corn, alfalfa, and vegetables (table 3). It is likely that, because these samples were predominately collected during the orchard dormant-spraying season, dry deposition contributed to the higher concentrations measured in samples collected from the funnel sampler compared to the autosampler. In addition, dacthal and chlorpyrifos are less water soluble and have greater log K_{oc} values compared to diazinon (table 15).

Bulk Wet-Deposition Sample Type

A total of 38 bulk-wet-deposition samples were collected during the 2002–04 study period (appendix 4). Table 18 compares wet and dry deposition amounts in mass per area (µg/m²) for compounds detected in both sample types collected from the Modesto Irrigation District rooftop autosampler (site 4) as discrete samples during the April 9–May 21, 2002, composite sample period. For some compounds (chlorpyrifos, simazine, and myclobutanil), the dry-deposition amount is between 3.4 and 6.8 times greater than deposition amounts measured in the wet-deposition sample type, whereas for the compound phosmet, the dry deposition was substantially greater (17 times greater). These differences could be a result of pesticide use during the sample period and the predominant deposition phase of the detected compounds.

The predominant deposition phase is related to the physical and chemical properties of a pesticide (table 18). Henry's Law describes partitioning properties with respect to a compound's potential to partition into the gas phase or the aqueous phase on the basis of vapor pressure. For compounds with low Henry's Law (H) constants and high water solubility, such as diazinon, rainfall was a more significant source of depositional loading to the ground than dry deposition. For compounds with high H constants and low water solubility, such as chlorpyrifos, dry deposition was more important.

Thirty-two compounds were detected in the samples collected as bulk-wet deposition. Detection frequencies for the bulk-wet-deposition samples ranged from one detection (2.6 percent) for eight compounds to 100 percent for dacthal (table 19). The most detected compounds in the San Joaquin Valley and Sacramento Valley were dacthal, diazinon, chlorpyrifos, and simazine (fig. 13). The median deposition amounts for the compounds dacthal and chlorpyrifos were comparable for the San Joaquin Valley and Sacramento Valley (table 17), but the median diazinon deposition amount was

Table 16.   Total number of samples analyzed and detection frequency for compounds analyzed in wet deposition and that had at least one detection during the 2002–04 study period in the Central Valley of California.

[Concentrations can be calculated from additional information listed in *appendix 3*]

Compound	Total number of samples analyzed	Detection frequency, in percent
Dacthal	184	97.8
Diazinon	184	97.3
Chlorpyrifos	184	92.4
Simazine	184	91.8
Pendimethalin	184	85.9
Trifluralin	184	64.7
Carbaryl	184	56.5
Metolachlor	184	56.5
Iprodione	184	49.5
Malathion	184	48.4
Myclobutanil	184	47.8
Chlorpyrifos oxon	184	46.2
1,4-Naphthoquinone	184	43.2
Carbofuran	37	43.2
cis-Propiconazole	37	43.2
trans-Propiconazole	37	43.2
Methidathion	37	37.5
Diazoxon	184	35.9
Oxyfluorfen	37	35.1
Prometryn	184	33.7
EPTC	38	29.7
Napropamide	38	29.7
3,4-Dichloroaniline	183	27.2
Malaoxon	183	27.2
3,5-Dichloroaniline	37	27.0
1-Naphthol	184	22.3
Propyzamide	184	17.9
Azinphos-methyl	184	16.8
Dichlorvos	184	16.3
Parathion-methyl	184	13.0
Phosmet	184	12.5
4-Chloro-2-methylphenol	184	10.9
2-(4-tert-Butylphenoxy)-cyclohexanol	37	8.1
p,p'-DDE	37	8.1
Metribuzin	184	5.4
4,4'-Dichlorobenzophenone	37	5.4
Cycloate	37	5.4
lambda-Cyhalothrin	37	5.4
cis-Permethrin	184	3.8
Alachlor	184	3.3
Dimethoate	184	3.3
Cyanazine	37	2.7
Molinate	37	2.7
Metalaxyl	184	2.2
Paraoxon-methyl	184	2.2
Atrazine	184	1.6
Phosmet oxon	184	1.2
Cyfluthrin	184	1.1
Terbufos-oxon-sulfone	184	1.1
Benfluralin	184	0.5
Cypermethrin	184	0.5
Fonofos oxon	184	0.5
Fonofos	184	0.5

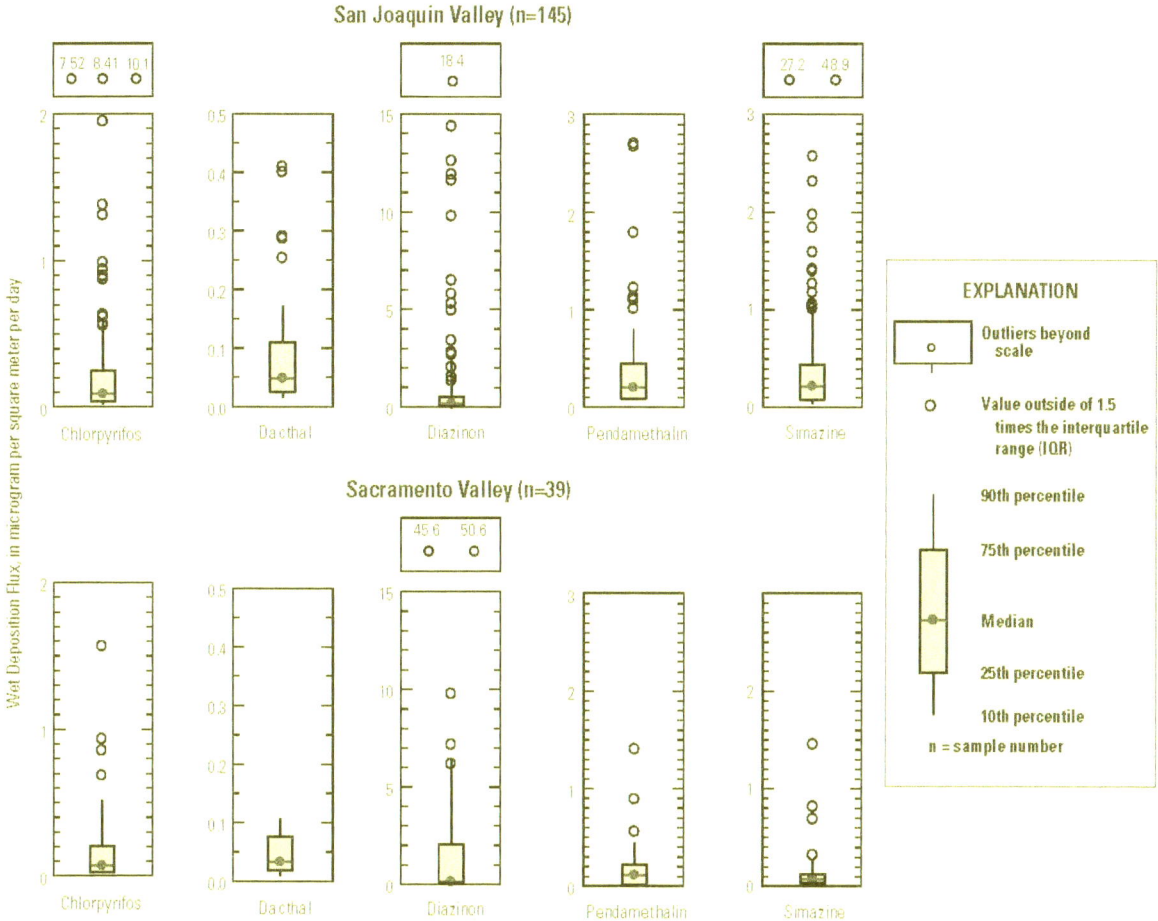

Figure 11. Distribution of calculated wet-deposition fluxes for the five most frequently detected compounds collected from the San Joaquin Valley and Sacramento Valley, California, sites during the 2002–04 study period.

Table 17. Detection frequency, and percentiles for the most frequently detected compounds collected as wet, bulk wet, dry, bulk dry deposition and soil box runoff collected in the San Joaquin Valley and Sacramento Valley, California, during the 2002–04 study period.

[For computation of percentiles, all non-detections were assumed to be less than the lowest detection. Abbreviations: E, estimated compound; NA, not applicable as there were not enough detections to calculate percentiles; μg/L, microgram per liter; μg/m²/day, microgram per square meter per day]

Compound	Detection frequency (in percent)	Maximum	90th percentile	75th percentile	50th percentile	25th percentile	10th percentile
Wet deposition, μg/m²/day							
San Joaquin Valley							
Chlorpyrifos	95.2	10.1	0.565	0.237	0.079	0.034	0.021
Dacthal	97.2	0.409	0.170	0.110	0.048	0.025	0.015
Diazinon	97.2	18.4	2.51	0.613	0.216	0.089	0.048
Pendimethalin	89.0	2.70	0.793	0.437	0.187	0.079	Non-detections
Simazine	95.9	48.9	1.04	0.449	0.219	0.079	0.045
Sacramento Valley							
Chlorpyrifos	82.1	1.56	0.508	0.200	0.066	0.023	Non-detections
Dacthal	100.0	0.149	0.109	0.078	0.035	0.021	0.013
Diazinon	97.4	50.6	6.39	2.06	0.134	0.051	0.034
Pendimethalin	74.4	1.40	0.433	0.212	0.103	0.013	Non-detections
Simazine	76.9	1.43	0.248	0.106	0.055	0.013	Non-detections
Bulk wet deposition, μg/m²/day							
San Joaquin Valley							
Chlorpyrifos	80.0	0.948	0.168	0.100	0.032	0.014	Non-detections
Dacthal	100.0	0.187	0.082	0.045	0.032	0.020	0.016
Diazinon	96.0	2.48	0.70	0.365	0.130	0.630	0.042
Simazine	100.0	7.46	1.16	0.724	0.295	0.145	0.083
Sacramento Valley							
Chlorpyrifos	76.9	3.73	0.12	0.058	0.025	0.002	Non-detections
Dacthal	100.0	0.207	0.091	0.058	0.042	0.020	0.012
Diazinon	76.9	8.63	0.308	0.179	0.064	0.003	Non-detections
Simazine	38.0	0.278	NA	NA	NA	NA	NA
¹Dry deposition, μg/m²/day							
San Joaquin Valley							
Carbaryl	64.7	E0.186	E0.036	E0.019	E0.004	Non-detections	Non-detections
Chlorpyrifos	73.5	E0.233	E0.052	E0.018	E0.004	E0.001	Non-detections
Dacthal	91.2	E0.009	E0.005	E0.002	E0.001	E0.001	Non-detections
Metolachlor	76.5	E0.153	E0.038	E0.011	E0.002	E0.001	Non-detections
Phosmet	61.8	E0.188	E0.056	E0.019	E0.009	Non-detections	Non-detections
¹Bulk dry deposition, μg/m²/day							
San Joaquin Valley							
Chlorpyrifos	95.0	E0.268	E0.220	E0.143	E0.011	E0.005	E0.004
Dacthal	95.0	E0.012	E0.009	E0.008	E0.006	E0.005	E0.003
Diazinon	75.0	E1.81	E1.56	E0.690	E0.008	E0.003	Non-detections
1-Naphthol	80.0	E0.011	E0.008	E0.004	E0.003	E0.002	Non-detections
Simazine	90.0	E0.099	E0.087	E0.071	E0.047	E0.024	E0.011
San Joaquin Valley soil-box runoff, μg/L							
Aqueous phase							
Dacthal	94.4	0.047	0.024	0.016	0.012	0.009	0.008
Diazinon	88.9	1.62	0.521	0.147	0.070	0.022	Non-detections
Metolachlor	94.4	0.309	0.228	0.140	0.048	0.021	0.010
Myclobutanil	88.9	1.02	0.127	0.073	0.051	0.030	Non-detections
Simazine	94.4	1.01	0.816	0.498	0.248	0.078	0.047
Suspended sediment							
Diazinon	50.0	E0.810	E0.622	E0.490	E0.078	Non-detections	Non-detections
Simazine	64.3	E0.930	E0.813	E0.679	E0.121	Non-detections	Non-detections

Table 17. Detection frequency, and percentiles for the most frequently detected compounds collected as wet, bulk-wet, dry, bulk-dry deposition and soil box runoff collected in the San Joaquin Valley and Sacramento Valley, California, during the 2002–04 study period.—Continued

[For computation of percentiles, all non-detections were assumed to be less than the lowest detection. Abbreviations: E, estimated compound; NA, not applicable as there were not enough detections to calculate percentiles; µg/L, microgram per liter; µg/m²/day, microgram per square meter per day]

Compound	Detection frequency (in percent)	Maximum	90th percentile	75th percentile	50th percentile	25th percentile	10th percentile
			Sacramento Valley soil-box runoff, µg/L				
			Aqueous phase				
Chlorpyrifos	100.0	0.057	0.041	0.026	0.012	0.005	0.004
Dacthal	100.0	0.014	0.010	0.009	0.006	0.004	0.003
Diazinon	100.0	1.75	0.730	0.091	0.046	0.020	0.010
Simazine	91.7	0.141	0.058	0.031	0.025	0.019	0.014
			Suspended sediment				
Chlorpyrifos	66.7	E0.570	E0.563	E0.492	E0.180	Non-detections	Non-detections
Dacthal	66.7	E0.49	E0.260	E0.253	E0.125	Non-detections	Non-detections
Simazine	75.0	E0.96	E0.729	E0.623	E0.250	E0.158	Non-detections

[1] Statistical summaries for the dry and bulk-dry deposition sample types are only presented for the San Joaquin Valley sites as there were too few samples collected in the Sacramento Valley. See *figures 14* and *15* for comparison of detections in the Sacramento Valley to compounds presented above as dry and bulk-dry deposition.

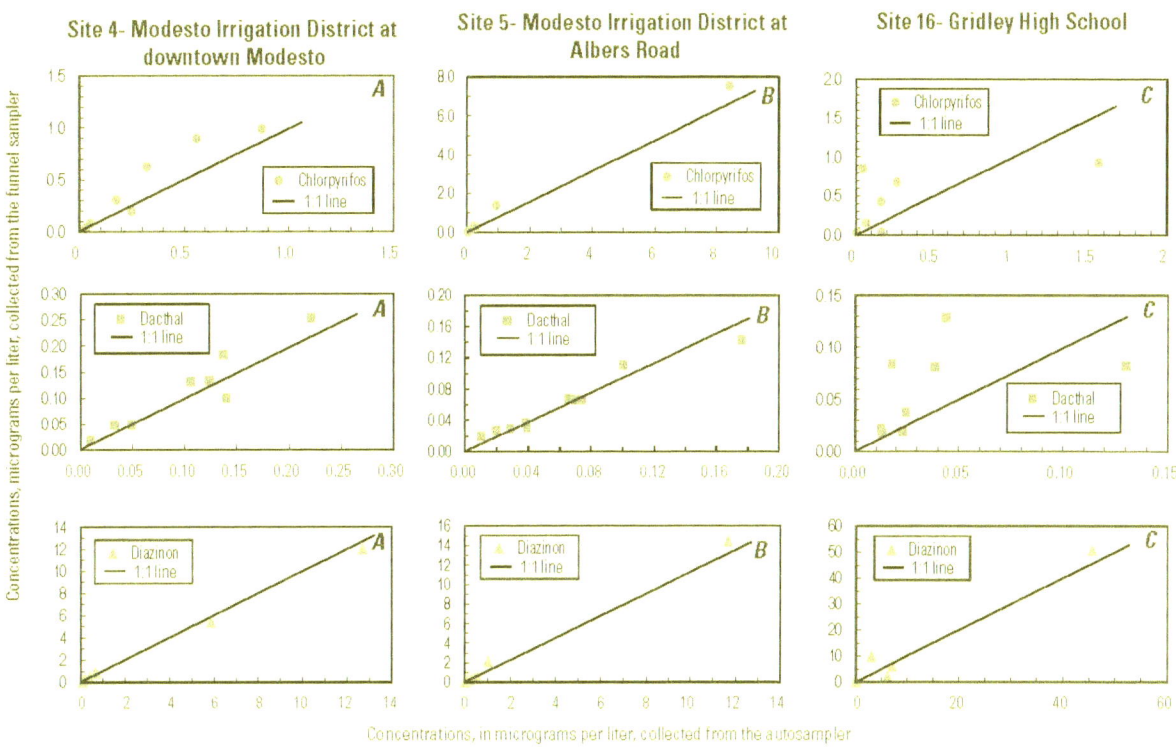

Figure 12. Comparison of concentrations of chlorpyrifos, dacthal, and diazinon collected from the autosampler and funnel sampler for the same sampling events at sites in the Central Valley of California during 2002–04: (*A*) site 4, MID at downtown Modesto; (*B*) site 5, MID at Albers Road; and (*C*) site 16, Gridley High School rooftop.

Table 18.   Comparison of wet and dry-deposition pesticide amounts detected in discrete samples collected from the autosampler at site 4 during the April 9–May 21, 2002, sampling event in the Central Valley of California and the physical and chemical properties of the pesticides.

[Abbreviations: E, estimated reported laboratory value; K_{OC}, soil-organic carbon partition coefficient; m^3/mol, cubic meters per mole; mg/L, milligram per liter; Pa, Pascals; $\mu g/m^2$, micrograms per square meter; –, not detected]

Compound	Wet deposition, ($\mu g/m^2$)	Dry deposition, ($\mu g/m^2$)	Henry's Law constant, H (Pa m^3/mol)	Water solubility, S_L (mg/L)	[10] Subcooled liquid vapor pressure, P_L (Pa)	Soil-organic carbon partition coefficient, log K_{OC}
Azinphos-methyl	E0.858	E1.82	[1]0.0003	[1]30	0.00009	[4]3.00
Carbaryl	E0.456	E0.586	[1]0.0012	[4]120	0.003	[8]2.36
Chlorpyrifos	0.132	E0.456	[1]1.753	[5]0.73	0.002	[4]3.78
Dacthal	0.078	E0.081	[2]0.2213	[7]0.5	0.0003	[7]3.75
Diazinon	_[9]	_[9]	[1]0.0641	[4]60	0.008	[8]2.76
Malathion	0.18	E0.042	[1]0.0022	[1]145	0.0006	[4]3.26
Metolachlor	1.14	E0.378	[3]0.0009	[6]530	0.002	[4]2.30
Myclobutanil	E0.174	E1.18	[1]0.0004	[7]142	0.074	[7]2.69
Pendimethilan	0.168	E0.042	[3]3.751	[7]0.275	0.008	[7]4.13
Phosmet	0.168	E2.86	[1]0.001	[1]25	0.0002	[4]2.91
Simazine	0.264	E1.39	[1]0.0003	[6]5	0.0009	[4]2.11

[1]Suntio and others, 1988.

[2]Wauchope and others, 1992.

[3]Worthing and Walker, 1987.

[4]Wauchope and others, 1992 (referenced from compilation in Mackay and others, 1997).

[5]Bowman and Sans, 1983 (referenced from compilation in Mackay and others, 1997).

[6]Worthing and Walker, 1987 (referenced from compilation in Mackay and others, 1997).

[7]U.S. Department of Agriculture, 2005.

[8]Howard, 1991 (referenced from compilation in Mackay and others, 1997).

[9]Diazinon was not detected in this sample but the physical and chemical properties are included for discussion purposes with respect to *figure 10*.

[10]Calculated using δ water solubility values of 56 joules per mole Kelvin from Suntio and others, 1988.

Table 19. Detection frequency for compounds analyzed as bulk-wet deposition with at least one detection during the 2002–04 study period in the Central Valley of California.

[Abbreviations: n, total number of samples]

Compound	Detection frequency in percent (n=38)	Compound	Detection frequency in percent (n=38)
Dacthal	100.0	Diazoxon	26.3
Diazinon	89.5	Prometryn	26.3
Chlorpyrifos	78.9	Propyzamide	21.1
Simazine	78.9	Azinphos-methyl	18.4
1-Naphthol	65.8	4-Chloro-2-methylphenol	15.8
Carbaryl	65.8	Phosmet	13.2
Metolachlor	65.8	Parathion-methyl	7.9
Pendimethalin	65.8	Metribuzin	7.9
Trifluralin	65.8	2,6-Diethylaniline	2.6
Myclobutanil	63.2	Alachlor	2.6
Iprodione	55.3	Atrazine	2.6
Malathion	42.1	Azinphos-methyl-oxon	2.6
Malaoxon	34.2	Cypermethrin	2.6
Methidathion	34.2	Dichlorvos	2.6
3,4-Dichloroaniline	31.6	Metalaxyl	2.6
Chlorpyrifos oxon	28.9	Paraoxon-methyl	2.6

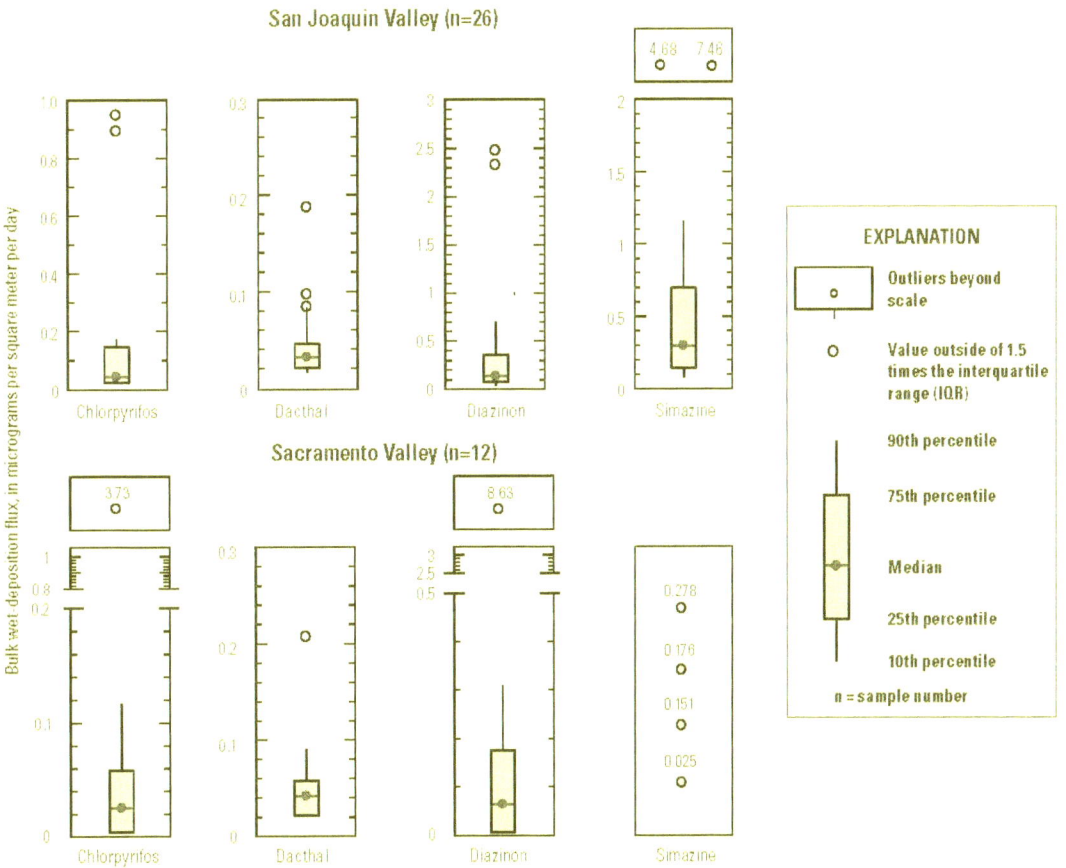

Figure 13. Distribution of calculated bulk wet-deposition fluxes for the most detected compounds collected from the San Joaquin Valley and Sacramento Valley sites during the 2002–04 study period.

twice as much in the San Joaquin Valley. Maximum bulk wet-deposition amounts in the San Joaquin Valley were found for the compounds iprodione (13.1 $\mu g/m^2/day$) and simazine (7.46 $\mu g/m^2/day$) at the Turlock Airport raingage (site 13; appendix 4). Dacthal was detected in all bulk-wet-deposition samples collected from the San Joaquin Valley and the Sacramento Valley sites. In the San Joaquin Valley, the median and maximum deposition amounts were 0.032 $\mu g/m^2/day$ and 0.187 $\mu g/m^2/day$, respectively. In the Sacramento Valley, the median and maximum deposition amounts were 0.042 $\mu g/m^2/day$ and 0.207 $\mu g/m^2/day$, respectively (table 17).

Dry-Deposition Sample Type

For the dry-deposition (and bulk dry) sample types, all detections were reported as estimated (E code preceding value) and have a greater, but unknown, degree of uncertainty. Additionally, the calculated dry-deposition and bulk dry-deposition flux values for 35 of the 64 compounds analyzed were biased low because the laboratory reagent spike recoveries were less than 50 percent for these compounds (table 7) and, therefore, could be present in the environment at higher concentrations than reported. A total of 37 dry-deposition samples were analyzed during the study period (appendix 5). Thirty-two compounds were detected with detection frequencies that ranged from 2.7 percent (representing one detection) for five compounds to 86.5 percent for the compound dacthal (table 20). Carbaryl, chlorpyrifos, dacthal, metolachlor, and phosmet were the most frequently detected compounds in the San Joaquin Valley (table 17) and were compared to the same compounds for the three samples collected from the Sacramento Valley at urban site 16 (fig. 14) because there were only three dry-deposition samples collected from site 16. Although there were few dry-deposition detections in the Sacramento Valley (chlorpyrifos, dacthal, and phosmet), these detections corresponded to the median flux values collected from the San Joaquin Valley for the same compounds (fig. 14 and table 17). In the San Joaquin Valley, the greatest dry-deposition amounts collected were for the compounds simazine (E 0.245 $\mu g/m^2/day$, site 5) and chlorpyrifos (E 0.233 $\mu g/m^2/day$, site 12; appendix 5). In the Sacramento Valley, the greatest dry-deposition amounts collected were for the compounds iprodione (E 0.078 $\mu g/m^2/day$) and simazine (E 0.025 $\mu g/m^2/day$; appendix 5).

Dry deposition computed for detections in both the funnel and autosampler from the San Joaquin Valley urban site 4, located in downtown Modesto, and the agricultural site 5, located in rural Modesto adjacent to agricultural land, were compared during two separate composite-sample periods: October–November 2002, and May–July 2003 (table 21). In general, the dry-deposition amounts collected from the funnel and autosampler were comparable for the respective sampling periods and sampling locations (urban vs. agricultural), but there were a few notable exceptions. During the October–November 2002, sampling period, chlorpyrifos at the urban site was more than twice the dry deposition at the agricultural

site, and, conversely, simazine was about three times greater at the agricultural site compared to the urban site. During the May–July 2003, sampling period, the dry deposition of chlorpyrifos was greater at the agricultural site (six times greater for the sample collected from the autosampler and more than twice for the funnel sampler), which was likely a result of its use on row crops during the summer. Detections of chlorpyrifos, diazinon, and dacthal were common to both sampling periods.

Bulk Dry-Deposition Sample Type

A total of 26 bulk-dry-deposition samples were analyzed for 64 compounds during the study period (appendix 6). Thirty-three compounds were detected with detection frequencies ranging from of 3.8 percent (representing one detection) for two compounds to 96.2 percent for dacthal (table 22). Chlorpyrifos, dacthal, diazinon, 1-napthol, and simazine were the most frequently detected compounds collected as bulk-dry deposition in the San Joaquin Valley (fig. 15) and were compared to the same compounds for the six samples collected from the Sacramento Valley at urban site 16. With the exception of simazine, the median bulk dry-deposition fluxes in the San Joaquin Valley were comparable to the median bulk-dry deposition collected from the Sacramento Valley (fig. 15) The compounds with the greatest bulk-dry deposition amounts collected in the San Joaquin Valley were iprodione (E 0.700 $\mu g/m^2/day$, site 5), methidathion (E 0.426 $\mu g/m^2/day$, site 4), and chlorpyrifos (E 0.268 $\mu g/m^2/day$, site 5; appendix 6). At the Gridley High School site (site 16) in the Sacramento Valley, diazinon and its oxon, diazoxon, had the greatest bulk dry-deposition amounts collected at E 5.26 $\mu g/m^2/day$ and E 0.203 $\mu g/m^2/day$, respectively (appendix 6). The Sacramento Valley bulk dry-deposition amounts were collected during the January 7–27, 2004, sampling event and coincided with the orchard dormant-spray season.

Soil-Box Runoff Sample Types (Aqueous Phase and Suspended Sediment)

Soil-box runoff samples (fig. 7) were submitted and analyzed as a filtered sample (aqueous phase) along with the corresponding suspended-sediment sample collected on one or more GFFs during filtration. A total of 30 aqueous-phase soil-box runoff samples (appendix 7) and 30 suspended-sediment samples (appendix 8) were submitted for analysis from the 2 sites in the San Joaquin Valley (urban site 4 and agricultural site 5) and one in the Sacramento Valley (urban site 16) during the 2002–04 study period.

In the San Joaquin Valley, 22 compounds were detected in the aqueous phase, and 11 compounds were detected in the suspended sediment samples (table 23). Detection frequencies for the aqueous phase in the San Joaquin Valley were greater than 88 percent for the top five most commonly detected compounds (dacthal, metolachlor, simazine, diazinon, and

Table 20. Detection frequency for compounds analyzed as dry deposition with at least one detection during the 2002–04 study period in the Central Valley of California.

[Abbreviations: n, total number of samples]

Compound	Detection frequency in percent (n=37)	Compound	Detection frequency in percent (n=37)
Dacthal	86.5	Paraoxon-methyl	24.3
Chlorpyrifos	70.3	Dimethoate	21.6
Metolachlor	70.3	Malathion	18.9
Simazine	64.9	Pendimethilan	18.9
Carbaryl	59.5	Phosmet oxon	18.9
Phosmet	59.5	Dichlorvos	16.2
Trifluralin	59.5	3,4-Dichloroaniline	13.5
Myclobutanil	56.8	Chlorpyrifos oxon	13.5
Diazinon	51.4	Diazoxon	10.8
Azinphos-methyl	43.2	Metribuzin	5.4
Parathion-methyl	43.2	Prometryn	5.4
Malaoxon	40.5	Alachlor	2.7
trans-Permethrin	40.5	Cypermethrin	2.7
1-Naphthol	37.8	Hexazinone	2.7
cis-Permethrin	35.1	Methidathion	2.7
Iprodione	32.4	Pronamide	2.7

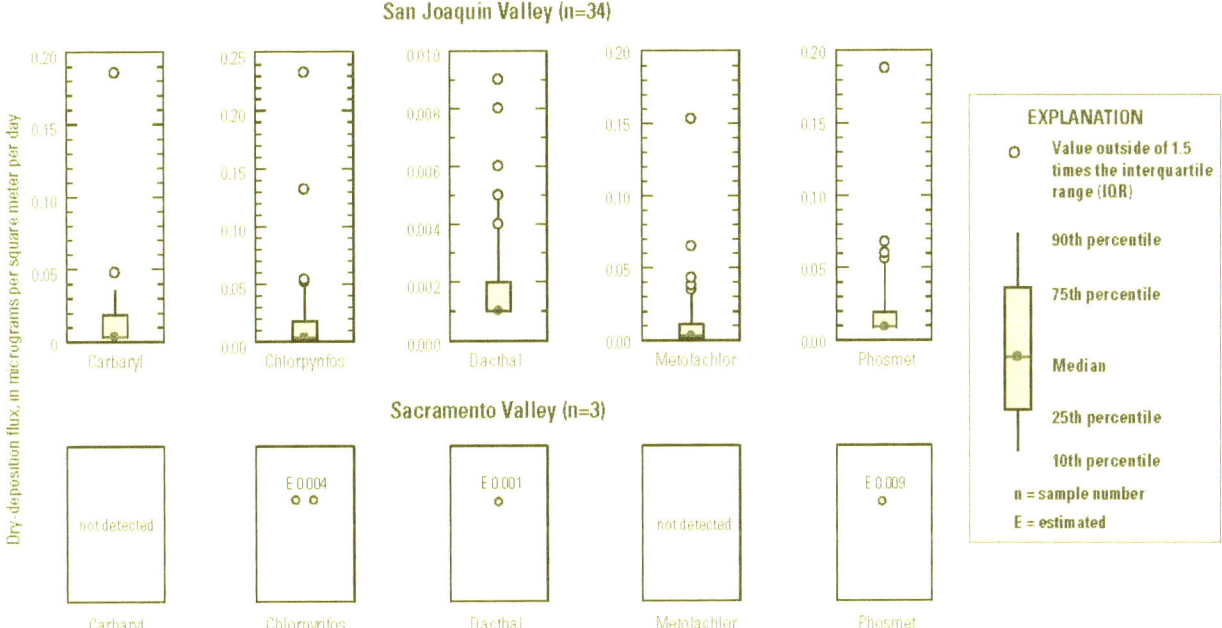

Figure 14. Distribution of calculated dry-deposition fluxes for the top five most detected compounds collected from the San Joaquin Valley compared to detections in the Sacramento Valley of California during the 2002–04 study period.

Table 21. Comparison of compounds detected in dry-deposition samples collected from funnel and autosamplers at San Joaquin Valley, California, urban site 4, and agricultural site 5 during two composite sampling periods in 2002 and 2003.

[Abbreviations: E, estimated reported laboratory value; µg/m²/day, microgram per square meter per day]

Site number	Site name	Composite sampling date	Sampler type	Chlorpyrifos (µg/m²/day)	Diazinon (µg/m²/day)	Dacthal (µg/m²/day)	Simazine (µg/m²/day)
4	Modesto Irrigation District gage rooftop at Modesto	October 4–November 6, 2002	Autosampler	E0.020	E0.006	E0.006	E0.079
		October 4–November 6, 2002	Funnel	E0.013	E0.016	E0.008	E0.077
5	Modesto Irrigation District gage rooftop at Albers Road	October 4–November 6, 2002	Autosampler	E0.009	E0.007	E0.009	E0.245
		October 4–November 6, 2002	Funnel	E0.006	E0.004	E0.005	E0.218

Site number	Site name	Composite sampling date	Sampler type	Chlorpyrifos (µg/m²/day)	Diazinon (µg/m²/day)	Dacthal (µg/m²/day)	Parathion-methyl (µg/m²/day)
4	Modesto Irrigation District gage rooftop at Modesto	May 5–July 22, 2003	Autosampler	E0.009	E0.001	E0.001	E0.009
		May 5–July 22, 2003	Funnel	E0.016	E0.002	E0.002	E0.007
5	Modesto Irrigation District gage rooftop at Albers Road	May 9–July 22, 2003	Autosampler	E0.054	E0.001	E0.001	E0.032
		May 9–July 22, 2003	Funnel	E0.039	E0.002	E0.001	E0.017

Site number	Site name	Composite sampling date	Sampler type	Paraoxon-methyl (µg/m²/day)	Metolachlor (µg/m²/day)	Dichlorvos (µg/m²/day)	cis-permethrin (µg/m²/day)	trans-permethrin (µg/m²/day)
4	Modesto Irrigation District gage rooftop at Modesto	May 5–July 22, 2003	Autosampler	E0.006	E0.002	E0.002	E0.004	E0.006
		May 5–July 22, 2003	Funnel	E0.006	E0.002	E0.004	E0.003	E0.005
5	Modesto Irrigation District gage rooftop at Albers Road	May 9–July 22, 2003	Autosampler	E0.017	E0.006	E0.006	E0.004	E0.006
		May 9–July 22, 2003	Funnel	E0.005	E0.011	E0.003	E0.006	E0.005

Table 22. Detection frequency for compounds analyzed as bulk-dry deposition with at least one detection during the 2002–04 study period in the Central Valley of California.

[Abbreviations: n, total number of samples]

Compound	Detection frequency in percent (n=26)	Compound	Detection frequency in percent (n=26)
Dacthal	96.2	Parathion-methyl	42.3
Chlorpyrifos	92.3	Diazoxon	34.6
Simazine	92.3	Malathion	34.6
1-Naphthol	84.6	Chlorpyrifos oxon	26.9
Diazinon	76.9	Prometryn	26.9
Metolachlor	73.1	Dichlorvos	19.2
3,4-Dichloroaniline	69.2	Phosmet	19.2
Iprodione	61.5	Dimethoate	15.4
Carbaryl	57.7	Azinphos-methyl	11.5
Methidathion	57.7	Pronamide	11.5
Pendimethilan	53.8	Metribuzin	7.7
Malaoxon	50.0	Paraoxon-methyl	7.7
Myclobutanil	50.0	Phosmet oxon	7.7
cis-Permethrin	50.0	Deethylatrazine, CIAT	3.8
trans-Permethrin	50.0	Cyfluthrin	3.8
Trifluralin	50.0	Fipronil sulfone	3.8
Hexazinone	42.3		

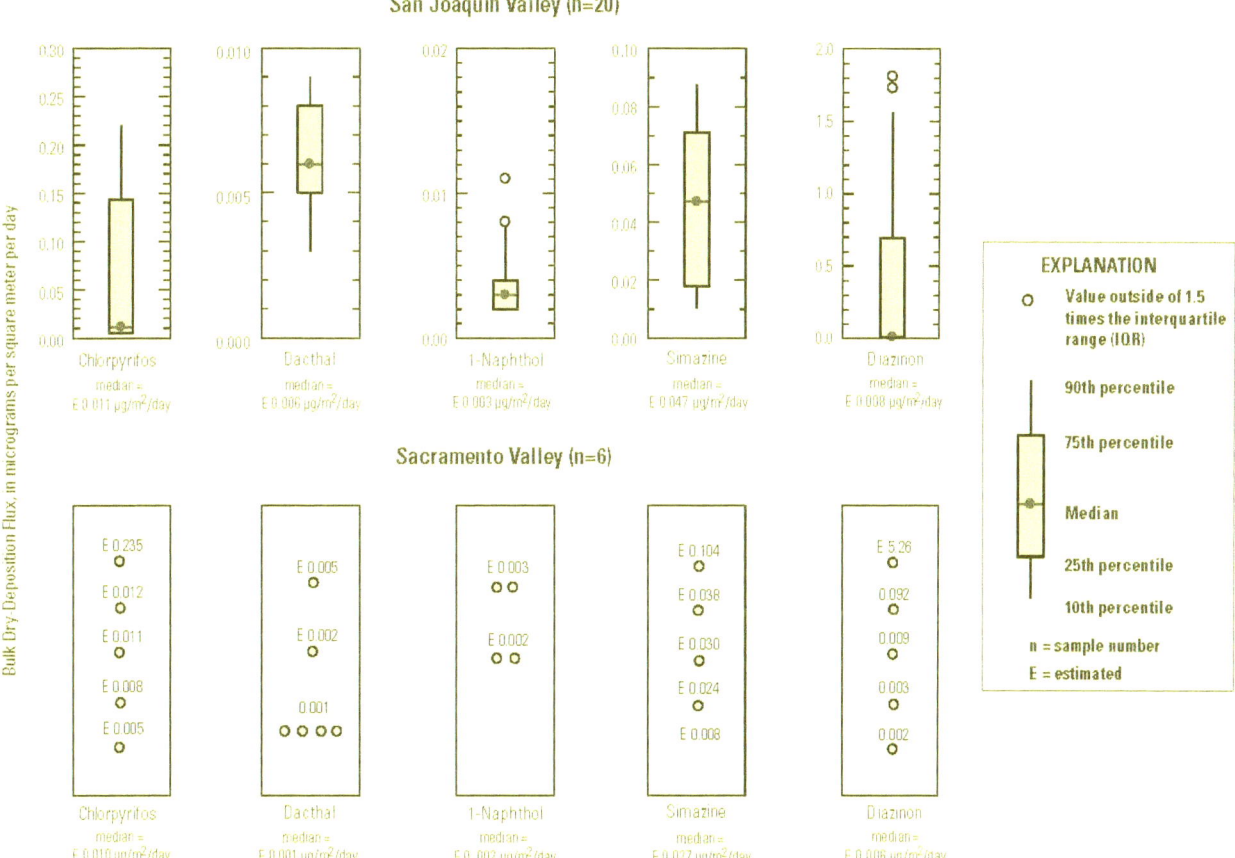

Figure 15. Distribution of calculated bulk dry-deposition fluxes for the top five most detected compounds collected from the San Joaquin Valley compared to detections in the Sacramento Valley, California, during the 2002–04 study period.

Table 23. Detection frequency for samples collected as soil-box runoff (aqueous-phase and suspended-sediment sample types) from the San Joaquin Valley and Sacramento Valley, California, soil-box sites during the 2002–04 study period.

[Abbreviations: n, total number of samples; (E), estimated compound because of poor performance by applied method (Childress and others, 1999; Sandstrom and others, 2001); NWQL, National Water Quality Laboratory]

San Joaquin Valley soil-box runoff (urban site 4 and agricultural site 5)		Sacramento Valley soil-box runoff (urban site 16)	
Compound	Aqueous-phase detection frequency, in percent (n=18)	Compound	Aqueous-phase detection frequency, in percent (n=12)
Dacthal	94.4	Chlorpyrifos	100.0
Metolachlor	94.4	Dacthal	100.0
Simazine	94.4	Diazinon	100.0
Diazinon	88.9	Simazine	91.7
Myclobutanil	[3]88.9	Pendimethalin	66.7
Chlorpyrifos	83.3	Carbaryl	58.3
3,4-Dichloroaniline	[3]82.4	Parathion-methyl	50.0
Pendimethalin	72.2	Prometon	[3]36.4
Methidathion	55.6	3,4-Dichloroaniline	33.3
Metribuzin	55.6	Diazoxon	33.3
Prometryn	[3]47.1	1-Naphthol (E)	[4]30
Diazoxon	38.9	Methidathion	25.0
Trifluralin	27.8	Myclobutanil	25.0
Carbaryl	22.2	Prometryn	25.0
Parathion-methyl	[3]22.2	Iprodione	16.7
Iprodione	[3]17.6	Metolachlor	16.7
Malathion	16.7	Azinphos-methyl	8.3
Pronamide	16.7	Hexazinone	8.3
Hexazinone	[6]14.3	Malathion	8.3
Azinphos-methyl	11.1	Malaoxon	8.3
Dimethoate	[3]5.9	Paraoxon-methyl	8.3
Prometon	[3]5.9		

Compound	Suspended-sediment detection frequency, in percent ([2]n=14)	Compound	Suspended-sediment detection frequency, in percent (n=12)
Simazine	64.3	Simazine	75.0
3,4-Dichloroaniline[1] (E)	50.0	Chlorpyrifos[1]	66.7
Diazinon	50.0	Dacthal	66.7
Myclobutanil	28.6	Carbaryl[1] (E)	50.0
Dacthal	21.4	Paraoxon-methyl	50.0
Pendimethalin	21.4	Pendimethalin	50.0
Chlorpyrifos[1]	14.3	Diazinon	33.3
Trifluralin	14.3	3,4-Dichloroaniline[1] (E)	25.0
1-Naphthol[1] (E)	7.1	Malaoxon1 (E)	[5]12.5
Fipronil	7.1	Fipronil	8.3
Iprodione[1] (E)	7.1	Iprodione[1] (E)	8.3
		Myclobutanil	8.3
		Trifluralin	8.3

[1]Compound is biased low because it is assumed that the analytical method might not adequately recover the compound because of matrix interference, adsorptive losses on the dry matrix, or other procedural problem for this sample type. See *table 11* and corresponding text for laboratory reagent spike recoveries.

[2]Four samples were ruined during laboratory preparation of these sample types, therefore the total number of samples does not equal the aqueous-phase sample type. See *appendix 8* for site specific samples.

[3] Total samples equal n–1, results of one sample deleted by NWQL because compound could not be determined because of matrix interference.

[4] Total samples equal n–2, results of two samples deleted by NWQL because compound could not be determined because of matrix interference.

[5] Total samples equal n–4, results of four samples deleted by NWQL because compound could not be determined because of matrix interference.

[6] Total samples equal n–11, results of eleven samples deleted by NWQL because compound could not be determined because of matrix interference.

myclobutanil), whereas only three compounds had detection frequencies greater than 50 percent for suspended sediment (simazine, 3,4-Dichloroaniline, and diazinon). Figure 16 shows the distribution of concentrations for the most detected compounds in the aqueous phase compared to the most detected compounds in suspended-sediment samples from the urban and agricultural soil-box sites in the San Joaquin Valley during the study period. The suspended-sediment sample type was analyzed by using a laboratory research method (NWQL research method LS750, appendix 1) to determine concentrations of pesticides, and in some cases, the LRL was above reported detections. For example, 3,4-Dichloroanaline was detected in 50 percent of the suspended-sediment sample types collected from the San Joaquin Valley; however, all the detections were below the LRL value of 5.9 µg/L and were reported as estimated concentrations and, therefore, assumed a greater degree of uncertainty (fig. 16 and appendix 8). With the exception of 3,4-Dichloroanaline and simazine collected in the San Joaquin Valley, the LRL for the most detected compounds collected from suspended sediment from the soil-box runoff, depicted in figures 16 and 17, was at or below the 50th percentile (median).

Nine compounds were detected in both the aqueous-phase and suspended-sediment sample types collected from the San Joaquin Valley soil-box sites (table 24). For compounds common to both sample types, concentrations of pesticides in the aqueous phase generally were detected in low concentrations and had few corresponding detections in the suspended-sediment sample type. Exceptions to this were the compounds simazine, 3,4-Dichloroaniline, and diazinon, for which concentrations in the aqueous phase were at least twice those in the suspended-sediment sample type. The highest concentrations detected in the aqueous phase for a given compound did not always correspond with the highest concentrations detected in corresponding suspended-sediment sample type because the physical and chemical properties of the pesticides vary (table 15), as did the timing and application of pesticides in the Central Valley during the study period. A soil-box runoff sample collected after the January 14–21, 2003, storm event from agricultural site 5 in the San Joaquin Valley had the highest concentrations of iprodione (E 3.52 µg/L), myclobutanil (1.02 µg/L), and simazine (1.01 µg/L) detected in the aqueous-phase sample type (appendix 7). The highest concentration of diazinon in the aqueous phase (1.62 µg/L) was collected from the same site after the previous storm event (January 9–14, 2003). The highest concentrations detected in suspended-sediment samples from the same site were for the compounds pendamethalin (0.96 µg/L) and myclobutanil (E 0.89 µg/L) following the November 7–10, 2002, storm event (appendix 8).

At site 16 at Gridley High School in the Sacramento Valley, 21 compounds were detected in the aqueous phase and 13 compounds were detected in the suspended-sediment sample type (table 23). Detection frequencies for the aqueous-phase sample type were greater than 91 percent for simazine, diazinon, dacthal, and chlorpyrifos, and only greater than 66 percent for simazine, chlorpyrifos, and dacthal for the suspended-sediment sample type. Figure 17 depicts the distribution of concentrations for the most detected compounds in the aqueous-phase sample type compared to the most detected compounds in suspended-sediment sample type collected from the urban and agricultural soil-box sites in the San Joaquin Valley during the study period. Concentrations of pesticides in the aqueous phase generally were detected in low concentrations, with the exception of diazinon at 1.75 µg/L (January 27– February 6, 2004 composite-sample time; table 24, appendix 7). Ten compounds were detected in both sample types collected from the Sacramento Valley soil box (table 24), and concentrations of pesticides in the aqueous phase generally were detected in low concentrations compared to the median concentrations detected in the suspended-sediment sample type for the same compound. In the Sacramento Valley at site 16, the highest concentrations of pesticides detected in suspended sediment were for the compounds methyl parathion (0.92 µg/L), chlorpyrifos (0.72 µg/L), and diazinon (0.52 µg/L; appendix 8).

When comparing concentration of compounds in the aqueous-phase sample type to concentrations in the rainfall (wet-deposition sample type collected from the autosampler and funnel sampler) collected during the same sampling event, pesticides present in the aqueous phase also were detected in the rainfall at similar concentrations, with the exception of metalachlor, myclobutanil, and simazine (table 25). In general, the aqueous-phase concentrations of metolachlor, myclobutanil, and simazine were an order of magnitude greater than concentrations detected in rainfall. The compound diazinon exhibited higher concentrations in rainfall than were detected in the aqueous phase for samples collected from the urban San Joaquin Valley soil-box at site 4 during the January 9–14, 2003, composite sample period (between 3 and 14 times greater). Conversely, the aqueous-phase concentration was twice the detected diazinon concentration in the rainfall at the San Joaquin Valley agricultural soil-box at site 5 collected during the same period (table 25). The differences were likely due to proximity of the sampler to the pesticide application areas as well as the timing and amount of pesticide used in conjunction with the physical and chemical properties of the pesticides.

Soil-Box Surficial-Soil Sample Type

Results for soil-box soil mixes prior to exposure (mix 1, 2, 3, and 4) and for samples collected from the soil boxes after exposure to the atmosphere during the study period are presented together for comparison purposes (appendix 9). Soil mix 1, 2, 3 was collected near specific sites in San Joaquin Valley (see appendix 9 header for locations), composited together, and placed in the San Joaquin Valley urban site 4 and

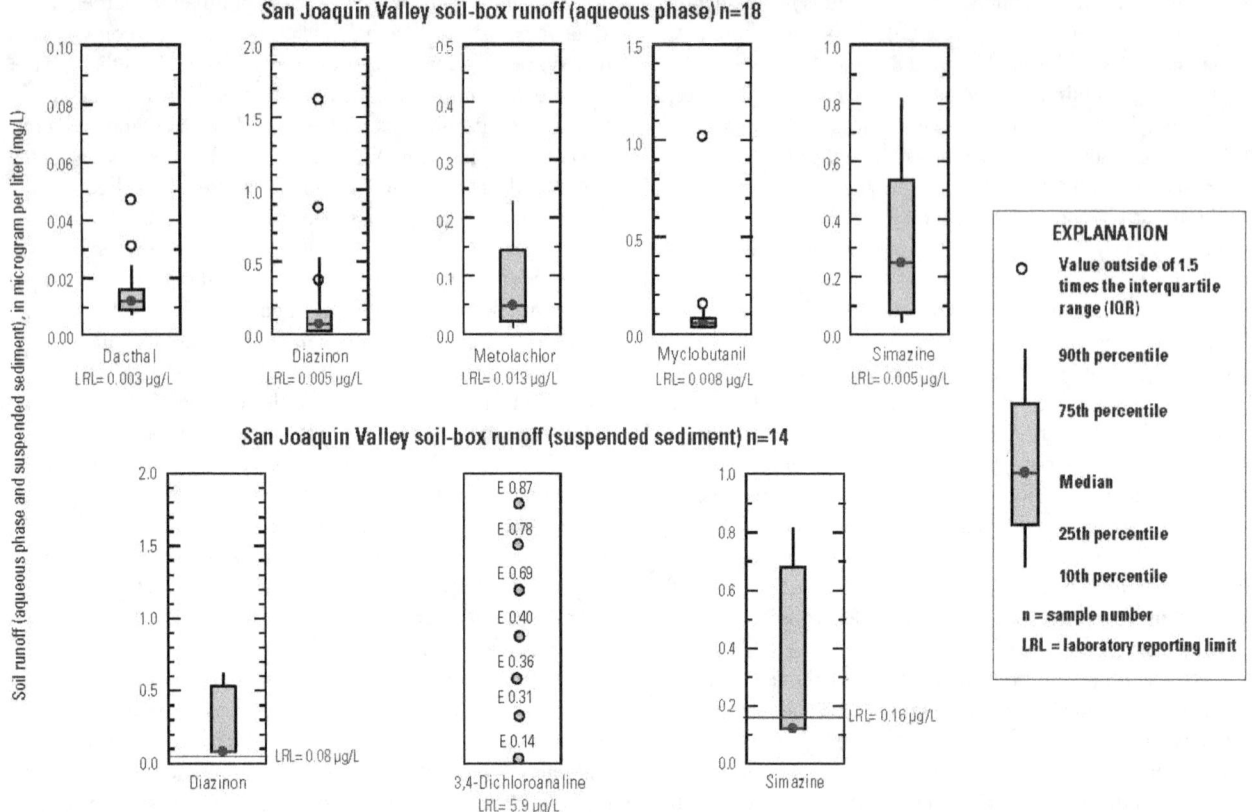

Figure 16.   Comparison of the distribution of concentrations of the most detected compounds collected from the San Joaquin Valley, California, soil-box sites 4 and 5 as aqueous-phase and suspended-sediment sample types during the 2002–04 study period.

Table 24. Comparison of detection frequencies and median concentrations of detected compounds common to aqueous-phase and suspended-sediment soil-box runoff samples collected from the San Joaquin Valley and Sacramento Valley, California, soil-box sites during the 2002–04 study period.

[Abbreviations: n, total number of samples; (E), estimated compound because of poor performance by applied method (Childress and others, 1999; Sandstrom and others, 2001); NWQL, National Water Quality Laboratory; µg/L, microgram per liter; <, less than]

Compound	San Joaquin Valley soil-box runoff (urban site 4 and agricultural site 5)			
	Suspended-sediment detection frequency, in percent ([2]n=14)	Aqueous-phase detection frequency, in percent (n=18)	Median suspended-sediment concentration, µg/L	Median aqueous-phase concentration, µg/L
Simazine	64.3	94.4	0.121	0.248
3,4-Dichloroaniline (E)	50.0	[3]82.4	[1]3.38	0.033
Diazinon	50.0	88.9	0.118	0.07
Myclobutanil	28.6	[3]88.9	[5]<0.08	0.052
Dacthal	21.4	94.4	[5]<0.08	0.012
Pendimethalin	21.4	72.2	[5]<0.11	0.033
Chlorpyrifos	14.3	83.3	[1,5]<0.08	0.017
Trifluralin	14.3	27.8	[5]<0.08	[5]<0.009
Iprodione (E)	7.1	[3]17.6	[1,5]<0.079	[5]<1.42

Compound	Sacramento Valley soil-box runoff (urban site 16)			
	Suspended-sediment detection frequency, in percent (n=12)	Aqueous-phase detection frequency, in percent (n=12)	Median suspended-sediment concentration, µg/L	Median aqueous-phase concentration, µg/L
Simazine	75.0	91.7	0.25	0.025
Chlorpyrifos	66.7	100.0	[1]0.18	0.012
Dacthal	66.7	100.0	0.13	0.006
Carbaryl (E)	50.0	58.3	[1,5]<0.16	[5]<0.041
Paraoxon-methyl	50.0	8.3	0.10	[5]<0.03
Diazinon	33.3	100.0	[5]<0.08	0.039
3,4-Dichloroaniline (E)	25.0	33.3	[1,5]<5.9	0.005
Malaoxon[1] (E)	[4]12.5	8.3	[1,5]<0.40	[5]<0.008
Iprodione[1] (E)	8.3	16.7	[1,5]<0.79	[5]<1.42
Myclobutanil	8.3	25.0	[5]<0.08	[5]<0.008

[1] Concentration result is biased low for given compound and sample type because it is assumed that the analytical method might not adequately recover the compound because of matrix interference, adsorptive losses on the dry matrix, or other procedural problem for this sample type. See *table 11* and corresponding text for LRS recoveries.

[2] Four samples were ruined during laboratory preparation of these sample types, therefore the total number of samples does not equal the aqueous phase sample type. See *appendix 5* for site specific samples.

[3] Total samples equal n–1, results of one sample deleted by NWQL because compound could not be determined because of matrix interference.

[4] Total samples equal n–4, results of four samples deleted by NWQL because compound could not be determined because of matrix interference.

[5] Concentration value given is the laboratory reporting limit for that compound and is considered a non-detection.

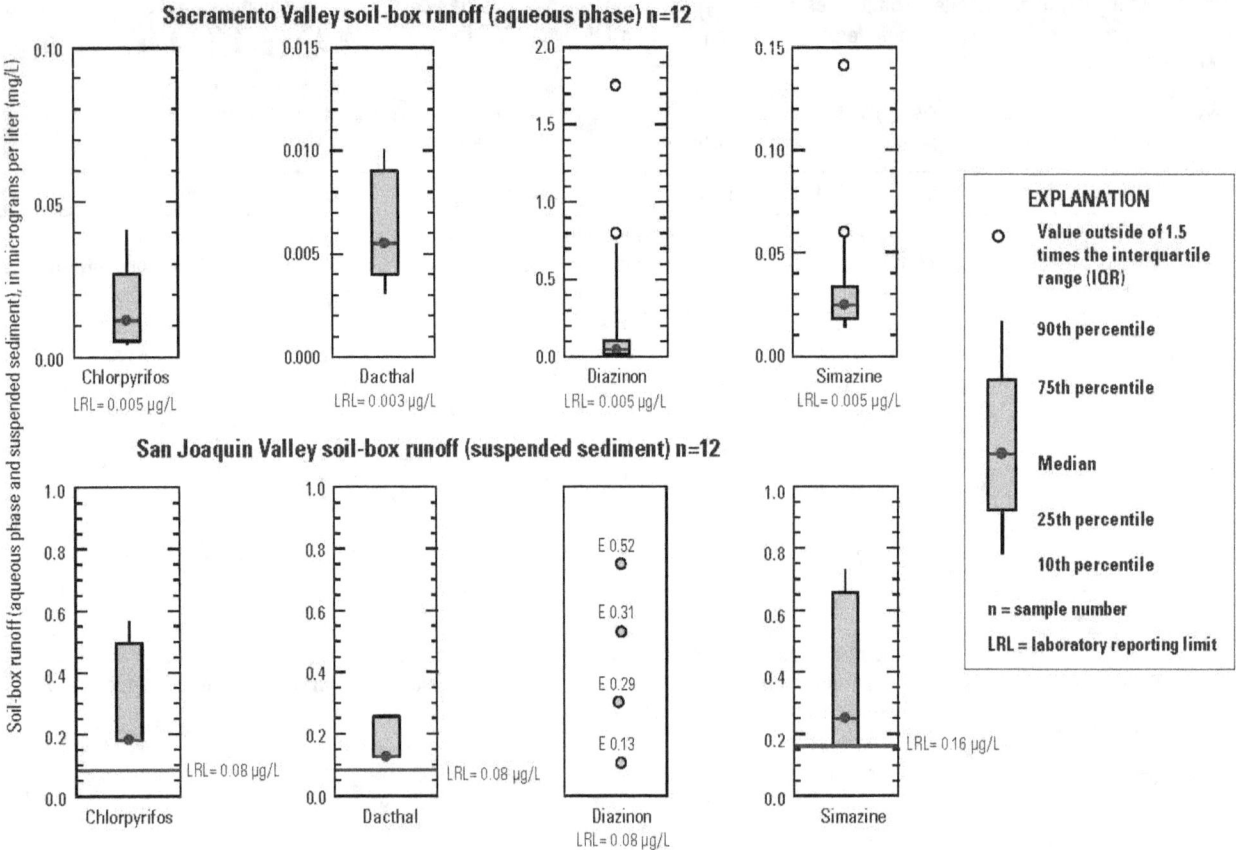

Figure 17.   Comparison of the distribution of concentrations of the most detected compounds collected from the Sacramento Valley, California, soil-box site 16 as aqueous-phase and suspended-sediment sample types during the 2002–04 study period.

Table 25. Comparison of concentrations of compounds detected in the aqueous-phase sample type to those detected in wet deposition for the same sampling periods during the 2002–04 study in the Central Valley of California.

[Compounds presented are for those with detection frequencies greater than 50 percent; all compounds were analyzed by gas chromatography/mass spectrometry (GCMS) at the U.S. Geological Survey (USGS) National Water Quality Laboratory (NWQL). Abbreviations: AS, autosampler; E, estimated reported laboratory value; (E), estimated compound due to known poor performance by applied method (Childress and others, 1999; Sandstrom and others, 2001); hh:mm, hour:minute; L, liters; mm/dd/yyyy, month/day/year; µg/L, micrograms per liter; <, less than laboratory reporting value]

Site name	Sample begin date (mm/dd/yyyy)	Start time (hh:mm)	Sample end date (mm/dd/yyyy)	End time (hh:mm)	Volume collected, (L)	Sample type	3,4-Dichloro-aniline (E), (µg/L)
Site 4							
Modesto Irrigation District gage rooftop at Modesto	11/07/2002	02:00	11/10/2002	03:10	0.422	Aqueous phase—soil-box runoff	E0.04
Modesto Irrigation District gage rooftop at Modesto	11/07/2002	02:00	11/10/2002	03:00	4.1	Wet deposition—funnel	<0.004
Modesto Irrigation District gage rooftop at Modesto	11/07/2002	02:00	11/10/2002	03:05	3.39	Wet deposition—AS	<0.004
Modesto Irrigation District gage rooftop at Modesto	12/23/2002	09:30	01/09/2003	15:35	0.7	Aqueous phase—soil-box runoff	E0.014
Modesto Irrigation District gage rooftop at Modesto	12/23/2002	09:30	01/09/2003	15:30	1.3	Wet deposition—funnel	E0.011
Modesto Irrigation District gage rooftop at Modesto	12/23/2002	09:30	01/09/2003	15:40	0.58	Wet deposition—AS	E0.017
Modesto Irrigation District gage rooftop at Modesto	01/09/2003	15:30	01/14/2003	10:40	0.66	Aqueous phase—soil-box runoff	E0.013
Modesto Irrigation District gage rooftop at Modesto	01/09/2003	15:30	01/14/2003	10:20	0.91	Wet deposition—funnel	E0.017
Modesto Irrigation District gage rooftop at Modesto	01/09/2003	15:30	01/14/2003	10:30	0.66	Wet deposition—AS	<0.025
Modesto Irrigation District gage rooftop at Modesto	12/11/2003	13:20	12/16/2003	15:20	0.12	Aqueous phase—soil-box runoff	<0.004
Modesto Irrigation District gage rooftop at Modesto	12/11/2003	13:10	12/16/2003	15:40	0.81	Wet deposition—funnel	E0.024
Modesto Irrigation District gage rooftop at Modesto	12/11/2003	13:00	12/16/2003	15:30	0.37	Wet deposition—AS	E0.082
Modesto Irrigation District gage rooftop at Modesto	12/23/2003	12:20	12/31/2003	09:40	0.325	Aqueous phase—soil-box runoff	<0.004
Modesto Irrigation District gage rooftop at Modesto	12/23/2003	13:00	12/31/2003	09:30	2.81	Wet deposition—funnel	<0.004
Site 5							
Modesto Irrigation District gage rooftop at Albers Road	11/07/2002	02:00	11/10/2002	03:10	0.389	Aqueous phase—soil-box runoff	E0.043
Modesto Irrigation District gage rooftop at Albers Road	11/07/2002	02:00	11/10/2002	03:00	4.18	Wet deposition—funnel	<0.004
Modesto Irrigation District gage rooftop at Albers Road	11/07/2002	02:00	11/10/2002	03:05	3.6	Wet deposition—AS	<0.004
Modesto Irrigation District gage rooftop at Albers Road	12/16/2002	12:00	12/23/2002	10:40	0.45	Aqueous phase—soil-box runoff	E0.041
Modesto Irrigation District gage rooftop at Albers Road	12/16/2002	12:00	12/23/2002	10:20	1.9	Wet deposition—funnel	E0.028
Modesto Irrigation District gage rooftop at Albers Road	12/16/2002	12:00	12/23/2002	10:30	1.44	Wet deposition—AS	E0.018
Modesto Irrigation District gage rooftop at Albers Road	01/09/2003	14:30	01/14/2003	11:50	1.07	Aqueous phase—soil-box runoff	E0.033
Modesto Irrigation District gage rooftop at Albers Road	01/09/2003	14:30	01/14/2003	11:30	1.2	Wet deposition—funnel	0.074
Modesto Irrigation District gage rooftop at Albers Road	01/09/2003	14:30	01/14/2003	11:40	0.86	Wet deposition—AS	0.038
Modesto Irrigation District gage rooftop at Albers Road	12/11/2003	14:00	12/16/2003	14:30	0.64	Aqueous phase—soil-box runoff	E0.027
Modesto Irrigation District gage rooftop at Albers Road	12/11/2003	14:05	12/16/2003	14:40	0.76	Wet deposition—funnel	E0.022
Modesto Irrigation District gage rooftop at Albers Road	12/11/2003	14:10	12/16/2003	14:35	0.67	Wet deposition—AS	<0.004

Table 25. Comparison of concentrations of compounds detected in the aqueous-phase sample type to those detected in wet deposition for the same sampling periods during the 2002–04 study in the Central Valley of California.—Continued

[Compounds presented are for those with detection frequencies greater than 50 percent; all compounds were analyzed by gas chromatography/mass spectrometry (GCMS) at the U.S. Geological Survey (USGS) National Water Quality Laboratory (NWQL). Abbreviations: AS, autosampler; E, estimated reported laboratory value; (E), estimated compound due to known poor performance by applied method (Childress and others, 1999; Sandstrom and others, 2001); hh:mm, hour:minute; L, liters; mm/dd/yyyy, month/day/year; µg/L, micrograms per liter; <, less than laboratory reporting value]

Site name	Sample begin date (mm/dd/yyyy)	Start time (hh:mm)	Sample end date (mm/dd/yyyy)	End time (hh:mm)	Metolachlor, (µg/L)	Myclobutanil (E), (µg/L)	Pendimethalin, (µg/L)	Simazine, (µg/L)
			Site 4					
Modesto Irrigation District gage rooftop at Modesto	11/07/2002	02:00	11/10/2002	03:10	0.309	E0.064	0.094	0.765
Modesto Irrigation District gage rooftop at Modesto	11/07/2002	02:00	11/10/2002	03:00	E0.006	<0.008	0.033	0.01
Modesto Irrigation District gage rooftop at Modesto	11/07/2002	02:00	11/10/2002	03:05	E0.005	E0.008	0.033	0.017
Modesto Irrigation District gage rooftop at Modesto	12/23/2002	09:30	01/09/2003	15:35	0.193	E0.047	0.029	0.297
Modesto Irrigation District gage rooftop at Modesto	12/23/2002	09:30	01/09/2003	15:30	E0.007	<0.008	0.071	0.04
Modesto Irrigation District gage rooftop at Modesto	12/23/2002	09:30	01/09/2003	15:40	0.01	<0.008	0.044	0.068
Modesto Irrigation District gage rooftop at Modesto	01/09/2003	15:30	01/14/2003	10:40	0.143	E0.049	0.03	0.246
Modesto Irrigation District gage rooftop at Modesto	01/09/2003	15:30	01/14/2003	10:20	E0.004	<0.008	0.048	0.022
Modesto Irrigation District gage rooftop at Modesto	01/09/2003	15:30	01/14/2003	10:30	E0.006	<0.008	0.062	0.025
Modesto Irrigation District gage rooftop at Modesto	12/11/2003	13:20	12/16/2003	15:20	0.05	<0.008	<0.095	<0.005
Modesto Irrigation District gage rooftop at Modesto	12/11/2003	13:10	12/16/2003	15:40	E0.006	<0.008	0.025	0.015
Modesto Irrigation District gage rooftop at Modesto	12/11/2003	13:00	12/16/2003	15:30	E0.011	<0.008	<0.022	0.037
Modesto Irrigation District gage rooftop at Modesto	12/23/2003	12:20	12/31/2003	09:40	E0.01	E0.026	0.039	0.027
Modesto Irrigation District gage rooftop at Modesto	12/23/2003	13:00	12/31/2003	09:30	<0.013	<0.008	0.043	0.014
			Site 5					
Modesto Irrigation District gage rooftop at Albers Road	11/07/2002	02:00	11/10/2002	03:10	0.132	E0.067	0.071	0.388
Modesto Irrigation District gage rooftop at Albers Road	11/07/2002	02:00	11/10/2002	03:00	E0.004	E0.006	E0.015	0.017
Modesto Irrigation District gage rooftop at Albers Road	11/07/2002	02:00	11/10/2002	03:05	E0.004	E0.006	E0.016	0.035
Modesto Irrigation District gage rooftop at Albers Road	12/16/2002	12:00	12/23/2002	10:40	0.154	E0.053	0.06	0.324
Modesto Irrigation District gage rooftop at Albers Road	12/16/2002	12:00	12/23/2002	10:20	<0.013	<0.008	0.028	0.012
Modesto Irrigation District gage rooftop at Albers Road	12/16/2002	12:00	12/23/2002	10:30	<0.013	<0.008	0.027	0.01
Modesto Irrigation District gage rooftop at Albers Road	01/09/2003	14:30	01/14/2003	11:50	0.066	E0.039	0.033	0.146
Modesto Irrigation District gage rooftop at Albers Road	01/09/2003	14:30	01/14/2003	11:30	<0.013	<0.008	0.035	0.015
Modesto Irrigation District gage rooftop at Albers Road	01/09/2003	14:30	01/14/2003	11:40	<0.013	<0.008	0.027	0.007
Modesto Irrigation District gage rooftop at Albers Road	12/11/2003	14:00	12/16/2003	14:30	0.023	E0.034	0.032	0.076
Modesto Irrigation District gage rooftop at Albers Road	12/11/2003	14:05	12/16/2003	14:40	<0.013	<0.008	0.035	0.033
Modesto Irrigation District gage rooftop at Albers Road	12/11/2003	14:10	12/16/2003	14:35	<0.013	<0.008	E0.019	0.019

Table 25. Comparison of concentrations of compounds detected in the aqueous-phase sample type to those detected in wet deposition for the same sampling periods during the 2002–04 study in the Central Valley of California.—Continued

[Compounds presented are for those with detection frequencies greater than 50 percent; all compounds were analyzed by gas chromatography/mass spectrometry (GCMS) at the U.S. Geological Survey (USGS) National Water Quality Laboratory (NWQL). Abbreviations: AS, autosampler; E, estimated reported laboratory value; (E), estimated compound due to known poor performance by applied method (Childress and others, 1999; Sandstrom and others, 2001); hh:mm, hour:minute; L, liters; mm/dd/yyyy, month/day/year; µg/L, micrograms per liter; <, less than laboratory reporting value]

Site name	Sample begin date (mm/dd/yyyy)	Start time (hh:mm)	Sample end date (mm/dd/yyyy)	End time (hh:mm)	Carbaryl (E), (µg/L)	Chlorpyrifos, (µg/L)	Dacthal, DCPA, (µg/L)	Diazinon, (µg/L)
Site 4								
Modesto Irrigation District gage rooftop at Modesto	11/07/2002	02:00	11/10/2002	03:10	E0.012	0.009	0.018	0.026
Modesto Irrigation District gage rooftop at Modesto	11/07/2002	02:00	11/10/2002	03:00	E0.026	0.007	0.009	0.03
Modesto Irrigation District gage rooftop at Modesto	11/07/2002	02:00	11/10/2002	03:05	E0.011	0.009	0.008	0.025
Modesto Irrigation District gage rooftop at Modesto	12/23/2002	09:30	01/09/2003	15:35	E0.007	0.016	0.009	0.154
Modesto Irrigation District gage rooftop at Modesto	12/23/2002	09:30	01/09/2003	15:30	E0.097	0.056	0.014	0.785
Modesto Irrigation District gage rooftop at Modesto	12/23/2002	09:30	01/09/2003	15:40	E0.087	0.109	0.019	2.22
Modesto Irrigation District gage rooftop at Modesto	01/09/2003	15:30	01/14/2003	10:40	<0.041	0.039	0.009	0.304
Modesto Irrigation District gage rooftop at Modesto	01/09/2003	15:30	01/14/2003	10:20	E0.01	0.072	0.008	0.962
Modesto Irrigation District gage rooftop at Modesto	01/09/2003	15:30	01/14/2003	10:30	E0.011	0.052	0.013	1.18
Modesto Irrigation District gage rooftop at Modesto	12/11/2003	13:20	12/16/2003	15:20	<0.041	0.053	0.031	0.101
Modesto Irrigation District gage rooftop at Modesto	12/11/2003	13:10	12/16/2003	15:40	<0.041	0.013	0.008	0.032
Modesto Irrigation District gage rooftop at Modesto	12/11/2003	13:00	12/16/2003	15:30	E0.052	0.023	0.008	0.064
Modesto Irrigation District gage rooftop at Modesto	12/23/2003	12:20	12/31/2003	09:40	<0.041	0.015	0.009	0.124
Modesto Irrigation District gage rooftop at Modesto	12/23/2003	13:00	12/31/2003	09:30	E0.009	0.019	0.004	0.074
Site 5								
Modesto Irrigation District gage rooftop at Albers Road	11/07/2002	02:00	11/10/2002	03:10	<0.041	0.011	0.009	0.012
Modesto Irrigation District gage rooftop at Albers Road	11/07/2002	02:00	11/10/2002	03:00	E0.005	0.009	0.005	0.013
Modesto Irrigation District gage rooftop at Albers Road	11/07/2002	02:00	11/10/2002	03:05	E0.006	0.01	0.006	0.016
Modesto Irrigation District gage rooftop at Albers Road	12/16/2002	12:00	12/23/2002	10:40	<0.041	0.028	0.01	0.043
Modesto Irrigation District gage rooftop at Albers Road	12/16/2002	12:00	12/23/2002	10:20	<0.041	0.035	0.007	0.024
Modesto Irrigation District gage rooftop at Albers Road	12/16/2002	12:00	12/23/2002	10:30	<0.041	0.037	0.01	0.029
Modesto Irrigation District gage rooftop at Albers Road	01/09/2003	14:30	01/14/2003	11:50	E0.004	0.544	0.005	1.62
Modesto Irrigation District gage rooftop at Albers Road	01/09/2003	14:30	01/14/2003	11:30	<0.041	0.458	0.004	0.879
Modesto Irrigation District gage rooftop at Albers Road	01/09/2003	14:30	01/14/2003	11:40	<0.041	0.601	0.005	0.834
Modesto Irrigation District gage rooftop at Albers Road	12/11/2003	14:00	12/16/2003	14:30	<0.041	0.015	0.011	0.096
Modesto Irrigation District gage rooftop at Albers Road	12/11/2003	14:05	12/16/2003	14:40	<0.041	0.015	0.007	0.06
Modesto Irrigation District gage rooftop at Albers Road	12/11/2003	14:10	12/16/2003	14:35	<0.041	0.012	0.007	0.059

Table 25. Comparison of concentrations of compounds detected in the aqueous-phase sample type to those detected in wet deposition for the same sampling periods during the 2002–04 study in the Central Valley of California.—Continued

[Compounds presented are for those with detection frequencies greater than 50 percent; all compounds were analyzed by gas chromatography/mass spectrometry (GCMS) at the U.S. Geological Survey (USGS) National Water Quality Laboratory (NWQL). Abbreviations: AS, autosampler; E, estimated reported laboratory value; (E), estimated compound due to known poor performance by applied method (Childress and others, 1999; Sandstrom and others, 2001); hh:mm, hour:minute; L, liters; mm/dd/yyyy, month/day/year; μg/L, micrograms per liter; <, less than laboratory reporting value]

Site name	Sample begin date (mm/dd/yyyy)	Start time (hh:mm)	Sample end date (mm/dd/yyyy)	End time (hh:mm)	Volume collected, (L)	Sample type	3,4-Dichloro-aniline (E), (μg/L)
			Site 16				
Gridley High School precipitation gage at Gridley	11/04/2003	10:15	11/12/2003	10:20	1.03	Aqueous phase—soil-box runoff	E0.01
Gridley High School precipitation gage at Gridley	11/04/2003	10:00	11/12/2003	10:15	2.05	Wet deposition—AS	<0.004
Gridley High School precipitation gage at Gridley	12/03/2003	12:30	12/12/2003	11:00	4.2	Aqueous phase—soil-box runoff	E0.008
Gridley High School precipitation gage at Gridley	12/03/2003	12:50	12/12/2003	11:20	3.885	Wet deposition—funnel	E0.006
Gridley High School precipitation gage at Gridley	12/03/2003	12:40	12/12/2003	11:10	2.13	Wet deposition—AS	<0.006
Gridley High School precipitation gage at Gridley	12/12/2003	11:00	12/18/2003	10:10	2	Aqueous phase—soil-box runoff	E0.009
Gridley High School precipitation gage at Gridley	12/12/2003	11:20	12/18/2003	10:30	1.19	Wet deposition—funnel	<0.004
Gridley High School precipitation gage at Gridley	12/12/2003	11:10	12/18/2003	10:20	0.57	Wet deposition—AS	<0.004
Gridley High School precipitation gage at Gridley	12/18/2003	10:00	12/23/2003	08:00	1.22	Aqueous phase—soil-box runoff	<0.004
Gridley High School precipitation gage at Gridley	12/18/2003	10:30	12/23/2003	08:10	1.72	Wet deposition—funnel	<0.004
Gridley High School precipitation gage at Gridley	12/23/2003	08:00	12/30/2003	09:45	3	Aqueous phase—soil-box runoff	<0.005
Gridley High School precipitation gage at Gridley	12/23/2003	08:10	12/30/2003	09:40	4.3	Wet deposition—funnel	<0.004
Gridley High School precipitation gage at Gridley	02/06/2004	10:30	02/17/2004	14:20	0.11	Aqueous phase—soil-box runoff	<0.004
Gridley High School precipitation gage at Gridley	02/06/2004	10:20	02/17/2004	14:00	3.4	Wet deposition—funnel	<0.004
Gridley High School precipitation gage at Gridley	02/17/2004	14:20	02/20/2004	10:30	1.28	Aqueous phase—soil-box runoff	<0.004
Gridley High School precipitation gage at Gridley	02/17/2004	14:00	02/20/2004	10:30	3.1	Wet deposition—funnel	<0.004
Gridley High School precipitation gage at Gridley	02/20/2004	10:30	03/02/2004	11:00	4.18	Aqueous phase—soil-box runoff	<0.004
Gridley High School precipitation gage at Gridley	02/20/2004	10:30	03/02/2004	10:40	4.28	Wet deposition—funnel	<0.004

Table 25. Comparison of concentrations of compounds detected in the aqueous-phase sample type to those detected in wet deposition for the same sampling periods during the 2002–04 study in the Central Valley of California.—Continued

[Compounds presented are for those with detection frequencies greater than 50 percent; all compounds were analyzed by gas chromatography/mass spectrometry (GCMS) at the U.S. Geological Survey (USGS) National Water Quality Laboratory (NWQL). Abbreviations: AS, autosampler; E, estimated reported laboratory value; (E), estimated compound due to known poor performance by applied method (Childress and others, 1999; Sandstrom and others, 2001); hh:mm, hour:minute; L, liters; mm/dd/yyyy, month/day/year; µg/L, micrograms per liter; <, less than laboratory reporting value]

Site name	Sample begin date (mm/dd/yyyy)	Start time (hh:mm)	Sample end date (mm/dd/yyyy)	End time (hh:mm)	Metolachlor, (µg/L)	Myclobutanil (E), (µg/L)	Pendimethalin, (µg/L)	Simazine, (µg/L)
			Site 16					
Gridley High School precipitation gage at Gridley	11/04/2003	10:15	11/12/2003	10:20	<0.013	<0.008	E0.014	0.025
Gridley High School precipitation gage at Gridley	11/04/2003	10:00	11/12/2003	10:15	<0.013	<0.008	<0.022	0.006
Gridley High School precipitation gage at Gridley	12/03/2003	12:30	12/12/2003	11:00	<0.013	<0.008	E0.013	0.019
Gridley High School precipitation gage at Gridley	12/03/2003	12:50	12/12/2003	11:20	<0.013	<0.008	E0.018	0.009
Gridley High School precipitation gage at Gridley	12/03/2003	12:40	12/12/2003	11:10	<0.013	<0.008	<0.022	0.011
Gridley High School precipitation gage at Gridley	12/12/2003	11:00	12/18/2003	10:10	<0.013	<0.008	0.023	0.026
Gridley High School precipitation gage at Gridley	12/12/2003	11:20	12/18/2003	10:30	E0.004	<0.008	0.026	0.009
Gridley High School precipitation gage at Gridley	12/12/2003	11:10	12/18/2003	10:20	<0.013	<0.008	E0.02	0.012
Gridley High School precipitation gage at Gridley	12/18/2003	10:00	12/23/2003	08:00	<0.013	<0.008	0.022	0.024
Gridley High School precipitation gage at Gridley	12/18/2003	10:30	12/23/2003	08:10	<0.013	<0.008	0.027	<0.01
Gridley High School precipitation gage at Gridley	12/23/2003	08:00	12/30/2003	09:45	<0.013	<0.008	E0.015	0.028
Gridley High School precipitation gage at Gridley	12/23/2003	08:10	12/30/2003	09:40	<0.013	<0.008	E0.017	0.009
Gridley High School precipitation gage at Gridley	02/06/2004	10:30	02/17/2004	14:20	<0.013	<0.008	<0.022	0.06
Gridley High School precipitation gage at Gridley	02/06/2004	10:20	02/17/2004	14:00	<0.013	<0.008	0.027	0.043
Gridley High School precipitation gage at Gridley	02/17/2004	14:20	02/20/2004	10:30	E0.002	E0.003	E0.022	0.014
Gridley High School precipitation gage at Gridley	02/17/2004	14:00	02/20/2004	10:30	<0.013	E0.006	E0.023	0.01
Gridley High School precipitation gage at Gridley	02/20/2004	10:30	03/02/2004	11:00	<0.013	E0.008	<0.022	0.021
Gridley High School precipitation gage at Gridley	02/20/2004	10:30	03/02/2004	10:40	<0.013	E0.022	E0.015	0.01

Table 25. Comparison of concentrations of compounds detected in the aqueous-phase sample type to those detected in wet deposition for the same sampling periods during the 2002–04 study in the Central Valley of California.—Continued

[Compounds presented are for those with detection frequencies greater than 50 percent; all compounds were analyzed by gas chromatography/mass spectrometry (GCMS) at the U.S. Geological Survey (USGS) National Water Quality Laboratory (NWQL). Abbreviations: AS, autosampler; E, estimated reported laboratory value; (E), estimated compound due to known poor performance by applied method (Childress and others, 1999; Sandstrom and others, 2001); hh:mm, hour:minute; L, liters; mm/dd/yyyy, month/day/year; µg/L, micrograms per liter; <, less than laboratory reporting value]

Site name	Sample begin date (mm/dd/yyyy)	Start time (hh:mm)	Sample end date (mm/dd/yyyy)	End time (hh:mm)	Carbaryl (E), (µg/L)	Chlorpyrifos, (µg/L)	Dacthal, DCPA, (µg/L)	Diazinon, (µg/L)
Site 16								
Gridley High School precipitation gage at Gridley	11/04/2003	10:15	11/12/2003	10:20	E0.023	0.005	0.008	0.012
Gridley High School precipitation gage at Gridley	11/04/2003	10:00	11/12/2003	10:15	E0.006	E0.005	0.006	0.009
Gridley High School precipitation gage at Gridley	12/03/2003	12:30	12/12/2003	11:00	E0.024	0.005	0.006	0.01
Gridley High School precipitation gage at Gridley	12/03/2003	12:50	12/12/2003	11:20	E0.011	0.006	0.005	0.007
Gridley High School precipitation gage at Gridley	12/03/2003	12:40	12/12/2003	11:10	<0.041	<0.005	0.005	0.007
Gridley High School precipitation gage at Gridley	12/12/2003	11:00	12/18/2003	10:10	<0.055	E0.004	0.004	0.013
Gridley High School precipitation gage at Gridley	12/12/2003	11:20	12/18/2003	10:30	E0.033	0.007	0.004	0.008
Gridley High School precipitation gage at Gridley	12/12/2003	11:10	12/18/2003	10:20	E0.021	0.054	0.004	0.036
Gridley High School precipitation gage at Gridley	12/18/2003	10:00	12/23/2003	08:00	E0.034	0.009	0.005	0.031
Gridley High School precipitation gage at Gridley	12/18/2003	10:30	12/23/2003	08:10	E0.007	0.011	0.003	0.037
Gridley High School precipitation gage at Gridley	12/23/2003	08:00	12/30/2003	09:45	E0.012	0.018	E0.003	0.022
Gridley High School precipitation gage at Gridley	12/23/2003	08:10	12/30/2003	09:40	E0.005	0.017	E0.003	0.008
Gridley High School precipitation gage at Gridley	02/06/2004	10:30	02/17/2004	14:20	E0.043	0.042	0.014	0.796
Gridley High School precipitation gage at Gridley	02/06/2004	10:20	02/17/2004	14:00	E0.007	0.03	0.005	0.311
Gridley High School precipitation gage at Gridley	02/17/2004	14:20	02/20/2004	10:30	<0.041	0.014	0.004	0.131
Gridley High School precipitation gage at Gridley	02/17/2004	14:00	02/20/2004	10:30	<0.041	0.018	0.005	0.097
Gridley High School precipitation gage at Gridley	02/20/2004	10:30	03/02/2004	11:00	<0.041	0.007	0.004	0.07
Gridley High School precipitation gage at Gridley	02/20/2004	10:30	03/02/2004	10:40	<0.041	0.011	0.004	0.051

agricultural site 5 soil-boxes. Soil mix 4 was collected from the Sacramento Valley near site 16 and placed in soil box at urban site 16 on the Gridley High School rooftop. Therefore, background concentrations of some compounds were detected only in the San Joaquin Valley mix, whereas others were only detected in the Sacramento Valley soil mix, and some in both. No soil-box runoff samples were collected from the side of the box from which the surficial-soil samples were collected.

In the San Joaquin Valley, 19 compounds were detected in the background concentrations of the composite soil mix placed in the urban and agricultural soil boxes (table 26). The highest background concentrations were detected for the compounds simazine (99 μg/Kg), metolachlor (98 μg/Kg), trifluralin (24.3 μg/Kg), and malathion (23.3 μg/Kg). For surficial-soil samples collected throughout the study period from the San Joaquin urban soil box, 16 of the 19 compounds detected in background concentrations were detected at lower concentrations following atmospheric exposure; of the exceptions were chlorpyrifos, dacthal, and iprodione (table 26). At the San Joaquin Valley agricultural soil-box site 5, carbaryl, chlorpyrifos, dacthal, iprodione, and myclobutinil also were detected at higher concentrations than the background concentrations of the composite soil mix, indicating atmospheric deposition of these compounds onto the soil-box surface. For chlorpyrifos, the concentration was twice that of background in at the urban soil-box site (1.6 μg/Kg compared to E 0.8 μg/Kg), whereas at the agricultural soil-box site, the concentration was nearly an order of magnitude more than background (6 μg/Kg compared to E 0.8 μg/Kg). The oxygen analog of chlopyrifos was not detected in the background concentrations of soil mix 1, 2, 3, but was detected at low concentrations in samples collected from both the urban and agricultural soil-box sites in July 2004. Detections of chlorpyrifos oxon could be attributed to the use of its parent compound, chlorpyrifos, on row crops during the summer in the San Joaquin Valley.

In the Sacramento Valley, six compounds were detected in background concentrations of soil mix 4 (table 26). These compounds also were detected in samples collected following atmospheric exposure at the same or slightly higher concentrations. Simazine had the highest concentration detected in both the background concentration and samples collected after atmospheric exposure at 10 μg/Kg and 11 μg/Kg, respectively. Chlorpyrifos, dacthal, iprodione, methyl parathion, and methyl paraoxon were detected in low concentrations in samples collected during the study period, but were not detected in the background concentration of soil mix 4, indicating atmospheric deposition of these compounds onto the soil-box surface.

Summary and Conclusions

2001 Study Period

A single storm event on January 25–26, 2001, was sampled, and concentrations in rainfall were measured at four urban and four agricultural sites in the Modesto, Calif., region, while simultaneously collecting hourly storm-drain runoff samples from a Modesto storm drain. Chlorpyrifos, dacthal, diazinon, pendimethalin, and simazine were detected in all rainfall samples collected from urban and agricultural sites. During a January 2001 storm event, the median concentrations of both chloryprifos and diazinon sampled at four urban (0.067 μg/L and 0.515 μg/L, respectively) and four agricultural sites (0.079 μg/L and 0.583 μg/L, respectively) in and around the Modesto, Calif., region were nearly identical, indicating that the overall atmospheric burden in the region could have been fairly evenly distributed during the sampling event. The median concentrations of chlorpyrifos, diazinon, and simazine collected from the agricultural sites (0.079 μg/L, 0.583 μg/L, and 0.051 μg/L, respectively) were greater than the median concentrations collected from the urban sites (0.067 μg/L, 0.515 μg/L, and 0.031 μg/L, respectively). Metolachlor and napropamide had higher detection frequencies in samples collected from the urban sites than those collected from the agricultural sites (100 percent compared to 50 percent). It is unknown how many, if any, urban applications there were during this study that could have contributed to the observed urban concentrations.

It is likely that rainfall can contribute significantly to pesticide concentrations in storm runoff for some compounds, depending on their physical and chemical properties. Many of the same pesticides detected in rainfall samples collected in the McHenry urban watershed also were detected in the storm-drain runoff samples from the urban watershed. When comparing the median concentrations in the rainfall to those in the McHenry storm drain runoff, it was found that for some compounds rainfall contributed a substantial percentage of the concentration found in the runoff, whereas for other compounds, the concentrations in rainfall were much greater than in the runoff. Diazinon is an example for which concentrations in rainfall likely contributed about 70 percent of the diazinon in the runoff, whereas the chlorpyrifos concentration in the rain was 1.8 times higher than in the runoff. Diazinon is much more water soluble than chlorpyrifos, whereas chlorpyrifos has a higher soil absorption coefficient. Since the runoff samples were filtered, and the filters were not analyzed, it can be inferred that most of the chlorpyrifos had been sorbed onto the suspended particulate matter, whereas diazinon stayed in solution.

Table 26.   Detection frequency and comparison of concentrations of compounds detected before and after atmospheric exposure for the surficial soil-box sample type collected from the San Joaquin Valley and Sacramento Valley, California, soil-box sites during the 2002–04 study period.

[See appendix 1 for background concentrations and sample weight of mix 1 collected near site 10 (Westly raingage), mix 2 collected near site 11 (Newman raingage), Mix 3 collected near site 5 (Modesto Irrigation District at Albers Road), and mix 4 collected near site 16 (Gridley High School); total area of soil box, 0.374 cubic meters. Abbreviations: E, estimated reported laboratory value; n, total number of samples analyzed for given sample type; NA, not applicable as there was only one detection for given compound; μg/Kg, micrograms per Kilogram; –, not detected]

Compound	[3]Surficial-soil average background concentration (μg/Kg) (n=3)	Surficial-soil detection frequency (in percent) (n=3)	Median surficial-soil concentration (μg/Kg)	Maximum surficial-soil concentration (μg/Kg)
San Joaquin Valley surficial soil (urban site 4)				
Chlorpyrifos	[6]E0.8	100	1.6	2
Dacthal[1]	E0.7	100	1	1
Dieldrin[1]	3.3	100	2.3	3
Fipronil sulfone	1.3	100	1	1
Desulfinylfipronil (E)	E0.7	100	E0.7	E0.8
Iprodione (E)	E7.3	100	E10	E15
Metolachlor	98	100	23	55
Myclobutanil (E)	E14.7	100	E9.6	E10
Pendimethalin	51	100	E32	E40
Simazine	99	100	22	53
Trifluralin	24.3	100	12	17
Carbaryl[2] (E)	E1.7	66.6	[6]E0.7	E0.8
Metribuzin[1]	E21	66.6	[6]E7	E12
3,4-Dichloroaniline[1] (E)	E5	33.3	[5]E2	NA
Chlorpyrifos oxon (E)	–	33.3	[5]E3	NA
Dimethoate	E22.3	33.3	[5]9	NA
Malathion	23.3	33.3	[5]10	NA
cis-Permethrin	[5]E11	–	–	–
trans-Permethrin	[5]6	–	–	–
Prometon (E)	E1.3	–	–	–
San Joaquin Valley surficial soil (agricultural site 5)				
Carbaryl[2] (E)	E1.7	100	E0.73	E1
Chlorpyrifos	[6]E0.8	100	6	10
Dacthal[1]	E0.7	100	E0.93	E1
Dieldrin[1]	3.3	100	2.3	3
Fipronil sulfone	1.30	100	1.3	2
Desulfinylfipronil (E)	E0.7	100	E0.73	E0.9
Iprodione (E)	E7.3	100	E18.6	E28
Metolachlor	98	100	28.3	66
Metribuzin[1]	E21	100	E8.5	E12
Myclobutanil (E)	E14.7	100	E16	E17
Pendimethalin	51	100	E39.6	E62
Simazine	99	100	34.6	72
Trifluralin	24.3	100	10.6	17
3,4-Dichloroaniline[1] (E)	E5	66.6	[6]E2.5	3
Chlorpyrifos oxon (E)	–	33.3	[5]E5	NA
Dimethoate	E22.3	33.3	[5]E6	NA
Malathion	23.3	33.3	[5]12	NA
Paraoxon-methyl	–	33.3	[5]E4	–
cis-Permethrin	[5]E11	–	–	–
trans-Permethrin	[5]6	–	–	–
Prometon (E)	E1.3	–	–	–

Table 26.   Detection frequency and comparison of concentrations of compounds detected before and after atmospheric exposure for the surficial soil-box sample type collected from the San Joaquin Valley and Sacramento Valley, California, soil-box sites during the 2002–04 study period.—Continued

[See appendix 1 for background concentrations and sample weight of mix 1 collected near site 10 (Westly raingage), mix 2 collected near site 11 (Newman raingage), Mix 3 collected near site 5 (Modesto Irrigation District at Albers Road), and mix 4 collected near site 16 (Gridley High School); total area of soil box, 0.374 cubic meters. Abbreviations: E, estimated reported laboratory value; n, total number of samples analyzed for given sample type; NA, not applicable as there was only one detection for given compound; µg/Kg, micrograms per Kilogram; --, not detected]

Compound	[3]Surficial-soil average background concentration (µg/Kg) (n=3)	Surficial-soil detection frequency (in percent) (n=3)	Median surficial-soil concentration (µg/Kg)	Maximum surficial-soil concentration (µg/Kg)
Sacramento Valley surficial soil (urban site 16)				
3,4-Dichloroaniline[1] (E)	E0.9	100.0	E1	E1
Chlorpyrifos	--	100.0	0.95	1
Dacthal[1]	--	100.0	E0.8	E0.8
Dieldrin[1]	E0.4	100.0	E0.5	0.6
Iprodione (E)	--	100.0	E0.95	1
Simazine	10	100.0	6.5	11
Trifluralin	E0.1	100.0	E0.45	E0.6
Carbaryl[2] (E)	E2	50.0	[5]E2	NA
Parathion-methyl	--	50.0	[5]E6	NA
Paraoxon-methyl	--	50.0	[5]E4	NA
Prometon (E)	E2	50.0	[5]E2	NA

[1]Compound is biased low (B-L) because of low laboratory matrix and reagent spike recoveries, see *table 11*. Compounds with non-detections are not presented in *appendix 9* but could be present in low concentrations in the environment.

[2]Compound is biased high (B-H) because laboratory matrix and reagent spike recoveries were greater than 150 percent, see *table 11*.

[3]Values represent the average of the background individual concentrations found in soil mix 1, soil mix 2, soil mix 3 (San Joaquin Valley) prior to exposure to the atmosphere and subsequent compositing of soil mixes for placement in the soil box.

[4]Values represent the background concentration found in soil mix 4 (Sacramento Valley) prior to exposure to the atmosphere and placement in the soil box.

[5]Value represents only one detection, n=1.

[6]Value represents two detections, n=2.

2002–04 Study Period

Throughout the 2.5-year study period, a wide variety of pesticides were detected in the wet, bulk wet, dry, and bulk dry atmospheric-deposition sample types. Each sample type was analyzed for 41 currently used pesticides and 23 pesticide degradates, including oxygen analogs (oxons) of 9 organophosphate insecticides. The suite of pesticides detected in the wet deposition during this study period was similar to that of the 2001 study. The differences in pesticide concentrations found in wet deposition (and bulk wet) compared to dry deposition (and bulk dry) appeared to be closely related to the Henry's Law value of each compound, although the mass deposited by dry deposition takes place over a much longer time frame. For those compounds characterized by low Henry's Law (H) values and high water solubility, such as diazinon, rainfall was a more significant source of depositional loading to the ground than dry deposition. For those compounds having high Henry's Law values and low water solubility, such as chlorpyrifos, dry deposition was more important.

Detection frequencies for the wet-deposition sample type collected from the San Joaquin Valley and Sacramento Valley were greater than 90 percent for the compounds dacthal, diazinon, chlorpyrifos, and simazine and greater than 78 percent for samples collected as bulk-wet deposition. For wet-deposition samples collected from the San Joaquin Valley and Sacramento Valley, the median wet-deposition flux values were comparable for the compounds dacthal (0.048 µg/L and 0.035 µg/L, respectively) and chlorpyrifos (0.079 µg/L and 0.066 µg/L, respectively). Dacthal concentrations also were comparable for bulk-wet deposition collected from the San Joaquin Valley and Sacramento Valley at 0.032 µg/L and 0.042 µg/L, respectively. In contrast, the median wet-deposition flux values for the compounds diazinon and simazine were 1.6 and 4.0 times greater, respectively, in samples collected from the San Joaquin Valley. The highest wet-deposition flux measured was for the compound diazinon at 50.64 µg/m^2/day and 45.61 µg/m^2/day, collected from the funnel and autosampler, respectively, in the Sacramento Valley at the Gridley High School rooftop site 16 during the 2003 orchard dormant-spray season.

The compounds most detected in the dry-deposition sample type were metolachlor, chlorpyrifos, and dacthal (detection frequencies greater than 70 percent). Detection frequencies were greater than 90 percent in samples collected as bulk-dry deposition for the compounds simazine, chlorpyrifos, and dacthal. Dry-deposition fluxes measured in the Sacramento Valley were two orders of magnitude less than many of the same compounds detected in the San Joaquin Valley. In the San Joaquin Valley, the largest dry deposition fluxes measured were for the herbicide simazine (E 0.244 µg/m^2/day) and insecticide chlorpyrifos (E 0.233 µg/m^2/day). The compounds with the highest bulk dry-deposition rates measured in the San Joaquin Valley were diazinon (E 1.81 µg/m^2/day) and iprodione (E 0.70 µg/m^2/day). During the January 2004 orchard dormant-spray season, diazinon and its oxygen analog, diazoxon, had the highest bulk dry-deposition rates measured at E 5.26 µg/m^2/day and E 0.203 µg/m^2/day, respectively, at the Gridley High School rooftop site 16 in the Sacramento Valley .

The soil-box results showed that many of the pesticides present in the soil-box runoff (aqueous phase) also were detected in the rainfall at similar concentrations, with the exception of the aqueous phase of metolachlor, myclobutanil, and simazine, which were found in concentrations an order of magnitude greater than those detected in rainfall. Compounds detected in aqueous-phase and suspended-sediment samples collected from the soil-box runoff were compared. In the San Joaquin Valley, detection frequencies for the aqueous-phase sample type were greater than 88 percent for the compounds myclobutanil, diazinon, simazine, metolachlor, and dacthal. In the Sacramento Valley, detection frequencies for the aqueous-phase sample type were 100 percent for the compounds diazinon, dacthal, and chlorypyrifos and 92 percent for simazine. In contrast, few pesticides were detected in the soil-box runoff suspended-sediment sample type collected from the San Joaquin Valley and Sacramento Valley, with the exception of the herbicide simazine (median concentration of E 0.121 µg/L and E 0.250 µg/L, respectively). Detection frequencies for simazine in the suspended-sediment sample type were 64 percent in the San Joaquin Valley and 75 percent in the Sacramento Valley.

Prior to atmospheric exposure, background concentrations for 19 compounds were measured in the composite soil mix placed in the San Joaquin Valley urban and agricultural soil boxes, and 6 compounds were detected in the soil mix placed in the urban Sacramento Valley soil box. Results for compounds detected after atmospheric exposure in the surficial-soil samples collected during the study period indicated that for some compounds concentration (µg/Kg) increased, which indicated atmospheric deposition of these compounds onto the soil-box surface. In the San Joaquin Valley, for samples collected from both the urban and agricultural soil-box sites, the compounds chlorpyrifos, dacthal, and iprodione were detected at higher concentrations than were found in the background samples. The remaining compounds also were detected in samples collected following exposure to the atmosphere during the study period, but at lower concentrations than the measured background concentrations. In the Sacramento Valley, the six compounds detected in the background concentration of soil-box mix were also detected throughout the study period, but at much lower concentrations. Conversely, chlorpyrifos, dacthal, iprodione, parathion-methyl, and its oxygen analog, paraoxon-methyl, were detected in low concentrations in samples collected during the study period, but were not detected in the background concentration of soil mix 4.

References Cited

Baker, L.W., Fitzell, D.L., Seiber, J.N., Parker, T.R., Shibamoto, T., Poore, M.W., Longley, K.E., Tomlin, R.P., Propper, R., and Duncan, D.W., 1996, Ambient air concentrations of pesticides in California: Environmental Science and Technology, v. 30, no. 4, p. 1365–1368.

Bowman, B.J., and Sans, W.W., 1983, Determination of octanol-water partitioning coefficients (Kow) of 61 organophosphorus and carbamate insecticides and their relationship to respective water solubility (S) values: Journal of Environmental Science and Health, v. 18, no. 6, p. 667–683.

Bertoldi, G.L., Johnston, R.H., and Evenson, K.D., 1991, Ground water in the Central Valley, California–A summary report: U.S. Geological Survey Professional Paper 1401-A, 44 p.

Childress, C.J.O., Foreman, W.T., Connor, B.F., and Maloney, T.J., 1999, New reporting procedures based on long-term method detection levels and some considerations for interpretations of water-quality data provided by the U.S. Geological Survey National Water Quality Laboratory: U.S. Geological Survey Open-File Report 99–193, 19 p.

Chiou, C.T., 2002, Partition and Adsorption of Organic Contaminants in Environmental Systems, John Wiley and Sons, Inc., 274 p.

Cliath, M.M., Spencer, W.F., Farmer, W.J., Shoup T.D., and Grover R., 1980, Volatilization of S-ethyl *N,N*-dipropyl-thiocarbamate from water and wet soil during and after flood irrigation of an alfalfa field: Journal of Agricultural and Food Chemistry, v. 28, no. 3, p. 610–613.

Daines, R.H., 1952, 2,4-D as an air pollutant and its effects on various species of plants, *in* McCabe, L.C., ed., Air Pollution: Proceeding of the United States Technical Conference on Air Pollution: McGraw-Hill Book Co., Inc., New York, p. 140–143.

Domalgalski, J.L., 1997a, Pesticides in surface and ground water of the San Joaquin-Tulare Basins, California: Analysis of available data, 1966 through 1992: U.S. Geological Survey Water-Supply Paper 2468, 74 p.

Domalgalski, J.L., 1997b, Results of a prototype surface water network design for pesticides developed for the San Joaquin River Basin, California: Journal of Hydrology, v. 192, p. 33–50.

Domalgalski, J.L., Dubrovsky, N.M., and Kratzer, C.R., 1997, Pesticides in the San Joaquin River, California: Inputs from dormant spray orchards: Journal of Environmental Quality, v. 26, no. 2, p. 454–465.

Domalgalski, J.L., and Munday, Cathy, 2003, Evaluation of diazinon and chlorpyrifos concentrations and loads, and other pesticides concentrations, at selected sites in the San Joaquin Valley, California, April to August, 2001: U.S. Geological Survey Water-Resources Investigations Report 03–4088, 60 p.

Epple, J., Maguhn, J., Spitzauer, P., and Kettrup, A., 2002, Input of pesticides by atmospheric deposition: Geoderma, v. 105, issue 3–4, p. 327–349.

Glotfelty, D.E., 1978, The atmosphere as a sink for applied pesticides: Journal of the Air Pollution Control Association, v. 28, no. 9, p. 917–920.

Glotfelty, D.E., Schomburg, C.J., McChesney, M.M., Segebiel, J.C., and Seiber, J.N., 1990, Studies of the distribution, drift and volatilization of diazinon resulting from spray application to a dormant peach orchard: Chemosphere, v. 21, no. 10–11, p. 1303–1304.

Gronberg, J.M., Dubrovsky, N.M., Kratzer, C.R., Domalgalski, J.L., Brown, L.R., and Burow, K.R.,1998, Environmental setting of the San Joaquin-Tulare Basins, California: U.S. Geological Survey Water-Resources Investigations Report 97–4205, 45 p.

Howard, P.H., Michalenko, E.M., Jarvis, W.F., Basu, D.K., Sage, G.W., Meylan, W.M., Beauman, J.A., and Gray, D.A., 1991, Handbook of environmental fate and exposure data for organic chemicals, Volume III Pesticides, Lewis Publishers: Chelsea, Mich.

Kratzer, C.R., 1998, Pesticides in storm runoff from agricultural and urban areas in the Tuolumne River Basin in the vicinity of Modesto, California: U.S. Geological Survey Water-Resources Investigations Report 98–4017, 17 p.

Kratzer, C.R., 1999, Transport of diazinon in the San Joaquin River Basin, California: Journal of American Water Resources Association, v. 35, no. 2, p. 379–395.

Kratzer, C.R., Zamora, Celia, and Knifong, D.L., 2002, Diazinon and chlorpyrifos loads in the San Joaquin River Basin, California, January and February 2000: U.S. Geological Survey Water-Resources Investigations Report 02–4103, 38 p.

Kratzer, C.R., Kent, R.H., Saleh, D.K., Knifong, D.L., Dileanis, P.D., and Orlando, J.L., 2011, Trends in nutrient concentrations, loads, and yields in streams in the Sacramento, San Joaquin, and Santa Ana Basins, California, 1975–2004: U.S. Geological Survey Scientific Investigations Report 2010–5228, 112 p.

Kuivila, K.M., and Foe, C.G., 1995, Concentrations, transport and biological effects of dormant spray pesticides in the San Francisco Estuary, California: Environmental Toxicology and Chemistry, v. 14, no. 7, p. 1141–1150.

Lindley, C.E., Stewart, J.T., and Sandstrom, M.W., 1996, Determination of low concentrations of acetochlor in water by automated solid-phase extraction and gas chromotagraphy/mass spectrometry: U.S. Geological Survey Open-File Report 02–462, 11 p.

Mackay, D., Shiu, W.Y., and Ma, K.C., 1997, Illustrated handbook of physical-chemical properties and environmental fate for organic chemicals–Pesticide chemicals, vol.V,: Boca Raton, Fla., Lewis Publishers, 812 p.

Madsen, J.E., Sandstrom, M.W., and Zaugg, S.D., 2003, Methods of analysis by the U.S. Geological Survey National Water Quality Laboratory—A method supplement for the determination of fipronil and degradates in water by gas chromatography/mass spectrometry: U.S. Geological Survey Open-File Report 02–462, 11 p.

Majewski, M.S., 1991, Sources, Movement, and fate of airborne pesticides, in Frehse, H., ed., Pesticide chemistry: Advances in international research, development, and legislation: Proceedings of the Seventh International Congress of Pesticide Chemistry (IUPAC), Hamburg, 1990: Weinheim, Germany, VCH Publishers, p. 307–317.

Majewski, M.S., and Baston, D.S., 2002, Atmospheric Transport of Pesticides in the Sacramento, California, Metropolitan Area, 1996-1997, U.S. Geological Survey Water Resources Investigation Report 02–4100, 56 p.

Majewski, M.S., and Capel, P.D., 1995, Pesticides in the atmosphere: Distrubution, trends and governing factors: Chelsea, Mich., Ann Arbor Press, Pesticides in the Hydrologic System series, v. 1, 214 p.

Majewski, M.S., Desjardins, R.L., Rochette, Philippe, Pattey, Elizabeth, Seiber, J.N., and Glotfelty, D.E., 1993, Field comparison of an eddy accumulation and an aerodynamic-gradient system for measuring pesticide volatilization fluxes: Environmental Science and Technology, v. 32, no. 23, p. 3689–3698.

Majewski, M.S., Zamora, Celia, Foreman, W.T., and Kratzer, C.R., 2005, Contribution of atmospheric deposition to pesticide loads in surface water runoff: U.S. Geological Survey Open-File Report 05–1307, 1 p.

Maloney, T.J., ed., 2005, Quality management system, U.S. Geological Survey National Water Quality Laboratory: U.S. Geological Survey Open-File Report 2005-1263, version 1.3, 9 November 2005, chapters and appendixes variously paged.

Mast, M.A., Foreman, W.T., and Skaates, S.V., 2006, Organo-chlorine compounds and current-use pesticides in snow and lake sediment in Rocky Mountain National Park, Colorado, and Glacier National Park, Montana, 2002–03: U.S Geological Survey Scientific Investigations Report 2006-5119, 54 p. Available electronically only at http://pubs.water.usgs.gov/sir2006-5119/

National Center for Atmospheric Research, 2003, DAYMET climate data, average precipitation and temperature data for 1980–1997. Accessed February 7, 2009 at URL http://www.daymet.org

Panshin, S.Y., Dubrovsky, N.M., Gronberg, J.M., and Domalgalski, J.L., 1998, Occurrence and distribution of dissolved pesticides in the San Joaquin River Basin, California: U.S. Geological Survey Water-Resources Investigations Report 98–4032, 88 p.

Risebrough, R.W., 1990, Beyond long-range transport: A model of a global gas chromatographic system, chap. 27 in Kurtz, D.A., ed., Long range transport of pesticides: Chelsea, Mich., Lewis Publishers, p. 417–426.

Sandstrom, M.W., 1995, Filtration of water-sediment samples for the determination of organic compounds: U.S. Geological Survey Water-Resources Investigations Report 95–4105, 13 p.

Sandstrom, M.W., Stroppel, M.E., Foreman, W.T., and Schroeder, M.P., 2001, Methods of analysis by the U.S. Geological Survey National Water Quality Laboratory–Determination of moderate-use pesticides and selected degradates in water by C-18 solid-phase extraction and capillary-column gas chromatography/mass spectrometry with selected-ion monitoring (methods 2002/2011): U.S. Geological Survey Water-Resources Investigations Report 01–4098, 70 p.

Seiber, J.N., and Woodrow, J.E., 1995, Origin and fate of pesticides in air, in Ragsdale, N., and other, eds., Proceedings of Eighth International Congress of Pesticide Chemistry: Options 2000, Washington, D.C., American Chemical Society Washington, D.C., p. 157–172.

Soil Conservation Service, 1993, State soil geographic data base (STATSGO): Data user guide: U.S. Department of Agriculture, Natural Resources Conservation Services, National Soil Survey Center, Miscellaneous Publication 1492, 88 p.

Suntio, L.R., Shiu, W.Y., Mackay, D., Seiber, J.N., and Glotfelty, D.E., 1988, Critical review of Henry's Law constants for pesticides: Reviews of Environmental Contamination and Toxicology, v. 103, p. 1–59.

U.S. Department of Agriculture, 2005, Pesticide Properties Database: U.S. Department of Agriculture-Agricultural Research Service, accessed November 15, 2011 at http://www.ars.usda.gov/Services/docs.htm?docid=14147

U.S. Environmental Protection Agency, 1992, Definition and procedure for the determination of the method detection limit, Revision 1.11, Appendix B of 40 CFR, Part 136, p. 565–567.

U.S. Environmental Protection Agency, 2005, Reregisration Eligibility Decision for Napropramide, Case No.2450, p. 16, accessed November 15, 2011, at *http://www.epa.gov/oppsrrd1/REDs/napropamide_red.pdf*

U.S. Geological Survey, 1998, Quality Control at the U.S. Geological Survey National Water Quality Laboratory: U.S. Geological Survey Fact Sheet FS-026-98, accessed October 31, 2007 at *http://www.nwql.cr.usgs.gov/Public/pubs/QC_Fact*

Wania, Frank, and Mackay, Donald, 1996, Tracking the distribution of persistent organic pollutants: Environmental Science and Technology, v. 30, no. 9, p. 390A–396A.

Wauchope, R.D., Buttler, T.M., Hornsby, A.G., Augustijn Beckers, P.W.M., and Burt, J.P., 1992, The SCS/ARS/CES pesticide properties database for environmental decision-making, *in* Ware, G.W., ed., Reviews of Environmental Contamination and Toxicology, v. 123, p. 1–155.

Wilde, F.D., Schertz, T.L., and Radtke, D.B., 1998, Quality control in national field manual for the collection of water quality data: U.S. Geological Survey Techniques of Water-Resources Investigations, book 9, chap. A4, 103 p.

Worthing, C.R., and Walker, S.B., 1987, The pesticide manual (8th ed.): The British Crop Protection Council, Thornton Heath, UK, 1,081 p.

Zamora, Celia, Kratzer, C.R., Majewski, M.S., Knifong, D.L., 2003, Diazinon and chlorpyrifos loads in precipitation and urban and agricultural storm runoff during January and February 2001 in the San Joaquin River basin, California: U.S. Geological Waters-Resources Investigations Report 03-4091, 56 p.

Zaugg, S.D., Sandstrom, M.W., Smith, S.G., and Fehlberg, K.M., 1995, Methods of analysis by the U.S. Geological Survey National Water Quality Laboratory–Determination of pesticides in water by C-18 solid phase extraction and capillary-column gas chromatography with selected-ion monitoring: U.S. Geological Survey Open-File Report 95–181, 49p.

Appendix 1.   Sample Preparation and Instrumental Analyses

Appendix 1.   Sample Preparation and Instrumental Analyses

This appendix provides more details about the sample preparation methods and instrumental analysis methods. Sample preparation methods varied by sample type; however, after the samples were prepared, all underwent the same instrumental analysis method.

Sample Preparation Methods

All samples were shipped to the National Water-Qulaity Laboratory (NWQL) on ice. Upon receipt by the laboratory, samples were stored at 4 degrees Celsius (°C) until prepared for analysis.

Dry and Bulk Dry-Deposition Samples (NWQL Research Method LS8054)

The methods for preparing and analyzing dry and bulk-dry-deposition samples were modified several times during the study in an effort to improve performance. All dry and bulk-dry samples were fortified with 100 nanograms (ng) of surrogate compounds α-HCH-d6 and diazinon-d10 before performing extractions at the NWQL. Samples collected from April 2002 through October 2003 were extracted by adding 250 milliliters (mL) of reagent water (if no water was already present) to aid in mixing and shaking the sample in a separatory funnel with two sequential 100-mL portions of 25-percent ethyl acetate in hexane. These extracts were combined, were dried with sodium sulfate, and then were concentrated as an azeotrope by using Kuderna-Danish (K-D) distillation at 85°C followed by nitrogen gas evaporation to about 1 mL. Samples collected from November 2003 through April 2004 were extracted by adding 250 mL of reagent water (if needed) and shaking with 50-mL of dichloromethane (DCM). The DCM extract was isolated from the water fraction by using a polytetrafluoroethylene membrane (DryDisk, Horizon Technologies, Inc., Salem, N.H.). The water fraction was returned to the sample bottle, shaken with another 50-mL aliquot of DCM, and this second DCM extract was isolated from the water fraction by using the membrane. The DCM extracts were combined, were dried further with sodium sulfate, and then were concentrated and solvent-exchanged to ethyl acetate by using Kuderna-Danish distillation at 65°C, followed by nitrogen gas evaporation to about 1 mL.

Extracts of samples collected in 2002 were cleaned by using a 0.5-gram (g) C-18 solid-phase extraction (SPE) column positioned over a 1-g Florisil® SPE column (International Sorbent Technologies; now Biotage, LLC, Charlotte, N.C.). Analytes were eluted from each of the columns with 6 mL of ethyl acetate, and the resulting eluents were reduced in volume and exchanged to toluene by nitrogen gas evaporation and then transferred to gas chromatography (GC) vials with a toluene rinse to achieve a 0.5 mL final volume for instrument analysis. The use of combined C-18/Florisil®

SPE columns provided cleaner extracts, but some analytes, especially diazinon oxon, chlorpyrifos oxon, and malathion oxon, had poor recoveries with this approach.

For samples collected in 2003 and 2004, each extract was introduced to a 0.5-g CarboPrep® 90 (Restek Corp., Belefonte, Pa.) graphitized-carbon SPE column, and the analytes were eluted with 13 mL of a 50-percent DCM in ethyl acetate solution. The resulting eluents were reduced to 1–2 mL by using micro-Kuderna-Danish distillation at 85°C. A procedural internal standard solution of perdeuterated polycyclic aromatic hydrocarbons was added to these extracts before final solvent reduction and exchange to toluene by using nitrogen gas evaporation. The extracts were transferred to vials with toluene rinse to achieve a 0.5-mL final volume for instrumental analysis.

Laboratory reagent blanks and spiked samples were prepared by using solvents and reagent water in amounts comparable to typical field samples for the set of prepared samples. Spiked samples were fortified with 100 ng of method analytes. These quality-control (QC) samples were prepared subsequent to and analyzed along with (and by using the same methods as) the field samples. Bulk-deposition samples collected at Modesto Irrigation District (MID) at MID, MID at Albers Road, Turlock Airport, and Turlock near Idaho Road from January 14 to February 14, 2003, were ruined during sample preparation.

Suspended Sediment Sample (NWQL Research Method LS7503)

The glass-fiber filters (GFFs) containing suspended sediment were extracted by using an Accelerated Solvent Extraction (ASE) system model 200 (ASE-200) or model 300 (ASE-300) instrument (Dionex, Sunnyvale, Calif.). The samples of suspended sediment collected on the GFFs were folded (particles inward), and each of the samples was placed inside a pre-weighed glass-fiber thimble (prebaked at 450 °C for 4 hours) and pre-positioned inside a 22-mL ASE-200 stainless-steel extraction cell. For samples with multiple GFFs, the GFFs were composited by placing the folded filters inside a 100-mL ASE model 300 extraction cell that also contained a pre-weighed 30-millimeter (mm) diameter GFF positioned as a support above the cell's bottom cap (thimbles were not used in the 100-mL cells). The composite samples collected on December 3, 2003, and February 20, 2004, from site 16 required 2 cells to extract the 9 GFFs and 6 cells for the 38 GFFs, respectively. All samples (including laboratory blanks and spiked samples) were fortified with 100 ng of α-HCH-d6 and diazinon-d10 surrogates by addition on top of the GFF. For the two samples requiring multiple cells (site 16), surrogate addition was proportioned between cells. All samples were extracted with 25-percent acetone in DCM at 100°C and 10,342 kilopascals (kPa) of pressure by using three 10-minute static cycles on the ASE model 200 or 300 instruments to isolate the analytes in organic solvent.

Quality-control samples extracted along with the field samples included laboratory reagent blanks and spiked samples prepared by using one GFF each for the ASE-200 and four GFFs each for the ASE-300 extractions. The GFFs for quality-control samples were moistened with reagent water prior to loading in extraction cells to simulate field samples. Spiked samples were fortified with 100 ng of all method analytes. These QC samples were subsequently prepared and analyzed along with (and by using the same methods as) the field samples. Suspended-material samples from site 4 on November 7, 2002, and site 5 on November 14, 2002, December 16, 2002, and January 14, 2004, were ruined during sample preparation at the NWQL.

Following extraction, the contents of each ASE cell were placed in beakers and dried at 120°C for 24 hours, then cooled in a desiccator for 1 hour, and finally weighed. The mass of suspended material isolated on the GFFs was calculated by subtracting the initial GFF and thimble or cap-filter tare weights from the post-ASE final weights. Masses ranged from a maximum of 21.8 g for the site 16 sample from February 20, 2004, to a minimum of negative or 0 g for samples with low particle loading, where either pieces of filter were lost or the initial GFF tare to the nearest 0.1 g was not precise enough to distinguish small mass loadings to the GFF.

For samples requiring multiple cells, ASE extracts were combined prior to further sample preparation. Residual water was removed from ASE extracts by reaction with sodium sulfate. Extracts were reduced in volume and exchanged to ethyl acetate by using Kuderna-Danish (K-D) distillation at 75°C and nitrogen gas evaporation to about 1 mL. Each extract was introduced to a 0.5-g Carboprep® 90 SPE column, and the analytes were eluted with 13 mL of a 50-percent DCM in ethyl acetate solution. Eluents were reduced to 1–2 mL by using micro-Kuderna-Danish distillation at 85°C. A procedural internal-standard solution of perdeuterated polycyclic aromatic hydrocarbons was added to the extracts before final solvent reduction and exchange to toluene by using nitrogen gas evaporation. The extracts were transferred to vials with toluene rinse to achieve a 0.5-mL final volume for instrumental analysis.

Surficial-Soil Samples (NWQL Research Method LS5503)

Each surficial soil sample was stirred by using a pre-cleaned stainless-stell spoon to break up any clumps, and its dry weight as a percentage of its original mass (ranging from 81– 8 percent) was determined on a high-temperature drying balance. Aliquots of 28 to 46-g dry-weight equivalent of the soil samples were mixed with up to 2 g of pelletized diatomaceous earth (Hydromatrix from Varian, Inc., now Agilent Technologies, Santa Clara, Calif.) to separate the particles in the soil matrix and bind any residual moisture. The soil

samples were then transferred to 33-mL extraction cells, followed by fortification with 100 ng of the surrogate compounds αHCH-d6 and diazinon-d10, and finally extracted with 25-percent acetone in DCM at 100°C and 10,342 kilopascals (kPa) by using three 10-min static cycles on the ASE-200 to isolate the analytes in organic solvent.

Quality-control samples extracted along with the field samples included laboratory reagent blanks and reagent spiked samples, which were both prepared by using dry Ottawa Sand. The spiked samples were fortified with 100 ng of all method analytes. A laboratory matrix spike sample was prepared by adding 200 ng of all method analytes to an aliquot of soil mix 2 sample. Duplicate aliquots were extracted and analyzed for the July 21, 2003, sample from Gridley High School, and for soil mix 1, 2 and 3 samples.

Residual water was removed from the ASE extracts by adding sodium sulfate. Extracts were reduced in volume, and the solvent-exchanged to ethyl acetate byusing K-D distillation at 75 °C and nitrogen gas evaporation to about 1 mL. Each extract was introduced to a 0.5-g CarboPrep® 90 graphitized-carbon SPE column, and the analytes were eluted with 13 mL of 50-percent DCM in ethyl acetate solution. Eluents were reduced by using micro-K-D distillation at 85°C and nitrogen gas evaporation, if needed, to about 2.5 mL. Analytes were further isolated from elemental sulfur and some unwanted coextractants by gel-permeation chromatography (GPC; Waters Corp., Milford, Mass.) of a 1.4-mL aliquot of the extract, with separation on a 300-mm long by 7.5-mm internal diameter GPC column containing 5-micrometer diameter particles of polymeric styrene-divinylbenzene with 50-angstrom pores (Polymer Laboratories Ltd., now Agilent Technologies) by using a 1-mL/minute ethyl acetate mobile phase. The 9-mL collected GPC fractions were reduced as hexane/ethyl acetate azeotropes by micro-K-D distillation at 75°C to 1–2 mL. A procedural internal-standard solution of perdeuterated polycyclic aromatic hydrocarbons was added to each of the extracts before final solvent reduction and exchange to toluene by using nitrogen gas evaporation. The extracts were transferred to GC vials with toluene rinses to achieve a 0.5-mL final volume. The extracts were transferred to vials with toluene rinse to achieve a 0.5-mL final volume for instrumental analysis.

Instrumental Analysis (All Sample Types)

Analytes of interest in the extracts were determined by capillary-column gas chromatography/electron-impact mass spectrometry (GC-MS) operated in the selected-ion monitoring mode as described in Sandstrom and others (2001). The analytes of interest included selected pesticides and degradates that typically are determined by the method described in Madsen and others (2003) and additional compounds determined by similar methods described by Zaugg and others (1995) and Madsen and others (2003).

References Cited

Madsen, J.E., Sandstrom, M.W., and Zaugg, S.D., 2003, Methods of analysis by the U.S. Geological Survey National Water Quality Laboratory—A method supplement for the determination of fipronil and degradates in water by gas chromatography/mass spectrometry: U.S. Geological Survey Open-File Report 02–462, 11 p.

Sandstrom, M.W., Stroppel, M.E., Foreman, W.T., and Schroeder, M.P., 2001, Methods of analysis by the U.S. Geological Survey National Water Quality Laboratory–Determination of moderate-use pesticides and selected degradates in water by C-18 solid-phase extraction and capillary-column gas chromatography/mass spectrometry with selected-ion monitoring (methods 2002/2011): U.S. Geological Survey Water-Resources Investigations Report 01–4098, 70 p.

Zaugg, S.D., Sandstrom, M.W., Smith, S.G., and Fehlberg, K.M., 1995, Methods of analysis by the U.S. Geological Survey National Water Quality Laboratory–Determination of pesticides in water by C-18 solid phase extraction and capillary-column gas chromatography with selected-ion monitoring: U.S. Geological Survey Open-File Report 95–181, 49 p.

Appendix 2.   Analytical Results for Wet Deposition (Rainfall) in Micrograms Per Liter (µg/L) for Samples With at Least One Detection Collected at Four Urban and Agricultural Sites, and Storm Run-Off Samples at the McHenry Storm Drain in Modesto, California, January 23–26, 2001

Appendix 2. Analytical results for wet deposition (rainfall) in micrograms per liter (μg/L) for samples with at least one detection collected at four urban and agricultural sites, and storm run-off samples at the McHenry storm drain in Modesto, California, January 23–26, 2001.

[All compounds were analyzed by gas chromatography/mass spectrometry (GC/MS) at the U.S. Geological Survey (USGS) National Water Quality Laboratory (NWQL). The Chemical Abstract Service Number is given below each compound name in brackets. The five digit number in parenthesis is a code used by the USGS to uniquely identify the given compound. Abbreviations: E, estimated reported laboratory value; hh:mm, hour:minute; Jan, January; [-], no Chemical Abstract Service Number available for given compound; <, less than laboratory reporting limit; (E), estimated compound due to known poor performance by applied method (Childress and others, 1999; Sandstrom and others, 2001)]

Site number	Site name	Site identification number	Date	Time (hh:mm)	Atrazine [1912-24-9] (39632)	Benfluralin [1861-40-1] (82673)	Carbaryl (E) [1563-66-2] (82680)	Chlorpyrifos [2921-88-2] (38933)	Dacthal [1861-32-1] (82682)	p,p'-DDE [72-55-9] (34653)
	Urban sites									
1	Waste Water Treatment Plant rooftop at Modesto	373637121004601	Jan 26, 2001	07:30	<0.007	<0.01	<0.041	0.086	0.011	<0.003
2	Cadoni Road lift station at Modesto	373725120543701	Jan 26, 2001	08:30	<0.007	<0.01	E0.074	0.071	0.009	<0.003
3	Bowen and Aloha Street at Modesto	374028120594301	Jan 26, 2001	08:00	<0.007	<0.01	<0.1	0.057	0.009	<0.003
4	Modesto Irrigation District gage rooftop at Modesto	373834121000601	Jan 26, 2001	08:30	<0.007	<0.01	E0.085	0.063	0.013	<0.003
	Agricultural sites									
5	Modesto Irrigation District gage rooftop at Albers Road	373841120504801	Jan 26, 2001	09:50	<0.007	<0.01	E0.087	0.148	0.005	<0.003
6	Modesto Irrigation District lateral #4	373750121092601	Jan 26, 2001	07:10	<0.007	<0.01	<0.041	0.034	0.013	<0.003
7	Turlock Irrigation District lateral #3	373228120551201	Jan 26, 2001	13:40	E0.003	<0.01	<0.09	0.052	0.009	E0.006
8	Tully Road near Modesto	374351121004701	Jan 26, 2001	09:10	<0.007	<0.01	E0.081	0.105	0.01	<0.003
	Urban storm drain									
9	McHenry storm drain at Bodem Street	373847120590801	Jan 23, 2001	23:50	<0.007	<0.01	E0.015	0.007	<0.003	<0.003
9	McHenry storm drain at Bodem Street	373847120590801	Jan 25, 2001	18:00	<0.007	<0.01	E0.411	<0.05	0.004	<0.003
9	McHenry storm drain at Bodem Street	373847120590801	Jan 25, 2001	19:00	<0.007	<0.01	E0.297	0.023	0.006	<0.003
9	McHenry storm drain at Bodem Street	373847120590801	Jan 25, 2001	20:00	<0.007	<0.01	E0.015	0.035	0.009	<0.003
9	McHenry storm drain at Bodem Street	373847120590801	Jan 25, 2001	21:00	<0.007	E0.004	E0.034	0.033	0.009	<0.003
9	McHenry storm drain at Bodem Street	373847120590801	Jan 25, 2001	22:00	<0.007	<0.01	E0.045	0.031	0.009	<0.003
9	McHenry storm drain at Bodem Street	373847120590801	Jan 25, 2001	23:00	<0.007	<0.01	E0.179	0.033	<0.003	<0.003
9	McHenry storm drain at Bodem Street	373847120590801	Jan 26, 2001	00:00	<0.007	<0.01	E0.181	0.018	<0.003	<0.003
9	McHenry storm drain at Bodem Street	373847120590801	Jan 26, 2001	01:00	<0.007	<0.01	E0.209	0.023	<0.003	<0.003
9	McHenry storm drain at Bodem Street	373847120590801	Jan 26, 2001	02:00	<0.007	<0.01	E0.207	0.033	<0.003	<0.003
	Equipment blank									
9	McHenry storm drain at Bodem Street	373847120590801	Jan 23, 2001	23:58	<0.007	<0.01	<0.041	<0.005	<0.003	<0.003
	Field blank									
9	McHenry storm drain at Bodem Street	373847120590801	Jan 25, 2001	13:08	<0.007	<0.01	<0.041	<0.005	<0.003	<0.003
9	McHenry storm drain at Bodem Street	373847120590801	Jan 25, 2001	21:08	<0.007	<0.01	<0.041	<0.005	<0.003	<0.003

Appendix 2. Analytical results for wet deposition (rainfall) in micrograms per liter (µg/L) for samples with at least one detection collected at four urban and agricultural sites, and storm run-off samples at the McHenry storm drain in Modesto, California, January 23–26, 2001.—Continued

[All compounds were analyzed by gas chromatography/mass spectrometry (GC/MS) at the U.S. Geological Survey (USGS) National Water Quality Laboratory (NWQL). The Chemical Abstract Service Number is given below each compound name in brackets. The five digit number in parenthesis is a code used by the USGS to uniquely identify the given compound. Abbreviations: E, estimated reported laboratory value; hh:mm, hour:minute; Jan, January; [-], no Chemical Abstract Service Number available for given compound; <, less than laboratory reporting limit; (E), estimated compound due to known poor performance by applied method (Childress and others, 1999; Sandstrom and others, 2001)]

Site number	Site name	Site identification number	Date	Time (hh:mm)	Diazinon [333-41-5] (39572)	Ethalfluralin [55283-68-6] (82663)	Malathion [121-75-5] (39532)	Metolachlor [51218-45-2] (39415)	Napropamide [15299-99-7] (82684)	Pendimethalin [40487-42-1] (82683)
						Urban sites—Continued				
1	Waste Water Treatment Plant rooftop at Modesto	373637121004601	Jan 26, 2001	07:30	0.908	<0.009	<0.025	E0.004	0.024	0.18
2	Cadoni Road lift station at Modesto	373725120543701	Jan 26, 2001	08:30	0.472	<0.009	<0.027	E0.004	0.018	E0.118
3	Bowen and Aloha Street at Modesto	374028120594301	Jan 26, 2001	08:00	0.486	<0.009	<0.027	E0.004	0.014	0.121
4	Modesto Irrigation District gage rooftop at Modesto	373834121000601	Jan 26, 2001	08:30	0.544	<0.009	<0.027	E0.004	0.008	E0.122
						Agricultural sites—Continued				
5	Modesto Irrigation District gage rooftop at Albers Road	373841120504801	Jan 26, 2001	09:50	0.87	<0.009	<0.027	<0.013	<0.007	0.083
6	Modesto Irrigation District lateral #4	373750121092601	Jan 26, 2001	07:10	0.188	<0.009	<0.027	<0.013	<0.007	0.02
7	Turlock Irrigation District lateral #3	373228120551201	Jan 26, 2001	13:40	0.491	E0.002	E0.015	E0.004	0.028	E0.292
8	Tully Road near Modesto	374351121004701	Jan 26, 2001	09:10	0.675	<0.009	<0.027	E0.005	0.037	E0.257
						Urban storm drain—Continued				
9	McHenry storm drain at Bodem Street	373847120590801	Jan 23, 2001	23:50	0.334	<0.009	<0.027	<0.013	0.019	0.034
9	McHenry storm drain at Bodem Street	373847120590801	Jan 25, 2001	18:00	0.506	<0.009	<0.15	E0.009	0.051	<0.075
9	McHenry storm drain at Bodem Street	373847120590801	Jan 25, 2001	19:00	0.947	<0.009	E0.065	E0.008	0.035	E0.105
9	McHenry storm drain at Bodem Street	373847120590801	Jan 25, 2001	20:00	0.922	<0.009	0.051	E0.005	0.02	0.092
9	McHenry storm drain at Bodem Street	373847120590801	Jan 25, 2001	21:00	0.708	<0.009	0.04	E0.004	0.017	0.097
9	McHenry storm drain at Bodem Street	373847120590801	Jan 25, 2001	22:00	0.691	<0.009	E0.022	E0.004	0.019	0.12
9	McHenry storm drain at Bodem Street	373847120590801	Jan 25, 2001	23:00	0.727	<0.009	<0.027	<0.013	0.011	0.061
9	McHenry storm drain at Bodem Street	373847120590801	Jan 26, 2001	00:00	0.756	<0.009	<0.027	<0.013	<0.02	0.033
9	McHenry storm drain at Bodem Street	373847120590801	Jan 26, 2001	01:00	0.761	<0.009	<0.027	<0.013	0.023	0.047
9	McHenry storm drain at Bodem Street	373847120590801	Jan 26, 2001	02:00	0.823	<0.009	<0.03	<0.013	<0.01	0.059
						Equipment blank—Continued				
9	McHenry storm drain at Bodem Street	373847120590801	Jan 23, 2001	23:58	<0.005	<0.009	<0.027	<0.013	<0.007	<0.01
						Field blank—Continued				
9	McHenry storm drain at Bodem Street	373847120590801	Jan 25, 2001	13:08	<0.005	<0.009	<0.027	<0.013	<0.007	<0.01
9	McHenry storm drain at Bodem Street	373847120590801	Jan 25, 2001	21:08	<0.005	<0.009	<0.027	<0.013	<0.007	<0.01

Appendix 2. Analytical results for wet deposition (rainfall) in micrograms per liter (µg/L) for samples with at least one detection collected at four urban and agricultural sites, and storm run-off samples at the McHenry storm drain in Modesto, California, January 23–26, 2001.—Continued

[All compounds were analyzed by gas chromatography/mass spectrometry (GC/MS) at the U.S. Geological Survey (USGS) National Water Quality Laboratory (NWQL). The Chemical Abstract Service Number is given below each compound name in brackets. The five digit number in parenthesis is a code used by the USGS to uniquely identify the given compound. Abbreviations: E, estimated reported laboratory value; hh:mm, hour:minute; Jan, January; [-], no Chemical Abstract Service Number available for given compound; <, less than laboratory reporting limit; (E), estimated compound due to known poor performance by applied method (Childress and others, 1999; Sandstrom and others, 2001)]

Site number	Site name	Site identification number	Date	Time (hh:mm)	cis-Permethrin [54774-45-7] (82687)	Propargite [2312-35-8] (82685)	Simazine [122-34-9] (04035)	Terbacil [5902-51-2] (82665)	Trifluralin [1582-09-8] (82661)
		Urban sites—Continued							
1	Waste Water Treatment Plant rooftop at Modesto	37363712100460 1	Jan 26, 2001	07:30	<0.006	<0.02	0.034	<0.034	<0.009
2	Cadoni Road lift station at Modesto	37372512054370 1	Jan 26, 2001	08:30	<0.006	<0.02	0.037	<0.034	0.048
3	Bowen and Aloha Street at Modesto	37402812059430 1	Jan 26, 2001	08:00	<0.006	<0.02	0.028	<0.034	0.025
4	Modesto Irrigation District gage rooftop at Modesto	37383412100060 1	Jan 26, 2001	08:30	<0.006	<0.02	0.025	E0.087	<0.009
		Agricultural sites—Continued							
5	Modesto Irrigation District gage rooftop at Albers Road	37384112050480 1	Jan 26, 2001	09:50	<0.006	<0.02	0.036	<0.034	E0.006
6	Modesto Irrigation District lateral #4	37375012109260 1	Jan 26, 2001	07:10	<0.006	<0.02	0.013	<0.034	E0.007
7	Turlock Irrigation District lateral #3	37322812055120 1	Jan 26, 2001	13:40	0.007	E0.01	0.066	E0.018	0.01
8	Tully Road near Modesto	37435112100470 1	Jan 26, 2001	09:10	<0.006	<0.02	0.172	<0.034	0.03
		Urban storm drain—Continued							
9	McHenry storm drain at Bodem Street	37384712059080 1	Jan 23, 2001	23:50	<0.006	<0.02	0.051	<0.034	E0.006
9	McHenry storm drain at Bodem Street	37384712059080 1	Jan 25, 2001	18:00	<0.006	<0.02	0.111	<0.034	<0.009
9	McHenry storm drain at Bodem Street	37384712059080 1	Jan 25, 2001	19:00	<0.006	<0.02	0.105	<0.034	<0.009
9	McHenry storm drain at Bodem Street	37384712059080 1	Jan 25, 2001	20:00	<0.006	<0.02	0.061	<0.034	E0.006
9	McHenry storm drain at Bodem Street	37384712059080 1	Jan 25, 2001	21:00	<0.006	<0.02	0.042	<0.034	E0.008
9	McHenry storm drain at Bodem Street	37384712059080 1	Jan 25, 2001	22:00	<0.006	<0.02	0.045	<0.034	E0.007
9	McHenry storm drain at Bodem Street	37384712059080 1	Jan 25, 2001	23:00	<0.006	<0.02	0.051	<0.034	E0.006
9	McHenry storm drain at Bodem Street	37384712059080 1	Jan 26, 2001	00:00	<0.006	<0.02	0.055	<0.034	<0.009
9	McHenry storm drain at Bodem Street	37384712059080 1	Jan 26, 2001	01:00	<0.006	<0.02	0.034	<0.034	<0.009
9	McHenry storm drain at Bodem Street	37384712059080 1	Jan 26, 2001	02:00	<0.006	<0.02	0.054	<0.034	0.005
		Equipment blank—Continued							
9	McHenry storm drain at Bodem Street	37384712059080 1	Jan 23, 2001	23:58	<0.006	<0.02	<0.011	<0.034	<0.009
		Field blank—Continued							
9	McHenry storm drain at Bodem Street	37384712059080 1	Jan 25, 2001	13:08	<0.006	<0.02	<0.011	<0.034	<0.009
9	McHenry storm drain at Bodem Street	37384712059080 1	Jan 25, 2001	21:08	<0.006	<0.02	<0.011	<0.034	<0.009

References Cited

Childress, C.J.O., Foreman, W.T., Connor, B.F., and Maloney, T.J., 1999, New reporting procedures based on long-term method detection levels and some considerations for interpretations of water-quality data provided by the U.S. Geological Survey National Water Quality Laboratory: U.S. Geological Survey Open-File Report 99–193, 19 p.

Sandstrom, M.W., Stroppel, M.E., Foreman, W.T., and Schroeder, M.P., 2001, Methods of analysis by the U.S. Geological Survey National Water Quality Laboratory–Determination of moderate-use pesticides and selected degradates in water by C-18 solid-phase extraction and capillary-column gas chromatography/mass spectrometry with selected-ion monitoring (methods 2002/2011): U.S. Geological Survey Water-Resources Investigations Report 01–4098, 70 p.

Appendix 3.   Calculated Flux Values Presented in Micrograms Per Square Meter Per Day (μg/m^2/day) for Wet-Deposition Samples With at Least One Detection Collected During the 2002–04 Study Period at Eight Sites in the Central Valley, California

Appendix 3. Calculated flux values presented as micrograms per square meter per day ($\mu g/m^2/day$) for wet-deposition samples with at least one detection collected during the 2002–04 study period at eight sites in the Central Valley, California.

[All compounds were analyzed by gas chromatography/mass spectrometry (GCMS) at the U.S. Geological Survey (USGS) National Water Quality Laboratory (NWQL). The Chemical Abstract Service Number is given below each compound name in brackets. The five-digit number in parenthesis is a code used by the USGS to uniquely identify the given compound. Flux calculated as the analytical result in $\mu g/L$ multiplied by the total sample volume collected in liters divided by the area of the respective sampler type (Funnel=0.0731 m^2, Autosampler (AS)= 0.0614 m^2) and sample composite time in days. Abbreviations: E, estimated reported laboratory value; na, not applicable; (E), estimated compound due to known poor performance by applied method (Childress and others, 1999; Sandstrom and others, 2001); hh:mm, hour:minute; –, compound not detected; [-], no Chemical Abstract Service Number available for given compound]

Site number	Site name	Site identification number	Start date	Start time (hh:mm)	End date	End time (hh:mm)	Composite sample time, days	Rain events during composite sample time, days	Sample type	Sampler type
11	Newman rain gage at wasteway levee near Draper Road	3717351210331201	Feb 9, 2002	09:00	Feb 11, 2002	10:45	2	1	wet	Funnel
11	Newman rain gage at wasteway levee near Draper Road	3717351210331201	Feb 15, 2002	11:15	Feb 18, 2002	12:30	3	2	wet	Funnel
11	Newman rain gage at wasteway levee near Draper Road	3717351210331201	Mar 5, 2002	10:15	Mar 8, 2002	10:15	3	2	wet	Funnel
11	Newman rain gage at wasteway levee near Draper Road	3717351210331201	Mar 15, 2002	10:00	Mar 19, 2002	10:20	4	3	wet	Funnel
11	Newman rain gage at wasteway levee near Draper Road	3717351210331201	Mar 22, 2002	09:40	Mar 24, 2002	13:20	2	1	wet	Funnel
11	Newman rain gage at wasteway levee near Draper Road	3717351210331201	Nov 7, 2002	02:00	Nov 10, 2002	03:00	3	2	wet	Funnel
11	Newman rain gage at wasteway levee near Draper Road	3717351210331201	Dec 9, 2002	10:15	Dec 16, 2002	09:50	7	5	wet	Funnel
11	Newman rain gage at wasteway levee near Draper Road	3717351210331201	Dec 16, 2002	09:50	Dec 22, 2002	13:00	6	6	wet	Funnel
11	Newman rain gage at wasteway levee near Draper Road	3717351210331201	Jan 8, 2003	14:00	Jan 14, 2003	10:10	6	1	wet	Funnel
11	Newman rain gage at wasteway levee near Draper Road	3717351210331201	Mar 13, 2003	10:20	Mar 18, 2003	10:15	5	3	wet	Funnel
11	Newman rain gage at wasteway levee near Draper Road	3717351210331201	Mar 18, 2003	10:15	Mar 27, 2003	10:20	9	2	wet	Funnel
11	Newman rain gage at wasteway levee near Draper Road	3717351210331201	Nov 5, 2003	10:50	Nov 10, 2003	10:30	5	4	wet	Funnel
11	Newman rain gage at wasteway levee near Draper Road	3717351210331201	Jan 29, 2004	14:00	Feb 4, 2004	12:30	6	2	wet	Funnel
11	Newman rain gage at wasteway levee near Draper Road	3717351210331201	Feb 4, 2004	12:30	Feb 19, 2004	13:00	15	5	wet	Funnel
11	Newman rain gage at wasteway levee near Draper Road	3717351210331201	Feb 19, 2004	13:00	Feb 24, 2004	14:30	5	3	wet	Funnel
11	Newman rain gage at wasteway levee near Draper Road	3717351210331201	Feb 24, 2004	14:30	Feb 27, 2004	14:50	3	3	wet	Funnel
12	Turlock rain gage near Idaho Road	3727131205334901	Jan 25, 2002	12:00	Jan 29, 2002	12:00	4	3	wet	Funnel
12	Turlock rain gage near Idaho Road	3727131205334901	Feb 7, 2002	09:30	Feb 9, 2002	13:00	2	2	wet	Funnel
12	Turlock rain gage near Idaho Road	3727131205334901	Feb 15, 2002	14:15	Feb 18, 2002	11:00	3	2	wet	Funnel
12	Turlock rain gage near Idaho Road	3727131205334901	Mar 5, 2002	16:00	Mar 8, 2002	10:45	3	2	wet	Funnel
12	Turlock rain gage near Idaho Road	3727131205334901	Mar 9, 2002	14:20	Mar 11, 2002	09:30	2	1	wet	Funnel
12	Turlock rain gage near Idaho Road	3727131205334901	Mar 17, 2002	07:00	Mar 18, 2002	13:00	1	1	wet	Funnel
12	Turlock rain gage near Idaho Road	3727131205334901	Mar 22, 2002	10:00	Mar 24, 2002	08:30	2	1	wet	Funnel
12	Turlock rain gage near Idaho Road	3727131205334901	Nov 7, 2002	02:00	Nov 10, 2002	03:00	3	2	wet	Funnel
12	Turlock rain gage near Idaho Road	3727131205334901	Dec 16, 2002	11:15	Dec 16, 2002	10:50	7	5	wet	Funnel
12	Turlock rain gage near Idaho Road	3727131205334901	Dec 16, 2002	10:40	Dec 22, 2002	13:30	6	6	wet	Funnel
12	Turlock rain gage near Idaho Road	3727131205334901	Jan 8, 2003	14:40	Jan 14, 2003	11:00	6	1	wet	Funnel
12	Turlock rain gage near Idaho Road	3727131205334901	Feb 14, 2003	09:20	Feb 21, 2003	11:00	7	3	wet	Funnel
12	Turlock rain gage near Idaho Road	3727131205334901	Mar 13, 2003	11:00	Mar 18, 2003	11:00	5	3	wet	Funnel
12	Turlock rain gage near Idaho Road	3727131205334901	Mar 18, 2003	11:00	Mar 27, 2003	11:20	9	2	wet	Funnel

Appendix 3. Calculated flux values presented as micrograms per square meter per day (µg/m²/day) for wet-deposition samples with at least one detection collected during the 2002–04 study period at eight sites in the Central Valley, California.—Continued

[All compounds were analyzed by gas chromatography mass spectrometry (GCMS) at the U.S. Geological Survey (USGS) National Water Quality Laboratory (NWQL). The Chemical Abstract Service Number is given below each compound name in brackets. The five-digit number in parenthesis is a code used by the USGS to uniquely identify the given compound. Flux calculated as the analytical result in µg/L multiplied by the total sample volume collected in liters divided by the area of the respective sampler type (Funnel=0.0731 m², Autosampler (AS)= 0.0614 m²) and sample composite time in days. Abbreviations: E, estimated reported laboratory value; na, not applicable; (E), estimated compound due to known poor performance by applied method (Childress and others, 1999; Sandstrom and others, 2001); hh:mm, hour:minute, —, compound not detected; [-], no Chemical Abstract Service Number available for given compound]

Site number	Site name	Site identification number	Start date	Start time (hh:mm)	End date	End time (hh:mm)	Composite sample time, days	Rain events during composite sample time, days	Sample type	Sampler type
12	Turlock rain gage near Idaho Road	3727131205349001	Apr 29, 2003	10:30	May 9, 2003	17:00	10	3	wet	Funnel
12	Turlock rain gage near Idaho Road	3727131205349001	Nov 5, 2003	11:30	Nov 10, 2003	11:10	5	4	wet	Funnel
12	Turlock rain gage near Idaho Road	3727131205349001	Jan 29, 2004	13:00	Feb 4, 2004	11:30	6	2	wet	Funnel
12	Turlock rain gage near Idaho Road	3727131205349001	Feb 4, 2004	11:30	Feb 19, 2004	11:50	15	5	wet	Funnel
12	Turlock rain gage near Idaho Road	3727131205349001	Feb 19, 2004	11:50	Feb 24, 2004	13:40	5	3	wet	Funnel
12	Turlock rain gage near Idaho Road	3727131205349001	Feb 24, 2004	13:40	Feb 27, 2004	14:00	3	3	wet	Funnel
12	Turlock rain gage near Idaho Road	3727131205349001	Mar 25, 2004	13:35	Mar 26, 2004	13:45	1	1	wet	Funnel
13	Turlock Airport rain gage	3728571204114001	Jan 25, 2002	12:00	Jan 29, 2002	09:40	4	3	wet	Funnel
13	Turlock Airport rain gage	3728571204114001	Feb 6, 2002	15:40	Feb 11, 2002	10:55	5	2	wet	Funnel
13	Turlock Airport rain gage	3728571204114001	Feb 15, 2002	12:00	Feb 18, 2002	13:50	3	2	wet	Funnel
13	Turlock Airport rain gage	3728571204114001	Mar 5, 2002	11:30	Mar 8, 2002	11:15	3	2	wet	Funnel
13	Turlock Airport rain gage	3728571204114001	Mar 8, 2002	11:15	Mar 19, 2002	11:25	3	2	wet	Funnel
13	Turlock Airport rain gage	3728571204114001	Mar 15, 2002	12:00	Mar 19, 2002	12:00	4	3	wet	Funnel
13	Turlock Airport rain gage	3728571204114001	Mar 22, 2002	10:30	Mar 24, 2002	10:30	2	1	wet	Funnel
13	Turlock Airport rain gage	3728571204114001	Nov 7, 2002	02:00	Nov 10, 2002	03:00	3	2	wet	Funnel
13	Turlock Airport rain gage	3728571204114001	Dec 9, 2002	12:00	Dec 16, 2002	11:30	7	5	wet	Funnel
13	Turlock Airport rain gage	3728571204114001	Dec 16, 2002	11:30	Dec 22, 2002	14:00	6	6	wet	Funnel
13	Turlock Airport rain gage	3728571204114001	Jan 9, 2003	13:50	Jan 14, 2003	11:40	5	1	wet	Funnel
13	Turlock Airport rain gage	3728571204114001	Feb 12, 2003	08:00	Feb 14, 2003	10:15	2	2	wet	Funnel
13	Turlock Airport rain gage	3728571204114001	Feb 14, 2003	10:00	Feb 21, 2003	11:40	7	3	wet	Funnel
13	Turlock Airport rain gage	3728571204114001	Mar 12, 2003	10:00	Mar 18, 2003	11:50	6	3	wet	Funnel
13	Turlock Airport rain gage	3728571204114001	Apr 26, 2003	13:00	Apr 29, 2003	11:15	3	1	wet	Funnel
13	Turlock Airport rain gage	3728571204114001	Apr 29, 2003	11:15	May 5, 2003	10:40	6	3	wet	Funnel
13	Turlock Airport rain gage	3728571204114001	Nov 5, 2003	12:10	Nov 10, 2003	11:40	5	4	wet	Funnel
13	Turlock Airport rain gage	3728571204114001	Jan 29, 2004	12:00	Feb 4, 2004	10:30	6	2	wet	Funnel
13	Turlock Airport rain gage	3728571204114001	Feb 19, 2004	10:10	Feb 24, 2004	11:40	5	3	wet	Funnel
13	Turlock Airport rain gage	3728571204114001	Feb 24, 2004	11:40	Feb 27, 2004	10:40	3	3	wet	Funnel
13	Turlock Airport rain gage	3728571204114001	Mar 25, 2004	12:30	Mar 26, 2004	13:00	1	1	wet	Funnel

Appendix 3. Calculated flux values presented as micrograms per square meter per day ($\mu g/m^2/day$) for wet-deposition samples with at least one detection collected during the 2002–04 study period at eight sites in the Central Valley, California.—Continued

[All compounds were analyzed by gas chromatography/mass spectrometry (GCMS) at the U.S. Geological Survey (USGS) National Water Quality Laboratory (NWQL). The Chemical Abstract Service Number is given below each compound name in brackets. The five-digit number in parenthesis is a code used by the USGS to uniquely identify the given compound. Flux calculated as the analytical result in $\mu g/L$ multiplied by the total sample volume collected in liters divided by the area of the respective sampler type (Funnel=0.0731 m^2, Autosampler (AS)= 0.0614 m^2) and sample composite time in days. Abbreviations: E, estimated reported laboratory value; na, not applicable; (E), estimated compound due to known poor performance by applied method (Childress and others, 1999; Sandstrom and others, 2001); hh:mm, hour:minute; –, compound not detected; [-], no Chemical Abstract Service Number available for given compound]

Site number	Site name	Site identification number	Start date	Start time (hh:mm)	End date	End time (hh:mm)	Composite sample time, days	Rain events during composite sample time, days	Sample type	Sampler type
10	Westley rain gage at pump building near lateral 6 North	373335121143001	Jan 25, 2002	09:45	Jan 29, 2002	11:20	4	3	wet	Funnel
10	Westley rain gage at pump building near lateral 6 North	373335121143001	Feb 15, 2002	10:30	Feb 18, 2002	11:45	3	2	wet	Funnel
10	Westley rain gage at pump building near lateral 6 North	373335121143001	Mar 5, 2002	09:30	Mar 8, 2002	09:30	3	2	wet	Funnel
10	Westley rain gage at pump building near lateral 6 North	373335121143001	Mar 8, 2002	09:30	Mar 11, 2002	09:45	3	2	wet	Funnel
10	Westley rain gage at pump building near lateral 6 North	373335121143001	Mar 15, 2002	09:30	Mar 19, 2002	09:30	4	3	wet	Funnel
10	Westley rain gage at pump building near lateral 6 North	373335121143001	Mar 22, 2002	08:45	Mar 24, 2002	12:45	2	1	wet	Funnel
10	Westley rain gage at pump building near lateral 6 North	373335121143001	Nov 7, 2002	02:00	Nov 10, 2002	03:00	3	2	wet	Funnel
10	Westley rain gage at pump building near lateral 6 North	373335121143001	Dec 9, 2002	09:15	Dec 16, 2002	09:00	7	5	wet	Funnel
10	Westley rain gage at pump building near lateral 6 North	373335121143001	Dec 16, 2002	09:00	Dec 22, 2002	12:00	6	6	wet	Funnel
10	Westley rain gage at pump building near lateral 6 North	373335121143001	Jan 8, 2003	13:15	Jan 14, 2003	09:20	6	1	wet	Funnel
10	Westley rain gage at pump building near lateral 6 North	373335121143001	Feb 15, 2003	11:45	Mar 1, 2003	11:00	14	4	wet	Funnel
10	Westley rain gage at pump building near lateral 6 North	373335121143001	Mar 13, 2003	09:30	Mar 18, 2003	09:30	5	3	wet	Funnel
10	Westley rain gage at pump building near lateral 6 North	373335121143001	Nov 6, 2003	12:30	Nov 11, 2003	10:30	5	4	wet	Funnel
10	Westley rain gage at pump building near lateral 6 North	373335121143001	Feb 19, 2004	14:30	Feb 24, 2004	15:30	5	3	wet	Funnel
10	Westley rain gage at pump building near lateral 6 North	373335121143001	Feb 24, 2004	15:30	Feb 27, 2004	16:00	3	3	wet	Funnel
4	Modesto Irrigation District gage rooftop at Modesto	373834121000601	Jan 25, 2002	12:00	Jan 29, 2002	14:00	4	3	wet	AS
4	Modesto Irrigation District gage rooftop at Modesto	373834121000601	Feb 7, 2002	14:00	Feb 11, 2002	12:50	4	2	wet	AS
4	Modesto Irrigation District gage rooftop at Modesto	373834121000601	Mar 5, 2002	13:00	Mar 8, 2002	12:30	3	2	wet	AS
4	Modesto Irrigation District gage rooftop at Modesto	373834121000601	Mar 8, 2002	12:30	Mar 11, 2002	12:30	3	2	wet	AS
4	Modesto Irrigation District gage rooftop at Modesto	373834121000601	Mar 15, 2002	11:00	Mar 19, 2002	12:30	4	3	wet	AS
4	Modesto Irrigation District gage rooftop at Modesto	373834121000601	Mar 22, 2002	12:00	Mar 24, 2002	09:45	2	1	wet	AS
4	Modesto Irrigation District gage rooftop at Modesto	373834121000601	Apr 9, 2002	10:30	May 21, 2002	13:20	42	6	wet	AS
4	Modesto Irrigation District gage rooftop at Modesto	373834121000601	Nov 7, 2002	02:00	Nov 10, 2002	03:00	3	2	wet	Funnel
4	Modesto Irrigation District gage rooftop at Modesto	373834121000601	Nov 7, 2002	02:00	Nov 10, 2002	03:05	3	2	wet	AS
4	Modesto Irrigation District gage rooftop at Modesto	373834121000601	Dec 9, 2002	14:00	Dec 16, 2002	12:50	7	5	wet	Funnel
4	Modesto Irrigation District gage rooftop at Modesto	373834121000601	Dec 9, 2002	14:00	Dec 16, 2002	13:00	7	5	wet	AS
4	Modesto Irrigation District gage rooftop at Modesto	373834121000601	Dec 16, 2002	12:50	Dec 23, 2002	09:30	7	6	wet	Funnel
4	Modesto Irrigation District gage rooftop at Modesto	373834121000601	Dec 16, 2002	12:50	Dec 23, 2002	09:40	7	6	wet	AS
4	Modesto Irrigation District gage rooftop at Modesto	373834121000601	Dec 23, 2002	09:30	Jan 9, 2003	15:40	17	6	wet	AS
4	Modesto Irrigation District gage rooftop at Modesto	373834121000601	Jan 9, 2003	15:30	Jan 14, 2003	10:20	5	1	wet	Funnel

Appendix 3. Calculated flux values presented as micrograms per square meter per day (µg/m²/day) for wet-deposition samples with at least one detection collected during the 2002–04 study period at eight sites in the Central Valley, California.—Continued

[All compounds were analyzed by gas chromatography/mass spectrometry (GCMS) at the U.S. Geological Survey (USGS) National Water Quality Laboratory (NWQL). The Chemical Abstract Service Number is given below each compound name in brackets. The five-digit number in parentheses is a code used by the USGS to uniquely identify the given compound. Flux calculated as the analytical result in µg/L multiplied by the total sample volume collected in liters divided by the area of the respective sampler type (Funnel=0.0731 m², Autosampler (AS)= 0.0614 m²) and sample composite time in days. Abbreviations: E, estimated reported laboratory value; na, not applicable; (E), estimated compound due to known poor performance by applied method (Childress and others, 1999; Sandstrom and others, 2001); hh:mm, hour:minute; –, compound not detected; [-], no Chemical Abstract Service Number available for given compound]

Site number	Site name	Site identification number	Start date	Start time (hh:mm)	End date	End time (hh:mm)	Composite sample time, days	Rain events during composite sample time, days	Sample type	Sampler type
4	Modesto Irrigation District gage rooftop at Modesto	373834121000601	Jan 9, 2003	15:30	Jan 14, 2003	10:30	5	1	wet	AS
4	Modesto Irrigation District gage rooftop at Modesto	373834121000601	Jan 14, 2003	10:30	Feb 14, 2003	14:10	31	6	wet	AS
4	Modesto Irrigation District gage rooftop at Modesto	373834121000601	Feb 14, 2003	14:00	Feb 21, 2003	14:40	7	3	wet	Funnel
4	Modesto Irrigation District gage rooftop at Modesto	373834121000601	Feb 14, 2003	14:10	Feb 21, 2003	14:30	7	3	wet	AS
4	Modesto Irrigation District gage rooftop at Modesto	373834121000601	Mar 13, 2003	11:40	Mar 18, 2003	16:15	5	3	wet	Funnel
4	Modesto Irrigation District gage rooftop at Modesto	373834121000601	Mar 13, 2003	11:40	Mar 18, 2003	16:25	5	3	wet	AS
4	Modesto Irrigation District gage rooftop at Modesto	373834121000601	Mar 18, 2003	16:25	Mar 27, 2003	15:00	9	2	wet	AS
4	Modesto Irrigation District gage rooftop at Modesto	373834121000601	Mar 27, 2003	15:00	Apr 8, 2003	13:15	12	3	wet	AS
4	Modesto Irrigation District gage rooftop at Modesto	373834121000601	Apr 29, 2003	14:30	May 5, 2003	09:15	6	3	wet	Funnel
4	Modesto Irrigation District gage rooftop at Modesto	373834121000601	Nov 5, 2003	14:20	Nov 10, 2003	13:15	5	4	wet	AS
4	Modesto Irrigation District gage rooftop at Modesto	373834121000601	Dec 11, 2003	13:00	Dec 16, 2003	15:30	5	5	wet	AS
4	Modesto Irrigation District gage rooftop at Modesto	373834121000601	Dec 11, 2003	13:10	Dec 16, 2003	15:40	5	5	wet	Funnel
4	Modesto Irrigation District gage rooftop at Modesto	373834121000601	Dec 16, 2003	15:40	Dec 23, 2003	13:00	7	3	wet	Funnel
4	Modesto Irrigation District gage rooftop at Modesto	373834121000601	Dec 23, 2003	13:00	Dec 31, 2003	09:30	8	5	wet	Funnel
4	Modesto Irrigation District gage rooftop at Modesto	373834121000601	Dec 31, 2003	09:30	Jan 6, 2004	08:40	6	3	wet	Funnel
4	Modesto Irrigation District gage rooftop at Modesto	373834121000601	Jan 29, 2004	09:55	Feb 4, 2004	08:00	6	2	wet	Funnel
4	Modesto Irrigation District gage rooftop at Modesto	373834121000601	Jan 29, 2004	10:00	Feb 4, 2004	08:10	6	2	wet	AS
4	Modesto Irrigation District gage rooftop at Modesto	373834121000601	Feb 19, 2004	08:10	Feb 24, 2004	09:10	5	3	wet	Funnel
4	Modesto Irrigation District gage rooftop at Modesto	373834121000601	Feb 24, 2004	09:10	Feb 27, 2004	09:15	3	3	wet	Funnel
4	Modesto Irrigation District gage rooftop at Modesto	373834121000601	Mar 25, 2004	10:00	Mar 26, 2004	10:20	1	1	wet	Funnel
5	Modesto Irrigation District gage rooftop at Albers Road	373841120504801	Jan 25, 2002	12:00	Jan 29, 2002	13:15	4	3	wet	AS
5	Modesto Irrigation District gage rooftop at Albers Road	373841120504801	Feb 7, 2002	13:00	Feb 11, 2002	12:15	4	2	wet	AS
5	Modesto Irrigation District gage rooftop at Albers Road	373841120504801	Feb 15, 2002	15:40	Feb 18, 2002	14:25	3	2	wet	AS
5	Modesto Irrigation District gage rooftop at Albers Road	373841120504801	Mar 5, 2002	12:00	Mar 8, 2002	11:45	3	2	wet	AS
5	Modesto Irrigation District gage rooftop at Albers Road	373841120504801	Mar 8, 2002	11:45	Mar 11, 2002	12:00	3	2	wet	AS
5	Modesto Irrigation District gage rooftop at Albers Road	373841120504801	Mar 22, 2002	11:00	Mar 24, 2002	10:30	2	1	wet	AS
5	Modesto Irrigation District gage rooftop at Albers Road	373841120504801	Nov 7, 2002	02:00	Nov 10, 2002	03:00	3	2	wet	Funnel
5	Modesto Irrigation District gage rooftop at Albers Road	373841120504801	Nov 7, 2002	02:00	Nov 10, 2002	03:05	3	2	wet	AS
5	Modesto Irrigation District gage rooftop at Albers Road	373841120504801	Dec 9, 2002	12:15	Dec 16, 2002	12:00	7	5	wet	Funnel
5	Modesto Irrigation District gage rooftop at Albers Road	373841120504801	Dec 9, 2002	12:15	Dec 16, 2002	12:10	7	5	wet	AS

Appendix 3. Calculated flux values presented as micrograms per square meter per day (μg/m²/day) for wet-deposition samples with at least one detection collected during the 2002–04 study period at eight sites in the Central Valley, California.—Continued

[All compounds were analyzed by gas chromatography/mass spectrometry (GCMS) at the U.S. Geological Survey (USGS) National Water Quality Laboratory (NWQL). The Chemical Abstract Service Number is given below each compound name in brackets. The five-digit number in parenthesis is a code used by the USGS to uniquely identify the given compound. Flux calculated as the analytical result in μg/L multiplied by the total sample volume collected in liters divided by the area of the respective sampler type (Funnel=0.0731 m², Autosampler (AS)= 0.0614 m²) and sample composite time in days. Abbreviations: E, estimated reported laboratory value; na, not applicable; (E), estimated compound due to known poor performance by applied method (Childress and others, 1999; Sandstrom and others, 2001); hh:mm, hour:minute; —, compound not detected; [-], no Chemical Abstract Service Number available for given compound]

Site number	Site name	Site identification number	Start date	Start time (hh:mm)	End date	End time (hh:mm)	Composite sample time, days	Rain events during composite sample time, days	Sample type	Sampler type
5	Modesto Irrigation District gage rooftop at Albers Road	37384112050480801	Dec 16, 2002	12:00	Dec 23, 2002	10:20	6	6	wet	Funnel
5	Modesto Irrigation District gage rooftop at Albers Road	37384112050480801	Dec 16, 2002	12:00	Dec 23, 2002	10:30	7	6	wet	AS
5	Modesto Irrigation District gage rooftop at Albers Road	37384112050480801	Dec 23, 2002	10:30	Jan 9, 2003	14:20	17	6	wet	AS
5	Modesto Irrigation District gage rooftop at Albers Road	37384112050480801	Jan 9, 2003	14:30	Jan 14, 2003	11:30	5	1	wet	Funnel
5	Modesto Irrigation District gage rooftop at Albers Road	37384112050480801	Jan 9, 2003	14:30	Jan 14, 2003	11:40	5	1	wet	AS
5	Modesto Irrigation District gage rooftop at Albers Road	37384112050480801	Jan 14, 2003	11:40	Feb 14, 2003	13:00	31	6	wet	Funnel
5	Modesto Irrigation District gage rooftop at Albers Road	37384112050480801	Feb 14, 2003	12:50	Feb 21, 2003	13:20	7	3	wet	AS
5	Modesto Irrigation District gage rooftop at Albers Road	37384112050480801	Feb 14, 2003	13:00	Feb 21, 2003	13:15	7	3	wet	Funnel
5	Modesto Irrigation District gage rooftop at Albers Road	37384112050480801	Mar 12, 2003	14:00	Mar 18, 2003	14:50	6	3	wet	Funnel
5	Modesto Irrigation District gage rooftop at Albers Road	37384112050480801	Mar 12, 2003	14:00	Mar 18, 2003	15:10	6	3	wet	AS
5	Modesto Irrigation District gage rooftop at Albers Road	37384112050480801	Mar 18, 2003	15:10	Apr 8, 2003	12:10	21	5	wet	AS
5	Modesto Irrigation District gage rooftop at Albers Road	37384112050480801	Apr 8, 2003	13:00	Apr 29, 2003	13:40	21	4	wet	AS
5	Modesto Irrigation District gage rooftop at Albers Road	37384112050480801	Apr 29, 2003	13:30	May 9, 2003	17:45	10	3	wet	Funnel
5	Modesto Irrigation District gage rooftop at Albers Road	37384112050480801	Apr 29, 2003	13:40	May 9, 2003	17:50	10	3	wet	AS
5	Modesto Irrigation District gage rooftop at Albers Road	37384112050480801	Nov 5, 2003	13:30	Nov 10, 2003	12:20	5	4	wet	Funnel
5	Modesto Irrigation District gage rooftop at Albers Road	37384112050480801	Nov 5, 2003	13:40	Nov 10, 2003	12:30	5	4	wet	AS
5	Modesto Irrigation District gage rooftop at Albers Road	37384112050480801	Dec 11, 2003	14:05	Dec 16, 2003	14:40	5	2	wet	Funnel
5	Modesto Irrigation District gage rooftop at Albers Road	37384112050480801	Dec 11, 2003	14:10	Dec 16, 2003	14:35	5	2	wet	AS
5	Modesto Irrigation District gage rooftop at Albers Road	37384112050480801	Dec 16, 2003	14:40	Dec 23, 2003	13:00	7	1	wet	Funnel
5	Modesto Irrigation District gage rooftop at Albers Road	37384112050480801	Dec 23, 2003	13:00	Dec 31, 2003	10:30	8	5	wet	Funnel
5	Modesto Irrigation District gage rooftop at Albers Road	37384112050480801	Dec 31, 2003	10:30	Jan 6, 2004	09:35	6	3	wet	Funnel
5	Modesto Irrigation District gage rooftop at Albers Road	37384112050480801	Jan 29, 2004	10:50	Feb 4, 2004	09:10	6	2	wet	Funnel
5	Modesto Irrigation District gage rooftop at Albers Road	37384112050480801	Jan 29, 2004	11:00	Feb 4, 2004	09:20	6	2	wet	AS
5	Modesto Irrigation District gage rooftop at Albers Road	37384112050480801	Feb 4, 2004	09:10	Feb 19, 2004	09:20	15	5	wet	Funnel
5	Modesto Irrigation District gage rooftop at Albers Road	37384112050480801	Feb 19, 2004	09:10	Feb 24, 2004	10:30	5	3	wet	Funnel
5	Modesto Irrigation District gage rooftop at Albers Road	37384112050480801	Feb 24, 2004	10:30	Feb 27, 2004	09:55	3	3	wet	Funnel
5	Modesto Irrigation District gage rooftop at Albers Road	37384112050480801	Mar 25, 2004	11:00	Mar 26, 2004	11:00	1	1	wet	Funnel
16	Gridley High School precipitation gage at Gridley	39220512141010201	Feb 7, 2003	09:15	Feb 14, 2003	11:50	7	2	wet	Funnel
16	Gridley High School precipitation gage at Gridley	39220512141010201	Feb 7, 2003	09:45	Feb 14, 2003	12:00	7	2	wet	AS
16	Gridley High School precipitation gage at Gridley	39220512141010201	Feb 14, 2003	11:50	Feb 20, 2003	14:00	6	3	wet	Funnel

Appendix 3. Calculated flux values presented as micrograms per square meter per day ($\mu g/m^2/day$) for wet-deposition samples with at least one detection collected during the 2002–04 study period at eight sites in the Central Valley, California.—Continued

[All compounds were analyzed by gas chromatography mass spectrometry (GCMS) at the U.S. Geological Survey (USGS) National Water Quality Laboratory (NWQL). The Chemical Abstract Service Number is given below each compound name in brackets. The five-digit number in parenthesis is a code used by the USGS to uniquely identify the given compound. Flux calculated as the analytical result in $\mu g/L$ multiplied by the total sample volume collected in liters divided by the area of the respective sampler type (Funnel=0.0731 m^2, Autosampler (AS)= 0.0614 m^2) and sample composite time in days. Abbreviations: E, estimated reported laboratory value; na, not applicable; (E), estimated compound due to known poor performance by applied method (Childress and others, 1999; Sandstrom and others, 2001); hh:mm, hour:minute. —, compound not detected; [-], no Chemical Abstract Service Number available for given compound]

Site number	Site name	Site identification number	Start date	Start time (hh:mm)	End date	End time (hh:mm)	Composite sample time, days	Rain events during composite sample time, days	Sample type	Sampler type
16	Gridley High School precipitation gage at Gridley	39220512141020l	Feb 14, 2003	11:50	Feb 20, 2003	14:10	6	3	wet	AS
16	Gridley High School precipitation gage at Gridley	39220512141020l	Mar 11, 2003	11:40	Mar 19, 2003	10:00	8	5	wet	AS
16	Gridley High School precipitation gage at Gridley	39220512141020l	Mar 11, 2003	11:40	Mar 19, 2003	10:10	8	5	wet	Funnel
16	Gridley High School precipitation gage at Gridley	39220512141020l	Apr 9, 2003	14:00	Apr 28, 2003	12:10	19	11	wet	Funnel
16	Gridley High School precipitation gage at Gridley	39220512141020l	Apr 28, 2003	12:10	May 9, 2003	10:30	11	6	wet	Funnel
16	Gridley High School precipitation gage at Gridley	39220512141020l	Apr 28, 2003	12:10	May 9, 2003	10:40	11	6	wet	AS
16	Gridley High School precipitation gage at Gridley	39220512141020l	Nov 4, 2003	10:00	Nov 12, 2003	10:15	8	3	wet	AS
16	Gridley High School precipitation gage at Gridley	39220512141020l	Nov 12, 2003	10:15	Nov 17, 2003	12:00	5	2	wet	AS
16	Gridley High School precipitation gage at Gridley	39220512141020l	Nov 17, 2003	12:00	Dec 3, 2003	12:40	16	4	wet	AS
16	Gridley High School precipitation gage at Gridley	39220512141020l	Dec 3, 2003	12:40	Dec 12, 2003	11:10	9	7	wet	AS
16	Gridley High School precipitation gage at Gridley	39220512141020l	Dec 3, 2003	12:50	Dec 12, 2003	11:20	9	7	wet	Funnel
16	Gridley High School precipitation gage at Gridley	39220512141020l	Dec 12, 2003	11:10	Dec 18, 2003	10:20	6	3	wet	AS
16	Gridley High School precipitation gage at Gridley	39220512141020l	Dec 12, 2003	11:20	Dec 18, 2003	10:30	6	3	wet	Funnel
16	Gridley High School precipitation gage at Gridley	39220512141020l	Dec 18, 2003	10:30	Dec 23, 2003	08:10	5	3	wet	Funnel
16	Gridley High School precipitation gage at Gridley	39220512141020l	Dec 23, 2003	08:10	Dec 30, 2003	09:40	7	5	wet	Funnel
16	Gridley High School precipitation gage at Gridley	39220512141020l	Dec 30, 2003	09:40	Jan 7, 2004	11:00	8	7	wet	Funnel
16	Gridley High School precipitation gage at Gridley	39220512141020l	Jan 27, 2004	10:00	Jan 30, 2004	09:30	3	3	wet	Funnel
16	Gridley High School precipitation gage at Gridley	39220512141020l	Jan 27, 2004	10:10	Jan 30, 2004	09:40	3	3	wet	AS
16	Gridley High School precipitation gage at Gridley	39220512141020l	Jan 30, 2004	09:30	Feb 6, 2004	10:20	7	4	wet	Funnel
16	Gridley High School precipitation gage at Gridley	39220512141020l	Jan 30, 2004	09:40	Feb 6, 2004	10:10	7	4	wet	AS
16	Gridley High School precipitation gage at Gridley	39220512141020l	Feb 6, 2004	10:20	Feb 17, 2004	14:00	11	3	wet	Funnel
16	Gridley High School precipitation gage at Gridley	39220512141020l	Feb 17, 2004	14:00	Feb 20, 2004	10:30	3	2	wet	Funnel
16	Gridley High School precipitation gage at Gridley	39220512141020l	Feb 20, 2004	10:30	Mar 2, 2004	10:40	11	7	wet	Funnel
15	Oroville Dam precipitation gage at spillway	39323412129270l	Dec 19, 2002	11:00	Dec 23, 2002	10:30	4	4	wet	Funnel
15	Oroville Dam precipitation gage at spillway	39323412129270l	Jan 7, 2003	10:30	Jan 13, 2003	09:50	6	5	wet	Funnel
15	Oroville Dam precipitation gage at spillway	39323412129270l	Jan 23, 2003	11:30	Jan 31, 2003	12:45	8	4	wet	Funnel
15	Oroville Dam precipitation gage at spillway	39323412129270l	Feb 14, 2003	10:30	Feb 20, 2003	15:20	6	3	wet	Funnel
15	Oroville Dam precipitation gage at spillway	39323412129270l	Mar 11, 2003	11:40	Mar 19, 2003	11:00	8	5	wet	Funnel
15	Oroville Dam precipitation gage at spillway	39323412129270l	Apr 9, 2003	12:30	Apr 28, 2003	13:10	19	11	wet	Funnel
15	Oroville Dam precipitation gage at spillway	39323412129270l	Apr 28, 2003	13:10	May 9, 2003	13:10	11	6	wet	Funnel

Appendix 3. Calculated flux values presented as micrograms per square meter per day (μg/m²/day) for wet-deposition samples with at least one detection collected during the 2002–04 study period at eight sites in the Central Valley, California.—Continued

[All compounds were analyzed by gas chromatography/mass spectrometry (GCMS) at the U.S. Geological Survey (USGS) National Water Quality Laboratory (NWQL). The Chemical Abstract Service Number is given below each compound name in parenthesis. The five-digit number in parenthesis is a code used by the USGS to uniquely identify the given compound. Flux calculated as the analytical result in μg/L multiplied by the total sample volume collected in liters divided by the area of the respective sampler type (Funnel=0.0731 m², Autosampler (AS)= 0.0614 m²) and sample composite time in days. Abbreviations: E, estimated reported laboratory value; na, not applicable; (E), estimated compound due to known poor performance by applied method (Childress and others, 1999; Sandstrom and others, 2001); hh:mm, hour:minute; –, compound not detected; [-], no Chemical Abstract Service Number available for given compound]

Site number	Site name	Site identification number	Start date	Start time (hh:mm)	End date	End time (hh:mm)	Composite sample time, days	Rain events during composite sample time, days	Sample type	Sampler type
15	Oroville Dam precipitation gage at spillway	393234121292701	Dec 3, 2003	12:00	Dec 18, 2003	11:40	15	9	wet	Funnel
15	Oroville Dam precipitation gage at spillway	393234121292701	Dec 30, 2003	10:40	Jan 7, 2004	12:30	8	7	wet	Funnel
15	Oroville Dam precipitation gage at spillway	393234121292701	Jan 27, 2004	11:15	Jan 30, 2004	10:40	3	3	wet	Funnel
15	Oroville Dam precipitation gage at spillway	393234121292701	Jan 30, 2004	10:40	Feb 6, 2004	11:50	7	4	wet	Funnel
15	Oroville Dam precipitation gage at spillway	393234121292701	Feb 17, 2004	15:00	Feb 20, 2004	11:30	3	2	wet	Funnel
15	Oroville Dam precipitation gage at spillway	393234121292701	Feb 20, 2004	11:30	Mar 2, 2004	11:40	11	7	wet	Funnel

Appendix 3. Calculated flux values presented as micrograms per square meter per day (µg/m²/day) for wet-deposition samples with at least one detection collected during the 2002–04 study period at eight sites in the Central Valley, California.—Continued

[All compounds were analyzed by gas chromatography/mass spectrometry (GCMS) at the U.S. Geological Survey (USGS) National Water Quality Laboratory (NWQL). The Chemical Abstract Service Number is given below each compound name in parenthesis. The five-digit number in parenthesis is a code used by the USGS to uniquely identify the given compound. Flux calculated as the analytical result in µg/L multiplied by the total sample volume collected in liters divided by the area of the respective sampler type (Funnel=0.0731 m², Autosampler= 0.0614 m²) and sample composite time in days. Abbreviations: E, estimated reported laboratory value; na, not applicable; (E), estimated compound due to known poor performance by applied method (Childress and others, 1999; Sandstrom and others, 2001); hh:mm, hour:minute; —, compound not detected. [-], no Chemical Abstract Service Number available for given compound]

Site number	Site name	Site identification number	Total sample volume collected, liters	1,4-Naphtho-quinone [130-15-4] (61611)	1-Naphthol (E) [90-15-3] (49295)	2-(4-*tert*-Butyl-phenoxy)-cy-clo-hexanol [1942-71-8] (61637)	3,4-Dichloro-aniline (E) [95-76-1] (61625)	3,5-Dichloro-aniline [626-43-7] (61627)	4,4'-Dichloro-benzo-phe-none [90-98-2] (61631)
11	Newman rain gage at wasteway levee near Draper Road	371735121031201	0.323	E0.13	—	—	—	—	—
11	Newman rain gage at wasteway levee near Draper Road	371735121031201	0.339	E0.02	—	—	—	—	—
11	Newman rain gage at wasteway levee near Draper Road	371735121031201	0.294	—	—	—	—	—	—
11	Newman rain gage at wasteway levee near Draper Road	371735121031201	0.895	—	—	—	—	—	—
11	Newman rain gage at wasteway levee near Draper Road	371735121031201	0.316	—	—	—	—	—	—
11	Newman rain gage at wasteway levee near Draper Road	371735121031201	3.52	na	—	na	—	na	na
11	Newman rain gage at wasteway levee near Draper Road	371735121031201	2.225	na	—	na	—	na	na
11	Newman rain gage at wasteway levee near Draper Road	371735121031201	3	na	—	na	E0.096	na	na
11	Newman rain gage at wasteway levee near Draper Road	371735121031201	0.87	na	—	na	E0.155	na	na
11	Newman rain gage at wasteway levee near Draper Road	371735121031201	1.39	na	—	na	—	na	na
11	Newman rain gage at wasteway levee near Draper Road	371735121031201	0.32	na	—	na	—	na	na
11	Newman rain gage at wasteway levee near Draper Road	371735121031201	0.52	na	—	na	—	na	na
11	Newman rain gage at wasteway levee near Draper Road	371735121031201	1	na	E0.02	na	E0.062	na	na
11	Newman rain gage at wasteway levee near Draper Road	371735121031201	3.1	na	—	na	E0.17	na	na
11	Newman rain gage at wasteway levee near Draper Road	371735121031201	0.16	na	—	na	E0.052	na	na
11	Newman rain gage at wasteway levee near Draper Road	371735121031201	2.2	na	—	na	—	na	na
12	Turlock rain gage near Idaho Road	372713120534901	0.855	E0.08	—	—	—	—	—
12	Turlock rain gage near Idaho Road	372713120534901	0.241	E0.05	—	—	—	—	—
12	Turlock rain gage near Idaho Road	372713120534901	0.636	E0.02	—	0.04	—	—	—
12	Turlock rain gage near Idaho Road	372713120534901	0.394	—	—	—	—	—	—
12	Turlock rain gage near Idaho Road	372713120534901	0.419	—	—	—	—	—	—
12	Turlock rain gage near Idaho Road	372713120534901	0.779	E0.11	—	—	—	—	—
12	Turlock rain gage near Idaho Road	372713120534901	0.465	—	—	—	—	—	—
12	Turlock rain gage near Idaho Road	372713120534901	2.78	na	—	na	—	na	na
12	Turlock rain gage near Idaho Road	372713120534901	3.6	na	—	na	—	na	na
12	Turlock rain gage near Idaho Road	372713120534901	1.8	na	—	na	E0.057	na	na
12	Turlock rain gage near Idaho Road	372713120534901	1.32	na	—	na	E0.271	na	na
12	Turlock rain gage near Idaho Road	372713120534901	0.44	na	E0.04	na	—	na	na
12	Turlock rain gage near Idaho Road	372713120534901	1.6	na	—	na	—	na	na

Appendix 3. Calculated flux values presented as micrograms per square meter per day ($\mu g/m^2/day$) for wet-deposition samples with at least one detection collected during the 2002–04 study period at eight sites in the Central Valley, California.—Continued

[All compounds were analyzed by gas chromatography/mass spectrometry (GCMS) at the U.S. Geological Survey (USGS) National Water Quality Laboratory (NWQL). The Chemical Abstract Service Number is given below each compound name in brackets. The five-digit number in parenthesis is a code used by the USGS to uniquely identify the given compound. Flux calculated as the analytical result in $\mu g/L$ multiplied by the total sample volume collected in liters divided by the area of the respective sampler type (Funnel=0.0731 m^2, Autosampler= 0.0614 m^2) and sample composite time in days. Abbreviations: E, estimated reported laboratory value; na, not applicable; (E), estimated compound due to known poor performance by applied method (Childress and others, 1999; Sandstrom and others, 2001); hh:mm, hour:minute; –, compound not detected; [–], no Chemical Abstract Service Number available for given compound]

Site number	Site name	Site identification number	Total sample volume collected, liters	1,4-Naphtho-quinone [130-15-4] (61611)	1-Naphthol (E) [90-15-3] (49295)	2-(4-*tert*-Butyl-phenoxy)-cyclo-hexanol [1942-71-8] (61637)	3,4-Dichloro-aniline (E) [95-76-1] (61625)	3,5-Dichloro-aniline [626-43-7] (61627)	4,4'-Dichloro-benzo-phenone [90-98-2] (61631)
12	Turlock rain gage near Idaho Road	372713120534901	0.4	na	E0.03	na	–	na	na
12	Turlock rain gage near Idaho Road	372713120534901	0.61	na	E0.03	na	–	na	na
12	Turlock rain gage near Idaho Road	372713120534901	0.66	na	–	na	–	na	na
12	Turlock rain gage near Idaho Road	372713120534901	1.34	na	E0.04	na	E0.049	na	na
12	Turlock rain gage near Idaho Road	372713120534901	2	na	E0.03	na	E0.010	na	na
12	Turlock rain gage near Idaho Road	372713120534901	0.165	na	E0.02	na	–	na	na
12	Turlock rain gage near Idaho Road	372713120534901	1.28	na	–	na	E0.057	na	na
12	Turlock rain gage near Idaho Road	372713120534901	0.22	na	–	na	–	na	na
13	Turlock Airport rain gage	372857120414001	1.958	E0.18	–	–	–	–	–
13	Turlock Airport rain gage	372857120414001	0.24	E0.03	–	–	–	–	–
13	Turlock Airport rain gage	372857120414001	0.692	E0.05	–	–	–	–	–
13	Turlock Airport rain gage	372857120414001	0.698	–	–	–	–	0.057	–
13	Turlock Airport rain gage	372857120414001	0.312	–	–	–	–	0.030	–
13	Turlock Airport rain gage	372857120414001	0.52	–	–	–	–	0.014	–
13	Turlock Airport rain gage	372857120414001	0.239	–	–	–	–	–	–
13	Turlock Airport rain gage	372857120414001	4.12	na	–	na	–	na	na
13	Turlock Airport rain gage	372857120414001	3.88	na	–	na	–	na	na
13	Turlock Airport rain gage	372857120414001	1.5	na	–	na	E0.099	na	na
13	Turlock Airport rain gage	372857120414001	1.2	na	–	na	E0.048	na	na
13	Turlock Airport rain gage	372857120414001	0.29	na	E0.02	na	–	na	na
13	Turlock Airport rain gage	372857120414001	0.42	na	–	na	–	na	na
13	Turlock Airport rain gage	372857120414001	1.59	na	E0.11	na	–	na	na
13	Turlock Airport rain gage	372857120414001	0.78	na	–	na	–	na	na
13	Turlock Airport rain gage	372857120414001	1.075	na	–	na	E0.029	na	na
13	Turlock Airport rain gage	372857120414001	0.715	na	E0.02	na	–	na	na
13	Turlock Airport rain gage	372857120414001	1.39	na	–	na	E0.013	na	na
13	Turlock Airport rain gage	372857120414001	0.465	na	–	na	–	na	na
13	Turlock Airport rain gage	372857120414001	2.22	na	–	na	–	na	na
13	Turlock Airport rain gage	372857120414001	0.85	na	E0.12	na	E0.105	na	na

Appendix 3. Calculated flux values presented as micrograms per square meter per day (μg/m²/day) for wet-deposition samples with at least one detection collected during the 2002–04 study period at eight sites in the Central Valley, California.—Continued

[All compounds were analyzed by gas chromatography/mass spectrometry (GCMS) at the U.S. Geological Survey (USGS) National Water Quality Laboratory (NWQL). The Chemical Abstract Service Number is given below each compound name in brackets. The five-digit number in parenthesis is a code used by the USGS to uniquely identify the given compound. Flux calculated as the analytical result in μg/L multiplied by the total sample volume collected in liters divided by the area of the respective sampler type (Funnel=0.0731 m², Autosampler= 0.0614 m²) and sample composite time in days. Abbreviations: E, estimated reported laboratory value; na, not applicable; (E), estimated compound due to known poor performance by applied method (Childress and others, 1999; Sandstrom and others, 2001); hh:mm, hour:minute; –, compound not detected. [-], no Chemical Abstract Service Number available for given compound]

Site number	Site name	Site identification number	Total sample volume collected, liters	1,4-Naphtho-quinone [130-15-4] (61611)	1-Naphthol (E) [90-15-3] (49295)	2-(4-tert-Butyl-phenoxy)-cyclo-hexanol [1942-71-8] (61637)	3,4-Dichloro-aniline (E) [95-76-1] (61625)	3,5-Dichloro-aniline [626-43-7] (61627)	4,4'-Dichloro-benzo-phe-none [90-98-2] (61631)
10	Westley rain gage at pump building near lateral 6 North	3733351211143001	0.371	E0.05	–	0.03	–	–	–
10	Westley rain gage at pump building near lateral 6 North	3733351211143001	0.759	E0.05	–	–	–	–	–
10	Westley rain gage at pump building near lateral 6 North	3733351211143001	0.152	–	–	–	–	0.061	–
10	Westley rain gage at pump building near lateral 6 North	3733351211143001	0.205	–	–	–	–	0.041	E0.006
10	Westley rain gage at pump building near lateral 6 North	3733351211143001	1.125	–	–	–	–	E0.02	–
10	Westley rain gage at pump building near lateral 6 North	3733351211143001	0.616	–	–	–	–	0.051	E0.025
10	Westley rain gage at pump building near lateral 6 North	3733351211143001	3.08	na	–	na	–	na	na
10	Westley rain gage at pump building near lateral 6 North	3733351211143001	2.94	na	–	na	E0.040	na	na
10	Westley rain gage at pump building near lateral 6 North	3733351211143001	1.82	na	–	na	–	na	na
10	Westley rain gage at pump building near lateral 6 North	3733351211143001	0.68	na	–	na	E0.214	na	na
10	Westley rain gage at pump building near lateral 6 North	3733351211143001	1.06	na	–	na	–	na	na
10	Westley rain gage at pump building near lateral 6 North	3733351211143001	1.76	na	–	na	–	na	na
10	Westley rain gage at pump building near lateral 6 North	3733351211143001	2.53	na	–	na	E0.041	na	na
10	Westley rain gage at pump building near lateral 6 North	3733351211143001	0.12	na	–	na	–	na	na
10	Westley rain gage at pump building near lateral 6 North	3733351211143001	2.52	na	–	na	–	na	na
4	Modesto Irrigation District gage rooftop at Modesto	3738341210000601	0.346	E0.06	–	0.06	–	–	–
4	Modesto Irrigation District gage rooftop at Modesto	3738341210000601	0.212	E0.02	–	–	–	–	–
4	Modesto Irrigation District gage rooftop at Modesto	3738341210000601	0.331	–	–	–	E0.024	–	–
4	Modesto Irrigation District gage rooftop at Modesto	3738341210000601	0.571	na	na	na	na	na	na
4	Modesto Irrigation District gage rooftop at Modesto	3738341210000601	0.719	–	–	–	–	–	–
4	Modesto Irrigation District gage rooftop at Modesto	3738341210000601	0.614	–	–	–	–	–	–
4	Modesto Irrigation District gage rooftop at Modesto	3738341210000601	0.21	–	–	–	–	–	–
4	Modesto Irrigation District gage rooftop at Modesto	3738341210000601	4.1	na	E0.28	na	–	na	na
4	Modesto Irrigation District gage rooftop at Modesto	3738341210000601	3.39	na	–	na	–	na	na
4	Modesto Irrigation District gage rooftop at Modesto	3738341210000601	4	na	–	na	–	na	na
4	Modesto Irrigation District gage rooftop at Modesto	3738341210000601	3.6	na	–	na	–	na	na
4	Modesto Irrigation District gage rooftop at Modesto	3738341210000601	1.9	na	–	na	–	na	na
4	Modesto Irrigation District gage rooftop at Modesto	3738341210000601	1.64	na	–	na	–	na	na
4	Modesto Irrigation District gage rooftop at Modesto	3738341210000601	0.58	na	E0.03	na	E0.027	na	na

Appendix 3. Calculated flux values presented as micrograms per square meter per day (µg/m^2/day) for wet-deposition samples with at least one detection collected during the 2002–04 study period at eight sites in the Central Valley, California.—Continued

[All compounds were analyzed by gas chromatography/mass spectrometry (GCMS) at the U.S. Geological Survey (USGS) National Water Quality Laboratory (NWQL). The Chemical Abstract Service Number is given below each compound name in brackets. The five-digit number in parenthesis is a code used by the USGS to uniquely identify the given compound. Flux calculated as the analytical result in µg/L multiplied by the total sample volume collected in liters divided by the area of the respective sampler type (Funnel=0.0731 m^2, Autosampler= 0.0614 m^2) and sample composite time in days. Abbreviations: E, estimated reported laboratory value; na, not applicable; (E), estimated compound due to known poor performance by applied method (Childress and others, 1999; Sandstrom and others, 2001); hh:mm, hour:minute; –, compound not detected; [-], no Chemical Abstract Service Number available for given compound]

Site number	Site name	Site identification number	Total sample volume collected, liters	1,4-Naphthoquinone [130-15-4] (61611)	1-Naphthol (E) [90-15-3] (49295)	2-(4-*tert*-Butylphenoxy)-cyclohexanol [1942-71-8] (61637)	3,4-Dichloroaniline (E) [95-76-1] (61625)	3,5-Dichloroaniline [626-43-7] (61627)	4,4'-Dichlorobenzophenone [90-98-2] (61631)
4	Modesto Irrigation District gage rooftop at Modesto	373834121000601	0.91	na	E0.37	na	E0.212	na	na
4	Modesto Irrigation District gage rooftop at Modesto	373834121000601	0.66	na	–	na	E0.035	na	na
4	Modesto Irrigation District gage rooftop at Modesto	373834121000601	0.56	na	E0.06	na	–	na	na
4	Modesto Irrigation District gage rooftop at Modesto	373834121000601	0.62	na	E0.03	na	–	na	na
4	Modesto Irrigation District gage rooftop at Modesto	373834121000601	0.54	na	E0.1	na	–	na	na
4	Modesto Irrigation District gage rooftop at Modesto	373834121000601	2.24	na	–	na	–	na	na
4	Modesto Irrigation District gage rooftop at Modesto	373834121000601	1.8	na	E0.03	na	–	na	na
4	Modesto Irrigation District gage rooftop at Modesto	373834121000601	0.35	na	–	na	–	na	na
4	Modesto Irrigation District gage rooftop at Modesto	373834121000601	0.62	na	–	na	–	na	na
4	Modesto Irrigation District gage rooftop at Modesto	373834121000601	0.56	na	–	na	E0.099	na	na
4	Modesto Irrigation District gage rooftop at Modesto	373834121000601	0.9	na	–	na	E0.053	na	na
4	Modesto Irrigation District gage rooftop at Modesto	373834121000601	0.37	na	–	na	–	na	na
4	Modesto Irrigation District gage rooftop at Modesto	373834121000601	0.81	na	–	na	–	na	na
4	Modesto Irrigation District gage rooftop at Modesto	373834121000601	2.81	na	–	na	–	na	na
4	Modesto Irrigation District gage rooftop at Modesto	373834121000601	2.81	na	E0.04	na	–	na	na
4	Modesto Irrigation District gage rooftop at Modesto	373834121000601	1.115	na	–	na	–	na	na
4	Modesto Irrigation District gage rooftop at Modesto	373834121000601	1.4	na	E0.04	na	–	na	na
4	Modesto Irrigation District gage rooftop at Modesto	373834121000601	1.26	na	E0.01	na	E0.025	na	na
4	Modesto Irrigation District gage rooftop at Modesto	373834121000601	0.3	na	E0.04	na	–	na	na
4	Modesto Irrigation District gage rooftop at Modesto	373834121000601	2.39	na	E0.08	na	E0.119	na	na
4	Modesto Irrigation District gage rooftop at Modesto	373834121000601	0.3	na	–	na	–	na	na
5	Modesto Irrigation District gage rooftop at Albers Road	373841120504801	1.184	E0.19	–	–	–	–	–
5	Modesto Irrigation District gage rooftop at Albers Road	373841120504801	0.229	E0.06	–	–	–	–	–
5	Modesto Irrigation District gage rooftop at Albers Road	373841120504801	0.781	E0.06	–	–	–	–	–
5	Modesto Irrigation District gage rooftop at Albers Road	373841120504801	0.15	–	–	–	–	0.094	–
5	Modesto Irrigation District gage rooftop at Albers Road	373841120504801	0.347	–	–	–	E0.051	0.045	–
5	Modesto Irrigation District gage rooftop at Albers Road	373841120504801	0.375	–	–	–	–	0.073	–
5	Modesto Irrigation District gage rooftop at Albers Road	373841120504801	4.18	na	–	na	na	na	na
5	Modesto Irrigation District gage rooftop at Albers Road	373841120504801	3.6	na	–	na	na	na	na

Appendix 3. Calculated flux values presented as micrograms per square meter per day (μg/m²/day) for wet-deposition samples with at least one detection collected during the 2002–04 study period at eight sites in the Central Valley, California.—Continued

[All compounds were analyzed by gas chromatography/mass spectrometry (GCMS) at the U.S. Geological Survey (USGS) National Water Quality Laboratory (NWQL). The Chemical Abstract Service Number is given below each compound name in brackets. The five-digit number in parenthesis is a code used by the USGS to uniquely identify the given compound. Flux calculated as the analytical result in μg/L multiplied by the total sample volume collected in liters divided by the area of the respective sampler type (Funnel=0.0731 m², Autosampler= 0.0614 m²) and sample composite time in days. Abbreviations: E, estimated reported laboratory value; na, not applicable; (E), estimated compound due to known poor performance by applied method (Childress and others, 1999; Sandstrom and others, 2001); hh:mm, hour:minute; —, compound not detected; [-], no Chemical Abstract Service Number available for given compound]

Site number	Site name	Site identification number	Total sample volume collected, liters	1,4-Naphtho-quinone [130-15-4] (61611)	1-Naphthol (E) [90-15-3] (49295)	2-(4-tert-Butyl-phenoxy)-cyclo-hexanol [1942-71-8] (61637)	3,4-Dichloro-aniline (E) [95-76-1] (61625)	3,5-Dichloro-aniline [626-43-7] (61627)	4,4'-Dichloro-benzo-phe-none [90-98-2] (61631)
5	Modesto Irrigation District gage rooftop at Albers Road	3738411205004801	4	na	—	na	E0.219	na	na
5	Modesto Irrigation District gage rooftop at Albers Road	3738411205004801	3.25	na	—	na	E0.074	na	na
5	Modesto Irrigation District gage rooftop at Albers Road	3738411205004801	1.9	na	—	na	E0.121	na	na
5	Modesto Irrigation District gage rooftop at Albers Road	3738411205004801	1.44	na	—	na	E0.070	na	na
5	Modesto Irrigation District gage rooftop at Albers Road	3738411205004801	0.7	na	—	na	E0.245	na	na
5	Modesto Irrigation District gage rooftop at Albers Road	3738411205004801	1.2	na	—	na	E1.215	na	na
5	Modesto Irrigation District gage rooftop at Albers Road	3738411205004801	0.86	na	—	na	0.532	na	na
5	Modesto Irrigation District gage rooftop at Albers Road	3738411205004801	0.34	na	—	na	—	na	na
5	Modesto Irrigation District gage rooftop at Albers Road	3738411205004801	0.42	na	—	na	E0.038	na	na
5	Modesto Irrigation District gage rooftop at Albers Road	3738411205004801	0.25	na	—	na	—	na	na
5	Modesto Irrigation District gage rooftop at Albers Road	3738411205004801	1.61	na	—	na	E0.066	na	na
5	Modesto Irrigation District gage rooftop at Albers Road	3738411205004801	1.42	na	—	na	—	na	na
5	Modesto Irrigation District gage rooftop at Albers Road	3738411205004801	1.4	na	—	na	—	na	na
5	Modesto Irrigation District gage rooftop at Albers Road	3738411205004801	1.64	na	—	na	—	na	na
5	Modesto Irrigation District gage rooftop at Albers Road	3738411205004801	0.71	na	—	na	—	na	na
5	Modesto Irrigation District gage rooftop at Albers Road	3738411205004801	0.38	na	—	na	—	na	na
5	Modesto Irrigation District gage rooftop at Albers Road	3738411205004801	0.64	na	—	na	E0.024	na	na
5	Modesto Irrigation District gage rooftop at Albers Road	3738411205004801	0.82	na	—	na	—	na	na
5	Modesto Irrigation District gage rooftop at Albers Road	3738411205004801	0.76	na	—	na	E0.114	na	na
5	Modesto Irrigation District gage rooftop at Albers Road	3738411205004801	0.67	na	—	na	—	na	na
5	Modesto Irrigation District gage rooftop at Albers Road	3738411205004801	0.89	na	—	na	—	na	na
5	Modesto Irrigation District gage rooftop at Albers Road	3738411205004801	2.82	na	—	na	—	na	na
5	Modesto Irrigation District gage rooftop at Albers Road	3738411205004801	0.24	na	—	na	—	na	na
5	Modesto Irrigation District gage rooftop at Albers Road	3738411205004801	0.98	na	—	na	E0.255	na	na
5	Modesto Irrigation District gage rooftop at Albers Road	3738411205004801	0.67	na	—	na	E0.142	na	na
5	Modesto Irrigation District gage rooftop at Albers Road	3738411205004801	1.7	na	E0.02	na	E0.745	na	na
5	Modesto Irrigation District gage rooftop at Albers Road	3738411205004801	0.42	na	E0.02	na	E0.556	na	na
5	Modesto Irrigation District gage rooftop at Albers Road	3738411205004801	1.5	na	—	na	E0.068	na	na
5	Modesto Irrigation District gage rooftop at Albers Road	3738411205004801	0.82	na	E0.11	na	E0.247	na	na

Appendix 3. Calculated flux values presented as micrograms per square meter per day (µg/m²/day) for wet-deposition samples with at least one detection collected during the 2002–04 study period at eight sites in the Central Valley, California.—Continued

[All compounds were analyzed by gas chromatography/mass spectrometry (GCMS) at the U.S. Geological Survey (USGS) National Water Quality Laboratory (NWQL). The Chemical Abstract Service Number is given below each compound name in brackets. The five-digit number in parenthesis is a code used by the USGS to uniquely identify the given compound. Flux calculated as the analytical result in µg/L multiplied by the total sample volume collected in liters divided by the area of the respective sampler type (Funnel=0.0731 m², Autosampler= 0.0614 m²) and sample composite time in days. Abbreviations: E, estimated reported laboratory value; na, not applicable; (E), estimated compound due to known poor performance by applied method (Childress and others, 1999; Sandstrom and others, 2001); hh:mm, hour:minute; –, compound not detected; [-], no Chemical Abstract Service Number available for given compound]

Site number	Site name	Site identification number	Total sample volume collected, liters	1,4-Naphtho-quinone [130-15-4] (61611)	1-Naphthol (E) [90-15-3] (49295)	2-(4-tert-Butyl-phenoxy)-cy-clo-hexanol [1942-71-8] (61637)	3,4-Dichloro-aniline (E) [95-76-1] (61625)	3,5-Dichloro-aniline [626-43-7] (61627)	4,4'-Dichloro-benzo-phe-none [90-98-2] (61631)
16	Gridley High School precipitation gage at Gridley	39220512141020l	4	na	E0.55	na	–	na	na
16	Gridley High School precipitation gage at Gridley	39220512141020l	4	na	–	na	–	na	na
16	Gridley High School precipitation gage at Gridley	39220512141020l	2.56	na	E0.12	na	–	na	na
16	Gridley High School precipitation gage at Gridley	39220512141020l	1.16	na	–	na		na	na
16	Gridley High School precipitation gage at Gridley	39220512141020l	0.85	na	E0.03	na		na	na
16	Gridley High School precipitation gage at Gridley	39220512141020l	2.96	na	E0.08	na		na	na
16	Gridley High School precipitation gage at Gridley	39220512141020l	4.24	na	E0.05	na		na	na
16	Gridley High School precipitation gage at Gridley	39220512141020l	3.7	na	E0.08	na		na	na
16	Gridley High School precipitation gage at Gridley	39220512141020l	1.09	na	–	na		na	na
16	Gridley High School precipitation gage at Gridley	39220512141020l	2.05	na	–	na	E0.039	na	na
16	Gridley High School precipitation gage at Gridley	39220512141020l	0.43	na	–	na	–	na	na
16	Gridley High School precipitation gage at Gridley	39220512141020l	1.67	na	–	na		na	na
16	Gridley High School precipitation gage at Gridley	39220512141020l	2.13	na	–	na		na	na
16	Gridley High School precipitation gage at Gridley	39220512141020l	3.885	na	–	na	E0.046	na	na
16	Gridley High School precipitation gage at Gridley	39220512141020l	0.57	na	E0.02	na		na	na
16	Gridley High School precipitation gage at Gridley	39220512141020l	1.19	na	–	na		na	na
16	Gridley High School precipitation gage at Gridley	39220512141020l	1.72	na	–	na		na	na
16	Gridley High School precipitation gage at Gridley	39220512141020l	4.3	na	–	na		na	na
16	Gridley High School precipitation gage at Gridley	39220512141020l	4.03	na	–	na	E0.017	na	na
16	Gridley High School precipitation gage at Gridley	39220512141020l	0.75	na	E0.02	na	E0.013	na	na
16	Gridley High School precipitation gage at Gridley	39220512141020l	0.48	na	E0.02	na		na	na
16	Gridley High School precipitation gage at Gridley	39220512141020l	1.43	na	E0.06	na		na	na
16	Gridley High School precipitation gage at Gridley	39220512141020l	1.42	na	–	na		na	na
16	Gridley High School precipitation gage at Gridley	39220512141020l	3.4	na	–	na		na	na
16	Gridley High School precipitation gage at Gridley	39220512141020l	3.1	na	–	na		na	na
16	Gridley High School precipitation gage at Gridley	39220512141020l	4.28	na	–	na		na	na
15	Oroville Dam precipitation gage at spillway	393234121292701	4	na	–	na		na	na
15	Oroville Dam precipitation gage at spillway	393234121292701	4	na	–	na		na	na
15	Oroville Dam precipitation gage at spillway	393234121292701	0.38	na	E0.03	na		na	na

Appendix 3. Calculated flux values presented as micrograms per square meter per day (µg/m²/day) for wet-deposition samples with at least one detection collected during the 2002–04 study period at eight sites in the Central Valley, California.—Continued

[All compounds were analyzed by gas chromatography mass spectrometry (GCMS) at the U.S. Geological Survey (USGS) National Water Quality Laboratory (NWQL). The Chemical Abstract Service Number is given below each compound name in brackets. The five-digit number in parenthesis is a code used by the USGS to uniquely identify the given compound. Flux calculated as the analytical result in µg/L multiplied by the total sample volume collected in liters divided by the area of the respective sampler type (Funnel=0.0731 m², Autosampler= 0.0614 m²) and sample composite time in days. Abbreviations: E, estimated reported laboratory value; na, not applicable; (E), estimated compound due to known poor performance by applied method (Childress and others, 1999; Sandstrom and others, 2001); hh:mm, hour:minute; –, compound not detected; [–], no Chemical Abstract Service Number available for given compound]

Site number	Site name	Site identification number	Total sample volume collected, liters	1,4-Naphtho-quinone [130-15-4] (61611)	1-Naphthol (E) [90-15-3] (49295)	2-(4-*tert*-Butyl-phenoxy)-cyclo-hexanol [1942-71-8] (61637)	3,4-Dichloro-aniline (E) [95-76-1] (61625)	3,5-Dichloro-aniline [626-43-7] (61627)	4,4'-Dichloro-benzo-phe-none [90-98-2] (61631)
15	Oroville Dam precipitation gage at spillway	393234121292701	1.78	na	E0.08	na	–	na	na
15	Oroville Dam precipitation gage at spillway	393234121292701	2.725	na	E0.07	na	–	na	na
15	Oroville Dam precipitation gage at spillway	393234121292701	4.27	na	E0.03	na	–	na	na
15	Oroville Dam precipitation gage at spillway	393234121292701	3.4	na	–	na	–	na	na
15	Oroville Dam precipitation gage at spillway	393234121292701	1.55	na	E0.01	na		na	na
15	Oroville Dam precipitation gage at spillway	393234121292701	2.8	na	–	na		na	na
15	Oroville Dam precipitation gage at spillway	393234121292701	0.91	na	E0.02	na	E0.021	na	na
15	Oroville Dam precipitation gage at spillway	393234121292701	2.535	na	–	na	E0.312	na	na
15	Oroville Dam precipitation gage at spillway	393234121292701	3.12	na	–	na	–	na	na
15	Oroville Dam precipitation gage at spillway	393234121292701	4.22	na	–	na		na	na

Appendix 3. Calculated flux values presented as micrograms per square meter per day (μg/m²/day) for wet-deposition samples with at least one detection collected during the 2002–04 study period at eight sites in the Central Valley, California.—Continued

[All compounds were analyzed by gas chromatography/mass spectrometry (GCMS) at the U.S. Geological Survey (USGS) National Water Quality Laboratory (NWQL). The Chemical Abstract Service Number is given below each compound name in parentheses. The five-digit number in parenthesis is a code used by the USGS to uniquely identify the given compound. Flux calculated as the analytical result in μg/L multiplied by the total sample volume collected in liters divided by the area of the respective sampler type (Funnel=0.0731 m², Autosampler= 0.0614 m²) and sample composite time in days. Abbreviations: E, estimated reported laboratory value; na, not applicable; (E), estimated compound due to known poor performance by applied method (Childress and others, 1999; Sandstrom and others, 2001); hh:mm, hour:minute; —, compound not detected; [-], no Chemical Abstract Service Number available for given compound]

Site number	Site name	Site identification number	4-Chloro-2-methyl-phenol (E) [1570-64-5] (61633)	Alachlor [15972-60-8] (46342)	Atrazine [1912-24-9] (39632)	Azinphos-methyl (E) [86-50-0] (82686)	Benfluralin [1861-40-1] (82673)	Carbaryl (E) [63-25-2] (82680)	Carbofuran [1563-66-2] (82674)
11	Newman rain gage at wasteway levee near Draper Road	37173512103120 1	—	—	—	—	—	E0.813	—
11	Newman rain gage at wasteway levee near Draper Road	37173512103120 1	—	—	—	—	—	—	—
11	Newman rain gage at wasteway levee near Draper Road	37173512103120 1	—	—	—	—	—	—	E0.053
11	Newman rain gage at wasteway levee near Draper Road	37173512103120 1	—	—	—	—	—	—	—
11	Newman rain gage at wasteway levee near Draper Road	37173512103120 1	—	—	—	E0.29	—	E0.120	na
11	Newman rain gage at wasteway levee near Draper Road	37173512103120 1	—	—	—	E0.05	—	E0.024	na
11	Newman rain gage at wasteway levee near Draper Road	37173512103120 1	—	—	—	—	—	—	na
11	Newman rain gage at wasteway levee near Draper Road	37173512103120 1	—	—	—	—	—	E0.048	na
11	Newman rain gage at wasteway levee near Draper Road	37173512103120 1	—	0.082	—	—	—	E0.032	na
11	Newman rain gage at wasteway levee near Draper Road	37173512103120 1	—	—	—	—	—	E0.050	na
11	Newman rain gage at wasteway levee near Draper Road	37173512103120 1	—	—	—	E0.03	—	E0.023	na
11	Newman rain gage at wasteway levee near Draper Road	37173512103120 1	E0.017	—	—	—	—	—	na
11	Newman rain gage at wasteway levee near Draper Road	37173512103120 1	—	—	—	—	—	E0.024	na
11	Newman rain gage at wasteway levee near Draper Road	37173512103120 1	—	E0.040	—	—	—	—	na
12	Turlock rain gage near Idaho Road	37271312053490 1	—	—	—	—	—	E0.062	—
12	Turlock rain gage near Idaho Road	37271312053490 1	—	—	—	—	—	E0.157	—
12	Turlock rain gage near Idaho Road	37271312053490 1	E0.07	—	—	—	—	—	—
12	Turlock rain gage near Idaho Road	37271312053490 1	E0.035	—	—	—	—	E0.035	E0.089
12	Turlock rain gage near Idaho Road	37271312053490 1	—	—	—	—	—	—	E0.310
12	Turlock rain gage near Idaho Road	37271312053490 1	—	—	—	—	—	—	E0.149
12	Turlock rain gage near Idaho Road	37271312053490 1	—	—	—	—	—	—	E0.465
12	Turlock rain gage near Idaho Road	37271312053490 1	—	—	—	E0.27	—	E0.247	na
12	Turlock rain gage near Idaho Road	37271312053490 1	—	—	—	E0.06	—	E0.049	na
12	Turlock rain gage near Idaho Road	37271312053490 1	—	—	—	E0.04	—	—	na
12	Turlock rain gage near Idaho Road	37271312053490 1	E0.036	—	—	—	—	E0.145	na
12	Turlock rain gage near Idaho Road	37271312053490 1	—	—	—	—	—	E0.026	na
12	Turlock rain gage near Idaho Road	37271312053490 1	E0.036	0.044	—	—	—	E0.044	na
12	Turlock rain gage near Idaho Road	37271312053490 1	—	—	—	—	—	E0.112	na

Appendix 3. Calculated flux values presented as micrograms per square meter per day (µg/m²/day) for wet-deposition samples with at least one detection collected during the 2002–04 study period at eight sites in the Central Valley, California.—Continued

[All compounds were analyzed by gas chromatography/mass spectrometry (GCMS) at the U.S. Geological Survey (USGS) National Water Quality Laboratory (NWQL). The Chemical Abstract Service Number is given below each compound name in brackets. The five-digit number in parenthesis is a code used by the USGS to uniquely identify the given compound. Flux calculated as the analytical result in µg/L multiplied by the total sample volume collected in liters divided by the area of the respective sampler type (Funnel=0.0731 m², Autosampler= 0.0614 m²) and sample composite time in days. Abbreviations: E, estimated compound due to known poor performance by applied method (Childress and others, 1999; Sandstrom and others, 2001); hh:mm, hour:minute; –, compound not detected; [-], no Chemical Abstract Service Number available for given compound]

Site number	Site name	Site identification number	4-Chloro-2-methyl-phenol (E) [1570-64-5] (61633)	Alachlor [15972-60-8] (46342)	Atrazine [1912-24-9] (39632)	Azinphos-methyl (E) [86-50-0] (82686)	Benfluralin [1861-40-1] (82673)	Carbaryl (E) [63-25-2] (82680)	Carbofuran [1563-66-2] (82674)
12	Turlock rain gage near Idaho Road	37271312053490 1	–	–	E0.017	E0.21	–	E0.359	na
12	Turlock rain gage near Idaho Road	37271312053490 1	–	–	–	E0.03	–	E0.029	na
12	Turlock rain gage near Idaho Road	37271312053490 1	E0.037	–	–	–	–	E0.064	na
12	Turlock rain gage near Idaho Road	37271312053490 1	E0.011	–	–	–	–	E0.104	na
12	Turlock rain gage near Idaho Road	37271312053490 1	E0.005	–	–	–	–	E0.039	na
12	Turlock rain gage near Idaho Road	37271312053490 1	–	–	–	–	–	–	na
12	Turlock rain gage near Idaho Road	37271312053490 1	–	–	–	–	–	–	na
13	Turlock Airport rain gage	37285712041400 1	–	–	–	–	–	–	–
13	Turlock Airport rain gage	37285712041400 1	–	–	–	–	–	E0.116	–
13	Turlock Airport rain gage	37285712041400 1	E0.052	–	–	–	–	E0.158	–
13	Turlock Airport rain gage	37285712041400 1	–	–	–	–	–	E0.071	–
13	Turlock Airport rain gage	37285712041400 1	–	–	–	–	–	E0.153	E2.109
13	Turlock Airport rain gage	37285712041400 1	–	–	–	–	–	–	E0.505
13	Turlock Airport rain gage	37285712041400 1	–	–	–	–	–	–	–
13	Turlock Airport rain gage	37285712041400 1	–	–	–	E0.39	–	E0.254	na
13	Turlock Airport rain gage	37285712041400 1	E0.042	–	–	E0.07	–	E0.042	na
13	Turlock Airport rain gage	37285712041400 1	–	–	–	–	–	–	na
13	Turlock Airport rain gage	37285712041400 1	E0.049	–	–	–	–	E0.082	na
13	Turlock Airport rain gage	37285712041400 1	E0.032	–	–	–	–	–	na
13	Turlock Airport rain gage	37285712041400 1	–	–	–	–	–	E0.015	na
13	Turlock Airport rain gage	37285712041400 1	–	–	–	–	–	E0.065	na
13	Turlock Airport rain gage	37285712041400 1	–	–	–	–	–	E0.673	na
13	Turlock Airport rain gage	37285712041400 1	–	–	–	E0.29	–	E0.270	na
13	Turlock Airport rain gage	37285712041400 1	–	–	–	E0.03	–	E0.034	na
13	Turlock Airport rain gage	37285712041400 1	–	–	–	–	–	E0.105	na
13	Turlock Airport rain gage	37285712041400 1	–	–	–	–	–	–	na
13	Turlock Airport rain gage	37285712041400 1	–	–	–	–	–	E0.030	na
13	Turlock Airport rain gage	37285712041400 1	E0.023	–	–	–	–	E0.151	na

Appendix 3. Calculated flux values presented as micrograms per square meter per day ($\mu g/m^2/day$) for wet-deposition samples with at least one detection collected during the 2002–04 study period at eight sites in the Central Valley, California.—Continued

[All compounds were analyzed by gas chromatography/mass spectrometry (GCMS) at the U.S. Geological Survey (USGS) National Water Quality Laboratory (NWQL). The Chemical Abstract Service Number is given below each compound name in brackets. The five-digit number in parentheses is a code used by the USGS to uniquely identify the given compound. Flux calculated as the analytical result in $\mu g/L$ multiplied by the total sample volume collected in liters divided by the area of the respective sampler type (Funnel=0.0731 m^2, Autosampler= 0.0614 m^2) and sample composite time in days. Abbreviations: E, estimated compound due to known poor performance by applied method (Childress and others, 1999; Sandstrom and others, 2001); hh:mm, hour:minute; na, not applicable value; na, not reported laboratory value; −, compound not detected; [−], no Chemical Abstract Service Number available for given compound]

Site number	Site name	Site identification number	4-Chloro-2-methyl-phenol (E) [1570-64-5] (61633)	Alachlor [15972-60-8] (46342)	Atrazine [1912-24-9] (39632)	Azinphos-methyl (E) [86-50-0] (82686)	Benfluralin [1861-40-1] (82673)	Carbaryl (E) [63-25-2] (82680)	Carbofuran [1563-66-2] (82674)
10	Westley rain gage at pump building near lateral 6 North	373335121143001	—	—	—	—	—	—	—
10	Westley rain gage at pump building near lateral 6 North	373335121143001	—	—	—	—	—	—	E10.173
10	Westley rain gage at pump building near lateral 6 North	373335121143001	—	—	—	—	—	E0.063	E5.205
10	Westley rain gage at pump building near lateral 6 North	373335121143001	—	—	—	—	—	E0.045	E0.082
10	Westley rain gage at pump building near lateral 6 North	373335121143001	—	—	—	—	—	E0.046	E0.169
10	Westley rain gage at pump building near lateral 6 North	373335121143001	—	—	—	—	—	E0.076	na
10	Westley rain gage at pump building near lateral 6 North	373335121143001	—	—	—	E0.38	—	E0.169	na
10	Westley rain gage at pump building near lateral 6 North	373335121143001	—	—	—	E0.06	—	—	na
10	Westley rain gage at pump building near lateral 6 North	373335121143001	—	—	—	—	—	E0.037	na
10	Westley rain gage at pump building near lateral 6 North	373335121143001	—	—	—	—	—	E0.074	na
10	Westley rain gage at pump building near lateral 6 North	373335121143001	—	—	—	—	—	E0.040	na
10	Westley rain gage at pump building near lateral 6 North	373335121143001	—	0.112	—	—	—	—	na
10	Westley rain gage at pump building near lateral 6 North	373335121143001	—	—	—	E0.09	E0.04	E0.061	na
10	Westley rain gage at pump building near lateral 6 North	373335121143001	—	E0.023	—	—	—	—	na
4	Modesto Irrigation District gage rooftop at Modesto	3738341210000601	—	—	—	—	—	E0.034	—
4	Modesto Irrigation District gage rooftop at Modesto	3738341210000601	—	—	—	—	—	E0.302	—
4	Modesto Irrigation District gage rooftop at Modesto	3738341210000601	—	—	—	—	—	—	E0.237
4	Modesto Irrigation District gage rooftop at Modesto	3738341210000601	—	—	—	—	—	E0.031	E0.995
4	Modesto Irrigation District gage rooftop at Modesto	3738341210000601	—	—	—	—	—	—	E1.330
4	Modesto Irrigation District gage rooftop at Modesto	3738341210000601	—	—	—	E0.14	—	E0.076	—
4	Modesto Irrigation District gage rooftop at Modesto	3738341210000601	—	—	—	E0.39	—	E0.729	na
4	Modesto Irrigation District gage rooftop at Modesto	3738341210000601	—	—	—	E0.41	—	E0.304	na
4	Modesto Irrigation District gage rooftop at Modesto	3738341210000601	—	—	—	E0.09	—	E0.077	na
4	Modesto Irrigation District gage rooftop at Modesto	3738341210000601	—	—	—	E0.07	—	—	na
4	Modesto Irrigation District gage rooftop at Modesto	3738341210000601	—	—	—	—	—	E0.035	na
4	Modesto Irrigation District gage rooftop at Modesto	3738341210000601	E0.028	—	—	—	—	E0.137	na
4	Modesto Irrigation District gage rooftop at Modesto	3738341210000601	E0.037	—	—	—	—	E0.125	na

Appendix 3. Calculated flux values presented as micrograms per square meter per day ($\mu g/m^2/day$) for wet-deposition samples with at least one detection collected during the 2002–04 study period at eight sites in the Central Valley, California.—Continued

[All compounds were analyzed by gas chromatography/mass spectrometry (GCMS) at the U.S. Geological Survey (USGS) National Water Quality Laboratory (NWQL). The Chemical Abstract Service Number is given below each compound name in brackets. The five-digit number in parenthesis is a code used by the USGS to uniquely identify the given compound. Flux calculated as the analytical result in $\mu g/L$ multiplied by the total sample volume collected in liters divided by the area of the respective sampler type (Funnel=0.0731 m², Autosampler= 0.0614 m²) and sample composite time in days. Abbreviations: E, estimated reported laboratory value; na, not applicable; (E), estimated compound due to known poor performance by applied method (Childress and others, 1999; Sandstrom and others, 2001); hh:mm, hour:minute; -, compound not detected; [-], no Chemical Abstract Service Number available for given compound]

Site number	Site name	Site identification number	4-Chloro-2-methyl-phenol (E) [1570-64-5] (61633)	Alachlor [15972-60-8] (46342)	Atrazine [1912-24-9] (39632)	Azinphos-methyl (E) [86-50-0] (82686)	Benfluralin [1861-40-1] (82673)	Carbaryl (E) [63-25-2] (82680)	Carbofuran [1563-66-2] (82674)
4	Modesto Irrigation District gage rooftop at Modesto	373834121000601	E0.054	-	-	-	-	E0.118	na
4	Modesto Irrigation District gage rooftop at Modesto	373834121000601	E0.012	-	-	-	-	-	na
4	Modesto Irrigation District gage rooftop at Modesto	373834121000601	-	-	-	-	-	E0.037	na
4	Modesto Irrigation District gage rooftop at Modesto	373834121000601	E0.012	-	-	-	-	-	na
4	Modesto Irrigation District gage rooftop at Modesto	373834121000601	-	-	-	-	-	E0.061	na
4	Modesto Irrigation District gage rooftop at Modesto	373834121000601	-	-	-	-	-	E0.029	na
4	Modesto Irrigation District gage rooftop at Modesto	373834121000601	-	-	-	-	-	E0.043	na
4	Modesto Irrigation District gage rooftop at Modesto	373834121000601	-	-	-	-	-	E0.377	na
4	Modesto Irrigation District gage rooftop at Modesto	373834121000601	-	-	-	E0.39	-	E0.623	na
4	Modesto Irrigation District gage rooftop at Modesto	373834121000601	-	-	-	E0.05	-	E0.051	na
4	Modesto Irrigation District gage rooftop at Modesto	373834121000601	-	-	-	-	-	E0.063	na
4	Modesto Irrigation District gage rooftop at Modesto	373834121000601	-	-	-	-	-	-	na
4	Modesto Irrigation District gage rooftop at Modesto	373834121000601	-	-	-	-	-	E0.282	na
4	Modesto Irrigation District gage rooftop at Modesto	373834121000601	-	-	-	-	-	E0.069	na
4	Modesto Irrigation District gage rooftop at Modesto	373834121000601	-	-	-	-	-	-	na
4	Modesto Irrigation District gage rooftop at Modesto	373834121000601	-	-	-	-	-	E0.086	na
4	Modesto Irrigation District gage rooftop at Modesto	373834121000601	-	-	-	-	-	E0.082	na
4	Modesto Irrigation District gage rooftop at Modesto	373834121000601	-	-	-	-	-	E0.031	na
4	Modesto Irrigation District gage rooftop at Modesto	373834121000601	-	-	-	-	-	E0.305	na
4	Modesto Irrigation District gage rooftop at Modesto	373834121000601	-	-	-	-	-	E0.209	na
5	Modesto Irrigation District gage rooftop at Albers Road	373841120504801	-	-	-	-	-	E0.154	-
5	Modesto Irrigation District gage rooftop at Albers Road	373841120504801	-	-	-	-	-	E0.179	-
5	Modesto Irrigation District gage rooftop at Albers Road	373841120504801	-	-	-	-	-	E0.127	-
5	Modesto Irrigation District gage rooftop at Albers Road	373841120504801	-	-	-	-	-	E0.109	E0.243
5	Modesto Irrigation District gage rooftop at Albers Road	373841120504801	-	-	-	-	-	-	E1.181
5	Modesto Irrigation District gage rooftop at Albers Road	373841120504801	-	-	-	-	-	-	-
5	Modesto Irrigation District gage rooftop at Albers Road	373841120504801	-	-	-	E0.49	-	E0.143	na
5	Modesto Irrigation District gage rooftop at Albers Road	373841120504801	-	-	-	E0.29	-	E0.176	na
5	Modesto Irrigation District gage rooftop at Albers Road	373841120504801	-	-	-	-	-	E0.044	na
5	Modesto Irrigation District gage rooftop at Albers Road	373841120504801	-	-	-	-	-	-	na

Appendix 3. Calculated flux values presented as micrograms per square meter per day (μg/m²/day) for wet-deposition samples with at least one detection collected during the 2002–04 study period at eight sites in the Central Valley, California.—Continued

[All compounds were analyzed by gas chromatography/mass spectrometry (GCMS) at the U.S. Geological Survey (USGS) National Water Quality Laboratory (NWQL). The Chemical Abstract Service Number is given below each compound name in brackets. The five-digit number in parentheses is a code used by the USGS to uniquely identify the given compound. Flux calculated as the analytical result in μg/L multiplied by the total sample volume collected in liters divided by the area of the respective sampler type (Funnel=0.0731 m², Autosampler= 0.0614 m²) and sample composite time in days. Abbreviations: E, estimated reported laboratory value; na, not applicable; (E), estimated compound due to known poor performance by applied method (Childress and others, 1999; Sandstrom and others, 2001); hh:mm, hour:minute; –, compound not detected; [-], no Chemical Abstract Service Number available for given compound]

Site number	Site name	Site identification number	4-Chloro-2-methyl-phenol (E) [1570-64-5] (61633)	Alachlor [15972-60-8] (46342)	Atrazine [1912-24-9] (39632)	Azinphos-methyl (E) [86-50-0] (82686)	Benfluralin [1861-40-1] (82673)	Carbaryl (E) [63-25-2] (82680)	Carbofuran [1563-66-2] (82674)
5	Modesto Irrigation District gage rooftop at Albers Road	373841120504801	–	–	–	–	–	–	na
5	Modesto Irrigation District gage rooftop at Albers Road	373841120504801	–	–	–	–	–	–	na
5	Modesto Irrigation District gage rooftop at Albers Road	373841120504801	–	–	–	–	–	E0.010	na
5	Modesto Irrigation District gage rooftop at Albers Road	373841120504801	–	–	–	–	–	–	na
5	Modesto Irrigation District gage rooftop at Albers Road	373841120504801	–	–	–	–	–	–	na
5	Modesto Irrigation District gage rooftop at Albers Road	373841120504801	–	–	–	–	–	–	na
5	Modesto Irrigation District gage rooftop at Albers Road	373841120504801	–	–	E0.019	–	–	–	na
5	Modesto Irrigation District gage rooftop at Albers Road	373841120504801	–	–	–	–	–	–	na
5	Modesto Irrigation District gage rooftop at Albers Road	373841120504801	–	–	–	–	–	–	na
5	Modesto Irrigation District gage rooftop at Albers Road	373841120504801	–	–	–	–	–	–	na
5	Modesto Irrigation District gage rooftop at Albers Road	373841120504801	–	0.033	–	–	–	E0.160	na
5	Modesto Irrigation District gage rooftop at Albers Road	373841120504801	–	–	–	E0.07	–	E0.301	na
5	Modesto Irrigation District gage rooftop at Albers Road	373841120504801	–	–	–	E0.41	–	E0.392	na
5	Modesto Irrigation District gage rooftop at Albers Road	373841120504801	–	–	–	–	–	–	na
5	Modesto Irrigation District gage rooftop at Albers Road	373841120504801	–	–	–	E0.02	–	E0.037	na
5	Modesto Irrigation District gage rooftop at Albers Road	373841120504801	–	–	–	E0.04	–	–	na
5	Modesto Irrigation District gage rooftop at Albers Road	373841120504801	–	–	–	–	–	–	na
5	Modesto Irrigation District gage rooftop at Albers Road	373841120504801	–	–	–	–	–	–	na
5	Modesto Irrigation District gage rooftop at Albers Road	373841120504801	–	–	–	–	–	E0.207	na
5	Modesto Irrigation District gage rooftop at Albers Road	373841120504801	–	–	–	–	–	–	na
5	Modesto Irrigation District gage rooftop at Albers Road	373841120504801	–	–	–	–	–	–	na
5	Modesto Irrigation District gage rooftop at Albers Road	373841120504801	–	–	–	–	–	E0.034	na
5	Modesto Irrigation District gage rooftop at Albers Road	373841120504801	E0.009	–	–	–	–	E0.033	na
5	Modesto Irrigation District gage rooftop at Albers Road	373841120504801	E0.004	–	–	–	–	E0.027	na
5	Modesto Irrigation District gage rooftop at Albers Road	373841120504801	–	–	–	–	–	–	na
5	Modesto Irrigation District gage rooftop at Albers Road	373841120504801	–	–	–	–	–	E0.157	na

Appendix 3. Calculated flux values presented as micrograms per square meter per day ($\mu g/m^2/day$) for wet-deposition samples with at least one detection collected during the 2002–04 study period at eight sites in the Central Valley, California.—Continued

[All compounds were analyzed by gas chromatography/mass spectrometry (GCMS) at the U.S. Geological Survey (USGS) National Water Quality Laboratory (NWQL). The Chemical Abstract Service Number is given below each compound name in brackets. The five-digit number in parenthesis is a code used by the USGS to uniquely identify the given compound. Flux calculated as the analytical result in $\mu g/L$ multiplied by the total sample volume collected in liters divided by the area of the respective sampler type (Funnel=0.0731 m^2, Autosampler= 0.0614 m^2) and sample composite time in days. Abbreviations: E, estimated reported laboratory value; na, not applicable; (E), estimated compound due to known poor performance by applied method (Childress and others, 1999; Sandstrom and others, 2001); hh:mm, hour:minute; –, compound not detected; [-], no Chemical Abstract Service Number available for given compound]

Site number	Site name	Site identification number	4-Chloro-2-methyl-phenol (E) [1570-64-5] (61633)	Alachlor [15972-60-8] (46342)	Atrazine [1912-24-9] (39632)	Azinphos-methyl (E) [86-50-0] (82686)	Benfluralin [1861-40-1] (82673)	Carbaryl (E) [63-25-2] (82680)	Carbofuran [1563-66-2] (82674)
16	Gridley High School precipitation gage at Gridley	39220512141020l	–	–	–		–	–	na
16	Gridley High School precipitation gage at Gridley	39220512141020l	–	–	–		–	–	na
16	Gridley High School precipitation gage at Gridley	39220512141020l	–	–	–		–	–	na
16	Gridley High School precipitation gage at Gridley	39220512141020l	–	–	–		–	–	na
16	Gridley High School precipitation gage at Gridley	39220512141020l	–	–	–		–	–	na
16	Gridley High School precipitation gage at Gridley	39220512141020l	–	–	–		–	–	na
16	Gridley High School precipitation gage at Gridley	39220512141020l	–	–	–	E0.72	–	E0.169	na
16	Gridley High School precipitation gage at Gridley	39220512141020l	–	–	–	E0.19	–	E0.135	na
16	Gridley High School precipitation gage at Gridley	39220512141020l	–	–	–		–	E0.024	na
16	Gridley High School precipitation gage at Gridley	39220512141020l	–	–	–		–	E0.067	na
16	Gridley High School precipitation gage at Gridley	39220512141020l	–	–	–		–	E0.046	na
16	Gridley High School precipitation gage at Gridley	39220512141020l	–	–	–		–	E0.048	na
16	Gridley High School precipitation gage at Gridley	39220512141020l	–	–	–		–	–	na
16	Gridley High School precipitation gage at Gridley	39220512141020l	–	–	–		–	E0.084	na
16	Gridley High School precipitation gage at Gridley	39220512141020l	–	–	–		–	E0.065	na
16	Gridley High School precipitation gage at Gridley	39220512141020l	–	–	–		–	E0.179	na
16	Gridley High School precipitation gage at Gridley	39220512141020l	–	–	–		–	E0.055	na
16	Gridley High School precipitation gage at Gridley	39220512141020l	–	–	–		–	E0.059	na
16	Gridley High School precipitation gage at Gridley	39220512141020l	–	–	–		–	–	na
16	Gridley High School precipitation gage at Gridley	39220512141020l	–	–	–		–	–	na
16	Gridley High School precipitation gage at Gridley	39220512141020l	–	–	–		–	–	na
16	Gridley High School precipitation gage at Gridley	39220512141020l	–	–	–		–	–	na
16	Gridley High School precipitation gage at Gridley	39220512141020l	–	–	–		–	E0.035	na
16	Gridley High School precipitation gage at Gridley	39220512141020l	–	–	–		–	E0.109	na
16	Gridley High School precipitation gage at Gridley	39220512141020l	–	–	–		–	–	na
16	Gridley High School precipitation gage at Gridley	39220512141020l	–	–	–		–	–	na

Appendix 3.   Calculated flux values presented as micrograms per square meter per day ($\mu g/m^2/day$) for wet-deposition samples with at least one detection collected during the 2002–04 study period at eight sites in the Central Valley, California.—Continued

[All compounds were analyzed by gas chromatography/mass spectrometry (GCMS) at the U.S. Geological Survey (USGS) National Water Quality Laboratory (NWQL). The Chemical Abstract Service Number is given below each compound name in parentheses. The five-digit number in brackets is a code used by the USGS to uniquely identify the given compound. Flux calculated as the analytical result in $\mu g/L$ multiplied by the total sample volume collected in liters divided by the area of the respective sampler type (Funnel=0.0731 m^2, Autosampler= 0.0614 m^2) and sample composite time in days. Abbreviations: E, estimated reported laboratory value; na, not applicable; (E), estimated compound due to known poor performance by applied method (Childress and others, 1999; Sandstrom and others, 2001); hh:mm, hour:minute; –, compound not detected; [-], no Chemical Abstract Service Number available for given compound]

Site number	Site name	Site identification number	4-Chloro-2-methyl-phenol (E) [1570-64-5] (61633)	Alachlor [15972-60-8] (46342)	Atrazine [1912-24-9] (39632)	Azinphos-methyl (E) [86-50-0] (82686)	Benfluralin [1861-40-1] (82673)	Carbaryl (E) [63-25-2] (82680)	Carbofuran [1563-66-2] (82674)
15	Oroville Dam precipitation gage at spillway	393234121292701	–	–	–	–	–	–	na
15	Oroville Dam precipitation gage at spillway	393234121292701	–	–	–	–	–	–	na
15	Oroville Dam precipitation gage at spillway	393234121292701	–	–	–	–	–	–	na
15	Oroville Dam precipitation gage at spillway	393234121292701	–	–	–	–	–	–	na
15	Oroville Dam precipitation gage at spillway	393234121292701	–	–	–	–	–	–	na
15	Oroville Dam precipitation gage at spillway	393234121292701	–	–	–	E0.13	–	E0.080	na
15	Oroville Dam precipitation gage at spillway	393234121292701	–	–	–	–	–	–	na
15	Oroville Dam precipitation gage at spillway	393234121292701	–	–	–	–	–	–	na
15	Oroville Dam precipitation gage at spillway	393234121292701	–	–	–	–	–	–	na
15	Oroville Dam precipitation gage at spillway	393234121292701	–	–	–	–	–	–	na
15	Oroville Dam precipitation gage at spillway	393234121292701	–	–	–	–	–	–	na
15	Oroville Dam precipitation gage at spillway	393234121292701	–	–	E0.085	–	–	–	na
15	Oroville Dam precipitation gage at spillway	393234121292701	–	–	–	–	–	E0.025	na

Appendix 3. Calculated flux values presented as micrograms per square meter per day (µg/m²/day) for wet-deposition samples with at least one detection collected during the 2002–04 study period at eight sites in the Central Valley, California.—Continued

[All compounds were analyzed by gas chromatography/mass spectrometry (GCMS) at the U.S. Geological Survey (USGS) National Water Quality Laboratory (NWQL). The Chemical Abstract Service Number is given below each compound name in brackets. The five-digit number in parenthesis is a code used by the USGS to uniquely identify the given compound. Flux calculated as the analytical result in µg/L multiplied by the total sample volume collected in liters divided by the area of the respective sampler type (Funnel=0.0731 m², Autosampler= 0.0614 m²) and sample composite time in days. Abbreviations: E, estimated reported laboratory value; na, not applicable; (E), estimated compound due to known poor performance by applied method (Childress and others, 1999; Sandstrom and others, 2001); hh:mm, hour:minute; —, compound not detected; [-], no Chemical Abstract Service Number available for given compound]

Site number	Site name	Site identification number	Chlorpyrifos [2921-88-2] (38933)	Chlorpyrifos oxon (E) [5598-15-2] (61636)	Cyanazine [21725-46-2] (04041)	Cycloate [1134-23-2] (04031)	Cyfluthrin (E) [68359-37-5] (61585)	Cypermethrin (E) [52315-07-8] (61586)	Dacthal (DCPA) [1861-32-1] (82682)
11	Newman rain gage at wasteway levee near Draper Road	371735121031201	0.040	—	—	—	—	—	0.119
11	Newman rain gage at wasteway levee near Draper Road	371735121031201	0.035	—	—	—	—	—	0.016
11	Newman rain gage at wasteway levee near Draper Road	371735121031201	0.062	—	—	0.107	—	—	0.048
11	Newman rain gage at wasteway levee near Draper Road	371735121031201	0.073	E0.04	—	—	—	—	—
11	Newman rain gage at wasteway levee near Draper Road	371735121031201	0.048	E0.09	—	0.095	—	—	0.195
11	Newman rain gage at wasteway levee near Draper Road	371735121031201	0.193	—	—	na	—	—	0.409
11	Newman rain gage at wasteway levee near Draper Road	371735121031201	—	—	na	na	—	—	0.091
11	Newman rain gage at wasteway levee near Draper Road	371735121031201	E0.027	—	na	na	—	—	0.075
11	Newman rain gage at wasteway levee near Draper Road	371735121031201	0.274	—	na	na	—	—	0.286
11	Newman rain gage at wasteway levee near Draper Road	371735121031201	0.159	E0.06	na	na	—	—	0.082
11	Newman rain gage at wasteway levee near Draper Road	371735121031201	0.020	—	na	na	—	—	0.018
11	Newman rain gage at wasteway levee near Draper Road	371735121031201	0.037	—	na	na	—	—	0.053
11	Newman rain gage at wasteway levee near Draper Road	371735121031201	0.041	E0.07	na	na	—	—	0.123
11	Newman rain gage at wasteway levee near Draper Road	371735121031201	0.085	E0.08	na	na	—	—	0.093
11	Newman rain gage at wasteway levee near Draper Road	371735121031201	0.015	—	na	na	—	—	0.015
11	Newman rain gage at wasteway levee near Draper Road	371735121031201	0.120	—	na	na	—	—	0.130
12	Turlock rain gage near Idaho Road	372713120534901	0.207	—	—	—	—	—	0.031
12	Turlock rain gage near Idaho Road	372713120534901	0.056	E0.07	—	—	—	—	0.030
12	Turlock rain gage near Idaho Road	372713120534901	0.044	E0.05	—	—	—	—	0.044
12	Turlock rain gage near Idaho Road	372713120534901	0.410	E0.06	—	—	—	—	0.032
12	Turlock rain gage near Idaho Road	372713120534901	0.430	E0.11	—	—	—	—	—
12	Turlock rain gage near Idaho Road	372713120534901	0.213	E0.06	—	—	—	—	—
12	Turlock rain gage near Idaho Road	372713120534901	0.121	—	—	—	—	—	0.102
12	Turlock rain gage near Idaho Road	372713120534901	0.133	—	na	na	—	E0.209	0.133
12	Turlock rain gage near Idaho Road	372713120534901	0.079	—	na	na	—	—	0.089
12	Turlock rain gage near Idaho Road	372713120534901	0.062	—	na	na	—	—	0.037
12	Turlock rain gage near Idaho Road	372713120534901	0.632	—	na	na	—	—	0.108
12	Turlock rain gage near Idaho Road	372713120534901	—	—	na	na	—	—	E0.016
12	Turlock rain gage near Idaho Road	372713120534901	0.569	E0.22	na	na	—	—	0.161
12	Turlock rain gage near Idaho Road	372713120534901	0.049	—	na	na	—	—	0.011
12	Turlock rain gage near Idaho Road	372713120534901	0.022	—	na	na	—	—	0.019

Appendix 3. Calculated flux values presented as micrograms per square meter per day (µg/m²/day) for wet-deposition samples with at least one detection collected during the 2002–04 study period at eight sites in the Central Valley, California.—Continued

[All compounds were analyzed by gas chromatography/mass spectrometry (GCMS) at the U.S. Geological Survey (USGS) National Water Quality Laboratory (NWQL). The Chemical Abstract Service Number is given below each compound name in brackets. The five-digit number in parentheses is a code used by the USGS to uniquely identify the given compound. Flux calculated as the analytical result in µg/L multiplied by the total sample volume collected in liters divided by the area of the respective sampler type (Funnel=0.0731 m², Autosampler= 0.0614 m²) and sample composite time in days. Abbreviations: E, estimated reported laboratory value; na, not applicable; (E), estimated compound due to known poor performance by applied method (Childress and others, 1999; Sandstrom and others, 2001); hh:mm, hour:minute; –, compound not detected; [-], no Chemical Abstract Service Number available for given compound]

Site number	Site name	Site identification number	Chlorpyrifos [2921-88-2] (38933)	Chlorpyrifos oxon (E) [5598-15-2] (61636)	Cyanazine [21725-46-2] (04041)	Cycloate [1134-23-2] (04031)	Cyfluthrin {E} [68359-37-5] (61585)	Cypermethrin {E} [52315-07-8] (61586)	Dacthal (DCPA) [1861-32-1] (82682)
12	Turlock rain gage near Idaho Road	372713120534901	0.020	–	na	na	–	–	0.056
12	Turlock rain gage near Idaho Road	372713120534901	0.504	E0.18	na	na	–	–	0.092
12	Turlock rain gage near Idaho Road	372713120534901	0.246	E0.16	na	na	–	–	0.038
12	Turlock rain gage near Idaho Road	372713120534901	0.047	E0.05	na	na	–	–	0.012
12	Turlock rain gage near Idaho Road	372713120534901	0.374	E0.06	na	na	–	–	0.035
12	Turlock rain gage near Idaho Road	372713120534901	0.066	–	na	na	–	–	0.078
13	Turlock Airport rain gage	372857120414001	0.232	E0.18	–	–	–	–	0.071
13	Turlock Airport rain gage	372857120414001	0.026	–	–	–	–	–	0.021
13	Turlock Airport rain gage	372857120414001	0.038	E0.05	–	–	–	–	0.047
13	Turlock Airport rain gage	372857120414001	0.115	–	–	–	–	–	0.110
13	Turlock Airport rain gage	372857120414001	0.062	–	–	–	–	–	0.019
13	Turlock Airport rain gage	372857120414001	0.026	E0.02	–	–	–	–	–
13	Turlock Airport rain gage	372857120414001	–	–	–	–	–	–	0.105
13	Turlock Airport rain gage	372857120414001	0.226	–	na	na	–	–	0.169
13	Turlock Airport rain gage	372857120414001	0.064	–	na	na	–	–	0.074
13	Turlock Airport rain gage	372857120414001	0.044	–	na	na	–	–	0.021
13	Turlock Airport rain gage	372857120414001	0.361	–	na	na	–	–	0.082
13	Turlock Airport rain gage	372857120414001	0.024	–	na	na	–	–	0.022
13	Turlock Airport rain gage	372857120414001	0.021	E0.02	na	na	–	–	0.036
13	Turlock Airport rain gage	372857120414001	0.225	E0.07	na	na	–	–	0.044
13	Turlock Airport rain gage	372857120414001	E0.021	–	na	na	–	–	0.171
13	Turlock Airport rain gage	372857120414001	0.029	–	na	na	–	–	0.025
13	Turlock Airport rain gage	372857120414001	0.076	–	na	na	–	–	0.054
13	Turlock Airport rain gage	372857120414001	0.618	E0.38	na	na	–	–	0.105
13	Turlock Airport rain gage	372857120414001	0.078	E0.06	na	na	–	–	0.025
13	Turlock Airport rain gage	372857120414001	0.132	E0.10	na	na	–	–	0.061
13	Turlock Airport rain gage	372857120414001	0.326	–	na	na	–	–	0.186
10	Westley rain gage at pump building near lateral 6 North	373335121143001	–	–	–	–	–	–	0.014
10	Westley rain gage at pump building near lateral 6 North	373335121143001	0.042	E0.05	–	–	–	–	0.187
10	Westley rain gage at pump building near lateral 6 North	373335121143001	0.028	E0.03	–	–	–	–	0.023

Appendix 3. Calculated flux values presented as micrograms per square meter per day (µg/m²/day) for wet-deposition samples with at least one detection collected during the 2002–04 study period at eight sites in the Central Valley, California.—Continued

[All compounds were analyzed by gas chromatography/mass spectrometry (GCMS) at the U.S. Geological Survey (USGS) National Water Quality Laboratory (NWQL). The Chemical Abstract Service Number is given below each compound name in brackets. The five-digit number in parenthesis is a code used by the USGS to uniquely identify the given compound. Flux calculated as the analytical result in µg/L multiplied by the total sample volume collected in liters divided by the area of the respective sampler type (Funnel=0.0731 m², Autosampler= 0.0614 m²) and sample composite time in days. Abbreviations: E, estimated reported laboratory value; na, not applicable; (E), estimated compound due to known poor performance by applied method (Childress and others, 1999; Sandstrom and others, 2001); hh:mm, hour:minute; –, compound not detected; [-], no Chemical Abstract Service Number available for given compound]

Site number	Site name	Site identification number	Chlorpyrifos [2921-88-2] (38933)	Chlorpyrifos oxon (E) [5598-15-2] (61636)	Cyanazine [21725-46-2] (04041)	Cycloate [1134-23-2] (04031)	Cyfluthrin (E) [68359-37-5] (61585)	Cypermethrin (E) [52315-07-8] (61586)	Dacthal (DCPA) [1861-32-1] (82682)
10	Westley rain gage at pump building near lateral 6 North	373335121143001	1.950	E0.41	0.06	–	–	–	0.022
10	Westley rain gage at pump building near lateral 6 North	373335121143001	0.144	E0.05	–	–	–	–	E0.015
10	Westley rain gage at pump building near lateral 6 North	373335121143001	0.177	E0.17	–	–	–	–	0.076
10	Westley rain gage at pump building near lateral 6 North	373335121143001	0.169	–	na	na	–	–	0.400
10	Westley rain gage at pump building near lateral 6 North	373335121143001	0.040	–	na	na	–	–	0.153
10	Westley rain gage at pump building near lateral 6 North	373335121143001	–	–	na	na	–	–	0.025
10	Westley rain gage at pump building near lateral 6 North	373335121143001	0.112	–	na	na	–	–	0.186
10	Westley rain gage at pump building near lateral 6 North	373335121143001	0.105	E0.04	na	na	–	–	0.036
10	Westley rain gage at pump building near lateral 6 North	373335121143001	0.241	E0.08	na	na	E0.136	–	0.136
10	Westley rain gage at pump building near lateral 6 North	373335121143001	0.164	–	na	na	E0.104	–	0.173
10	Westley rain gage at pump building near lateral 6 North	373335121143001	0.014	E0.03	na	na	–	–	0.015
10	Westley rain gage at pump building near lateral 6 North	373335121143001	0.092	E0.11	na	na	–	–	0.172
4	Modesto Irrigation District gage rooftop at Modesto	373834121000601	0.079	–	–	–	–	–	0.021
4	Modesto Irrigation District gage rooftop at Modesto	373834121000601	0.022	–	–	–	–	–	0.033
4	Modesto Irrigation District gage rooftop at Modesto	373834121000601	0.043	–	–	–	–	–	0.030
4	Modesto Irrigation District gage rooftop at Modesto	373834121000601	0.326	–	–	–	–	–	0.042
4	Modesto Irrigation District gage rooftop at Modesto	373834121000601	0.148	E0.12	–	–	–	–	E0.012
4	Modesto Irrigation District gage rooftop at Modesto	373834121000601	–	E0.2	–	–	–	–	0.290
4	Modesto Irrigation District gage rooftop at Modesto	373834121000601	0.022	E0.1	–	–	–	–	0.013
4	Modesto Irrigation District gage rooftop at Modesto	373834121000601	0.196	–	na	na	–	–	0.253
4	Modesto Irrigation District gage rooftop at Modesto	373834121000601	0.248	–	na	na	–	–	0.221
4	Modesto Irrigation District gage rooftop at Modesto	373834121000601	0.307	–	na	na	–	–	0.131
4	Modesto Irrigation District gage rooftop at Modesto	373834121000601	0.176	–	na	na	–	–	0.106
4	Modesto Irrigation District gage rooftop at Modesto	373834121000601	0.078	E0.04	na	na	–	–	0.048
4	Modesto Irrigation District gage rooftop at Modesto	373834121000601	0.053	–	na	na	–	–	0.049
4	Modesto Irrigation District gage rooftop at Modesto	373834121000601	0.172	E0.06	na	na	–	–	0.030
4	Modesto Irrigation District gage rooftop at Modesto	373834121000601	0.897	E0.12	na	na	–	–	0.100
4	Modesto Irrigation District gage rooftop at Modesto	373834121000601	0.559	–	na	na	–	–	0.140
4	Modesto Irrigation District gage rooftop at Modesto	373834121000601	0.041	–	na	na	–	–	0.014
4	Modesto Irrigation District gage rooftop at Modesto	373834121000601	0.034	E0.03	na	na	–	–	0.048
4	Modesto Irrigation District gage rooftop at Modesto	373834121000601	0.029	–	na	na	–	–	0.032

Appendix 3. Calculated flux values presented as micrograms per square meter per day (µg/m²/day) for wet-deposition samples with at least one detection collected during the 2002–04 study period at eight sites in the Central Valley, California.—Continued

[All compounds were analyzed by gas chromatography/mass spectrometry (GCMS) at the U.S. Geological Survey (USGS) National Water Quality Laboratory (NWQL). The Chemical Abstract Service Number is given below each compound name in brackets. The five-digit number in parenthesis is a code used by the USGS to uniquely identify the given compound. Flux calculated as the analytical result in µg/L multiplied by the total sample volume collected in liters divided by the area of the respective sampler type (Funnel=0.0731 m², Autosampler= 0.0614 m²) and sample composite time in days. Abbreviations: E, estimated reported laboratory value; na, not applicable; (E), estimated compound due to known poor performance by applied method (Childress and others, 1999; Sandstrom and others, 2001); hh:mm, hour:minute; —, compound not detected; [-], no Chemical Abstract Service Number available for given compound]

Site number	Site name	Site identification number	Chlorpyrifos [2921-88-2] (38833)	Chlorpyrifos oxon (E) [5598-15-2] (61636)	Cyanazine [21725-46-2] (04041)	Cycloate [1134-23-2] (04031)	Cyfluthrin (E) [68359-37-5] (61585)	Cypermethrin (E) [52315-07-8] (61586)	Dacthal (DCPA) [1861-32-1] (82682)
4	Modesto Irrigation District gage rooftop at Modesto	37383412000601	0.623	E0.31	na	na	—	—	0.184
4	Modesto Irrigation District gage rooftop at Modesto	37383412000601	0.322	E0.10	na	na	—	—	0.137
4	Modesto Irrigation District gage rooftop at Modesto	37383412000601	0.020	—	na	na	—	—	0.029
4	Modesto Irrigation District gage rooftop at Modesto	37383412000601	0.040	—	na	na	—	—	0.024
4	Modesto Irrigation District gage rooftop at Modesto	37383412000601	0.020	E0.03	na	na	—	—	0.033
4	Modesto Irrigation District gage rooftop at Modesto	37383412000601	0.033	—	na	na	—	—	0.059
4	Modesto Irrigation District gage rooftop at Modesto	37383412000601	0.028	—	na	na	—	—	0.010
4	Modesto Irrigation District gage rooftop at Modesto	37383412000601	0.029	—	na	na	—	—	0.018
4	Modesto Irrigation District gage rooftop at Modesto	37383412000601	0.526	—	na	na	—	—	0.115
4	Modesto Irrigation District gage rooftop at Modesto	37383412000601	0.146	—	na	na	—	—	0.031
4	Modesto Irrigation District gage rooftop at Modesto	37383412000601	0.056	—	na	na	—	—	0.025
4	Modesto Irrigation District gage rooftop at Modesto	37383412000601	0.987	E0.19	na	na	—	—	0.134
4	Modesto Irrigation District gage rooftop at Modesto	37383412000601	0.872	E0.31	na	na	—	—	0.123
4	Modesto Irrigation District gage rooftop at Modesto	37383412000601	0.034	E0.05	na	na	—	—	0.018
4	Modesto Irrigation District gage rooftop at Modesto	37383412000601	0.207	E0.11	na	na	—	—	0.076
4	Modesto Irrigation District gage rooftop at Modesto	37383412000601	0.062	—	na	na	—	—	0.066
5	Modesto Irrigation District gage rooftop at Albers Road	37384112050804801	0.270	—	—	—	—	—	0.032
5	Modesto Irrigation District gage rooftop at Albers Road	37384112050804801	0.024	E0.06	—	—	—	—	0.034
5	Modesto Irrigation District gage rooftop at Albers Road	37384112050804801	0.197	E0.13	—	—	—	—	0.159
5	Modesto Irrigation District gage rooftop at Albers Road	37384112050804801	0.034	E0.06	—	—	—	—	0.037
5	Modesto Irrigation District gage rooftop at Albers Road	37384112050804801	0.057	E0.06	—	—	—	—	0.048
5	Modesto Irrigation District gage rooftop at Albers Road	37384112050804801	0.110	E0.86	—	—	—	—	0.110
5	Modesto Irrigation District gage rooftop at Albers Road	37384112050804801	0.257	—	na	na	—	—	0.143
5	Modesto Irrigation District gage rooftop at Albers Road	37384112050804801	0.293	—	na	na	—	—	0.176
5	Modesto Irrigation District gage rooftop at Albers Road	37384112050804801	0.142	—	na	na	—	—	0.066
5	Modesto Irrigation District gage rooftop at Albers Road	37384112050804801	0.138	—	na	na	—	—	0.074
5	Modesto Irrigation District gage rooftop at Albers Road	37384112050804801	0.152	—	na	na	—	—	0.030
5	Modesto Irrigation District gage rooftop at Albers Road	37384112050804801	0.145	—	na	na	—	—	0.039
5	Modesto Irrigation District gage rooftop at Albers Road	37384112050804801	0.893	E0.04	na	na	—	—	0.017
5	Modesto Irrigation District gage rooftop at Albers Road	37384112050804801	7.522	E0.33	na	na	—	—	0.066
5	Modesto Irrigation District gage rooftop at Albers Road	37384112050804801	8.418	E0.14	na	na	—	—	0.070

Appendix 3. Calculated flux values presented as micrograms per square meter per day (µg/m²/day) for wet-deposition samples with at least one detection collected during the 2002–04 study period at eight sites in the Central Valley, California.—Continued

[All compounds were analyzed by gas chromatography mass spectrometry (GCMS) at the U.S. Geological Survey (USGS) National Water Quality Laboratory (NWQL). The Chemical Abstract Service Number is given below each compound name in parenthesis. The five-digit number in parentheses is a code used by the USGS to uniquely identify the given compound. Flux calculated as the analytical result in µg/L multiplied by the total sample volume collected in liters divided by the area of the respective sampler type (Funnel=0.0731 m², Autosampler= 0.0614 m²) and sample composite time in days. Abbreviations: E, estimated reported laboratory value; na, not applicable; (E), estimated compound due to known poor performance by applied method (Childress and others, 1999; Sandstrom and others, 2001); hh:mm, hour:minute; —, compound not detected; [-], no Chemical Abstract Service Number available for given compound]

Site number	Site name	Site identification number	Chlorpyrifos [2921-88-2] (38933)	Chlorpyrifos oxon (E) [5598-15-2] (61636)	Cyanazine [21725-46-2] (04041)	Cycloate [1134-23-2] (04031)	Cyfluthrin (E) [68359-37-5] (61585)	Cypermethrin (E) [52315-07-8] (61586)	Dacthal (DCPA) [1861-32-1] (82682)
5	Modesto Irrigation District gage rooftop at Albers Road	37384112050480	0.041	—	na	na	—	—	0.006
5	Modesto Irrigation District gage rooftop at Albers Road	37384112050480	0.048	—	na	na	—	—	0.029
5	Modesto Irrigation District gage rooftop at Albers Road	37384112050480	E0.018	—	na	na	—	—	0.029
5	Modesto Irrigation District gage rooftop at Albers Road	37384112050480	0.301	E0.15	na	na	—	—	0.110
5	Modesto Irrigation District gage rooftop at Albers Road	37384112050480	0.224	E0.08	na	na	—	—	0.100
5	Modesto Irrigation District gage rooftop at Albers Road	37384112050480	0.064	—	na	na	—	—	0.032
5	Modesto Irrigation District gage rooftop at Albers Road	37384112050480	E0.033	E0.07	na	na	—	—	0.114
5	Modesto Irrigation District gage rooftop at Albers Road	37384112050480	0.023	E0.03	na	na	—	—	0.019
5	Modesto Irrigation District gage rooftop at Albers Road	37384112050480	—	—	na	na	—	—	0.010
5	Modesto Irrigation District gage rooftop at Albers Road	37384112050480	0.020	—	na	na	—	—	0.026
5	Modesto Irrigation District gage rooftop at Albers Road	37384112050480	0.033	—	na	na	—	—	0.020
5	Modesto Irrigation District gage rooftop at Albers Road	37384112050480	0.078	—	na	na	—	—	0.036
5	Modesto Irrigation District gage rooftop at Albers Road	37384112050480	0.065	—	na	na	—	—	0.038
5	Modesto Irrigation District gage rooftop at Albers Road	37384112050480	10.122	E1.83	na	na	—	—	0.110
5	Modesto Irrigation District gage rooftop at Albers Road	37384112050480	1.312	E0.23	na	na	—	—	0.031
5	Modesto Irrigation District gage rooftop at Albers Road	37384112050480	0.090	E0.03	na	na	—	—	0.005
5	Modesto Irrigation District gage rooftop at Albers Road	37384112050480	1.381	E0.47	na	na	—	—	0.067
5	Modesto Irrigation District gage rooftop at Albers Road	37384112050480	0.938	E0.11	na	na	—	—	0.065
5	Modesto Irrigation District gage rooftop at Albers Road	37384112050480	0.554	E0.28	na	na	—	—	0.023
5	Modesto Irrigation District gage rooftop at Albers Road	37384112050480	0.084	E0.11	na	na	—	—	0.021
5	Modesto Irrigation District gage rooftop at Albers Road	37384112050480	0.219	E0.14	na	na	—	—	0.055
5	Modesto Irrigation District gage rooftop at Albers Road	37384112050480	0.303	—	na	na	—	—	0.123
16	Gridley High School precipitation gage at Gridley	39220512141020	0.931	E0.27	na	na	—	—	0.082
16	Gridley High School precipitation gage at Gridley	39220512141020	1.564	E0.65	na	na	—	—	0.130
16	Gridley High School precipitation gage at Gridley	39220512141020	0.853	—	na	na	—	—	0.128
16	Gridley High School precipitation gage at Gridley	39220512141020	0.044	E0.06	na	na	—	—	0.044
16	Gridley High School precipitation gage at Gridley	39220512141020	0.066	E0.03	na	na	—	—	0.039
16	Gridley High School precipitation gage at Gridley	39220512141020	0.154	E0.08	na	na	—	—	0.081
16	Gridley High School precipitation gage at Gridley	39220512141020	E0.021	—	na	na	—	—	0.032
16	Gridley High School precipitation gage at Gridley	39220512141020	E0.034	—	na	na	—	—	0.084
16	Gridley High School precipitation gage at Gridley	39220512141020	E0.006	—	na	na	—	—	0.018

Appendix 3. Calculated flux values presented as micrograms per square meter per day (µg/m²/day) for wet-deposition samples with at least one detection collected during the 2002–04 study period at eight sites in the Central Valley, California.—Continued

[All compounds were analyzed by gas chromatography/mass spectrometry (GCMS) at the U.S. Geological Survey (USGS) National Water Quality Laboratory (NWQL). The Chemical Abstract Service Number is given below each compound name in brackets. The five-digit number in parenthesis is a code used by the USGS to uniquely identify the given compound. Flux calculated as the analytical result in µg/L multiplied by the total sample volume collected in liters divided by the area of the respective sampler type (Funnel=0.0731 m², Autosampler= 0.0614 m²) and sample composite time in days. Abbreviations: E, estimated reported laboratory value; na, not applicable; (E), estimated compound due to known poor performance by applied method (Childress and others, 1999; Sandstrom and others, 2001); hh:mm, hour:minute; –, compound not detected; [-], no Chemical Abstract Service Number available for given compound]

Site number	Site name	Site identification number	Chlorpyrifos [2921-88-2] (38933)	Chlorpyrifos oxon (E) [5598-15-2] (61636)	Cyanazine [21725-46-2] (04041)	Cycloate [1134-23-2] (04031)	Cyfluthrin (E) [68359-37-5] (61585)	Cypermethrin (E) [52315-07-8] (61586)	Dacthal (DCPA) [1861-32-1] (82682)
16	Gridley High School precipitation gage at Gridley	3922051214102 01	E0.056	–	na	na	–	–	0.067
16	Gridley High School precipitation gage at Gridley	3922051214102 01	0.032	–	na	na	–	–	0.021
16	Gridley High School precipitation gage at Gridley	3922051214102 01	–	–	na	na	–	–	0.020
16	Gridley High School precipitation gage at Gridley	3922051214102 01	0.046	–	na	na	–	–	0.025
16	Gridley High School precipitation gage at Gridley	3922051214102 01	0.167	E0.06	na	na	–	–	0.038
16	Gridley High School precipitation gage at Gridley	3922051214102 01	0.038	–	na	na	–	–	0.012
16	Gridley High School precipitation gage at Gridley	3922051214102 01	0.086	E0.08	na	na	–	–	0.022
16	Gridley High School precipitation gage at Gridley	3922051214102 01	0.200	E0.12	na	na	–	–	0.024
16	Gridley High School precipitation gage at Gridley	3922051214102 01	0.276	E0.08	na	na	–	–	E0.035
16	Gridley High School precipitation gage at Gridley	3922051214102 01	0.681	E0.10	na	na	–	–	0.032
16	Gridley High School precipitation gage at Gridley	3922051214102 01	0.266	E0.08	na	na	–	–	0.017
16	Gridley High School precipitation gage at Gridley	3922051214102 01	0.431	E0.10	na	na	–	–	0.013
16	Gridley High School precipitation gage at Gridley	3922051214102 01	0.162	E0.17	na	na	–	–	0.020
16	Gridley High School precipitation gage at Gridley	3922051214102 01	0.465	E0.31	na	na	–	–	0.023
16	Gridley High School precipitation gage at Gridley	3922051214102 01	0.382	–	na	na	–	–	0.078
16	Gridley High School precipitation gage at Gridley	3922051214102 01	0.092	E0.08	na	na	–	–	0.106
16	Gridley High School precipitation gage at Gridley	3922051214102 01	–	–	na	na	–	–	0.033
15	Oroville Dam precipitation gage at spillway	3932341212927 01	–	–	na	na	–	–	0.068
15	Oroville Dam precipitation gage at spillway	3932341212927 01	0.009	–	na	na	–	–	0.044
15	Oroville Dam precipitation gage at spillway	3932341212927 01	0.073	–	na	na	–	–	E0.003
15	Oroville Dam precipitation gage at spillway	3932341212927 01	0.157	E0.07	na	na	–	–	0.122
15	Oroville Dam precipitation gage at spillway	3932341212927 01	–	–	na	na	–	–	0.082
15	Oroville Dam precipitation gage at spillway	3932341212927 01	–	–	na	na	–	–	0.037
15	Oroville Dam precipitation gage at spillway	3932341212927 01	–	–	na	na	–	–	0.070
15	Oroville Dam precipitation gage at spillway	3932341212927 01	0.038	–	na	na	–	–	E0.007
15	Oroville Dam precipitation gage at spillway	3932341212927 01	0.029	E0.04	na	na	–	–	0.027
15	Oroville Dam precipitation gage at spillway	3932341212927 01	0.087	E0.09	na	na	–	–	E0.012
15	Oroville Dam precipitation gage at spillway	3932341212927 01	0.107	–	na	na	–	–	0.026
15	Oroville Dam precipitation gage at spillway	3932341212927 01	E0.025	E0.08	na	na	–	–	0.149
15	Oroville Dam precipitation gage at spillway	3932341212927 01	–	–	na	na	–	–	0.050

Appendix 3. Calculated flux values presented as micrograms per square meter per day (µg/m²/day) for wet-deposition samples with at least one detection collected during the 2002–04 study period at eight sites in the Central Valley, California.—Continued

[All compounds were analyzed by gas chromatography/mass spectrometry (GCMS) at the U.S. Geological Survey (USGS) National Water Quality Laboratory (NWQL). The Chemical Abstract Service Number is given below each compound name in brackets. The five-digit number in parenthesis is a code used by the USGS to uniquely identify the given compound. Flux calculated as the analytical result in µg/L multiplied by the total sample volume collected in liters divided by the area of the respective sampler type (Funnel=0.0731 m², Autosampler= 0.0614 m²) and sample composite time in days. Abbreviations: E, estimated reported laboratory value; na, not applicable; (E), estimated compound due to known poor performance by applied method (Childress and others, 1999; Sandstrom and others, 2001); hh:mm, hour:minute; –, compound not detected; [-], no Chemical Abstract Service Number available for given compound]

Site number	Site name	Site identification number	p,p'-DDE [72-55-9] (34653)	Diazinon [333-41-5] (39572)	Diazoxon (E) [962-58-3] (61638)	Dichlorvos (E) [62-73-7] (38775)	Dimethoate [60-51-5] (82662)	EPTC [759-94-4] (82668)	Fonofos [944-22-9] (04095)
11	Newman rain gage at wasteway levee near Draper Road	371735121031201	–	0.199	na	–	–	–	–
11	Newman rain gage at wasteway levee near Draper Road	371735121031201	–	0.160	na	–	–	–	–
11	Newman rain gage at wasteway levee near Draper Road	371735121031201	–	0.074	na	–	–	–	–
11	Newman rain gage at wasteway levee near Draper Road	371735121031201	–	0.065	na	–	–	–	–
11	Newman rain gage at wasteway levee near Draper Road	371735121031201	–	0.747	na	E0.06	E0.819	na	–
11	Newman rain gage at wasteway levee near Draper Road	371735121031201	na	0.049	na	–	E0.067	na	–
11	Newman rain gage at wasteway levee near Draper Road	371735121031201	na	–	–	–	–	na	–
11	Newman rain gage at wasteway levee near Draper Road	371735121031201	na	0.988	E0.06	–	–	na	–
11	Newman rain gage at wasteway levee near Draper Road	371735121031201	na	0.140	–	–	E0.074	na	–
11	Newman rain gage at wasteway levee near Draper Road	371735121031201	na	0.037	–	E0.04	E0.027	na	–
11	Newman rain gage at wasteway levee near Draper Road	371735121031201	na	0.048	–	–	–	na	–
11	Newman rain gage at wasteway levee near Draper Road	371735121031201	na	1.28	E0.41	–	–	na	–
11	Newman rain gage at wasteway levee near Draper Road	371735121031201	na	0.526	E0.17	–	–	na	–
11	Newman rain gage at wasteway levee near Draper Road	371735121031201	na	0.064	E0.04	–	–	na	–
11	Newman rain gage at wasteway levee near Draper Road	371735121031201	na	0.462	E0.20	E0.10	–	na	–
12	Turlock rain gage near Idaho Road	372713120534901	–	6.55	na	–	–	–	–
12	Turlock rain gage near Idaho Road	372713120534901	–	0.744	na	–	–	–	–
12	Turlock rain gage near Idaho Road	372713120534901	–	0.487	na	–	–	0.039	–
12	Turlock rain gage near Idaho Road	372713120534901	–	0.089	na	–	–	0.027	–
12	Turlock rain gage near Idaho Road	372713120534901	–	0.080	na	–	–	0.132	–
12	Turlock rain gage near Idaho Road	372713120534901	–	0.064	na	E0.04	–	–	–
12	Turlock rain gage near Idaho Road	372713120534901	–	0.102	na	–	–	0.064	–
12	Turlock rain gage near Idaho Road	372713120534901	na	0.266	na	E0.19	–	na	–
12	Turlock rain gage near Idaho Road	372713120534901	na	0.325	na	–	–	na	–
12	Turlock rain gage near Idaho Road	372713120534901	na	0.033	–	–	–	na	–
12	Turlock rain gage near Idaho Road	372713120534901	na	0.921	–	–	–	na	–
12	Turlock rain gage near Idaho Road	372713120534901	na	0.151	–	na	–	na	–
12	Turlock rain gage near Idaho Road	372713120534901	na	0.307	–	–	E0.041	na	–
12	Turlock rain gage near Idaho Road	372713120534901	na	0.126	–	–	–	na	–
12	Turlock rain gage near Idaho Road	372713120534901	na	0.031	–	–	–	na	–

Appendix 3. Calculated flux values presented as micrograms per square meter per day (µg/m²/day) for wet-deposition samples with at least one detection collected during the 2002–04 study period at eight sites in the Central Valley, California.—Continued

[All compounds were analyzed by gas chromatography/mass spectrometry (GCMS) at the U.S. Geological Survey (USGS) National Water Quality Laboratory (NWQL). The Chemical Abstract Service Number is given below each compound name in brackets. The five-digit number in parenthesis is a code used by the USGS to uniquely identify the given compound. Flux calculated as the analytical result in µg/L multiplied by the total sample volume collected in liters divided by the area of the respective sampler type (Funnel=0.0731 m², Autosampler= 0.0614 m²) and sample composite time in days. Abbreviations: E, estimated reported laboratory value; na, not applicable; (E), estimated compound due to known poor performance by applied method (Childress and others, 1999; Sandstrom and others, 2001); hh:mm, hour:minute; –, compound not detected; [–], no Chemical Abstract Service Number available for given compound]

Site number	Site name	Site identification number	p,p'-DDE [72-55-9] (34653)	Diazinon [333-41-5] (39572)	Diazoxon (E) [962-58-3] (61638)	Dichlorvos (E) [62-73-7] (38775)	Dimethoate [60-51-5] (82662)	EPTC [759-94-4] (82668)	Fonofos [944-22-9] (04095)
12	Turlock rain gage near Idaho Road	372713120534901	na	0.099	–	E0.02	–	na	–
12	Turlock rain gage near Idaho Road	372713120534901	na	2.77	E0.64	E0.04	–	na	–
12	Turlock rain gage near Idaho Road	372713120534901	na	0.345	E0.11	–	–	na	–
12	Turlock rain gage near Idaho Road	372713120534901	na	0.090	E0.05	–	–	na	–
12	Turlock rain gage near Idaho Road	372713120534901	na	0.146	E0.06	E0.01	–	na	–
12	Turlock rain gage near Idaho Road	372713120534901	na	0.120	E0.15	E0.03	–	na	–
13	Turlock Airport rain gage	372857120414001	–	5.01	na	–	–	–	–
13	Turlock Airport rain gage	372857120414001	–	0.376	na	–	–	–	–
13	Turlock Airport rain gage	372857120414001	–	0.421	na	–	–	0.086	–
13	Turlock Airport rain gage	372857120414001	–	0.272	na	–	–	0.034	–
13	Turlock Airport rain gage	372857120414001	–	0.060	na	–	–	–	–
13	Turlock Airport rain gage	372857120414001	–	0.021	na	–	–	na	–
13	Turlock Airport rain gage	372857120414001	–	0.111	na	–	–	na	–
13	Turlock Airport rain gage	372857120414001	na	0.536	na	–	–	na	–
13	Turlock Airport rain gage	372857120414001	na	0.096	–	–	–	na	–
13	Turlock Airport rain gage	372857120414001	na	0.065	–	–	–	na	–
13	Turlock Airport rain gage	372857120414001	na	9.85	E0.99	–	–	na	–
13	Turlock Airport rain gage	372857120414001	na	0.617	–	–	–	na	–
13	Turlock Airport rain gage	372857120414001	na	0.240	E0.07	–	–	na	–
13	Turlock Airport rain gage	372857120414001	na	0.326	–	–	–	na	–
13	Turlock Airport rain gage	372857120414001	na	0.117	–	–	–	na	–
13	Turlock Airport rain gage	372857120414001	na	0.054	–	–	–	na	–
13	Turlock Airport rain gage	372857120414001	na	0.046	–	–	–	na	–
13	Turlock Airport rain gage	372857120414001	na	1.62	E0.38	–	–	na	–
13	Turlock Airport rain gage	372857120414001	na	0.129	E0.06	–	–	na	–
13	Turlock Airport rain gage	372857120414001	na	0.192	E0.10	–	–	na	–
13	Turlock Airport rain gage	372857120414001	na	0.081	0.233	E0.04	–	na	–
10	Westley rain gage at pump building near lateral 6 North	373335121143001	–	0.530	na	–	–	–	–
10	Westley rain gage at pump building near lateral 6 North	373335121143001	–	1.29	na	E0.16	–	0.057	–
10	Westley rain gage at pump building near lateral 6 North	373335121143001	–	0.086	na	–	–	–	–

Appendix 3. Calculated flux values presented as micrograms per square meter per day ($\mu g/m^2/day$) for wet-deposition samples with at least one detection collected during the 2002–04 study period at eight sites in the Central Valley, California.—Continued

[All compounds were analyzed by gas chromatography mass spectrometry (GCMS) at the U.S. Geological Survey (USGS) National Water Quality Laboratory (NWQL). The Chemical Abstract Service Number is given below each compound name in brackets. The five-digit number in parenthesis is a code used by the USGS to uniquely identify the given compound. Flux calculated as the analytical result in $\mu g/L$ multiplied by the total sample volume collected in liters divided by the area of the respective sampler type (Funnel=0.0731 m^2, Autosampler= 0.0614 m^2) and sample composite time in days. Abbreviations: E, estimated reported laboratory value; na, not applicable; (E), estimated compound due to known poor performance by applied method (Childress and others, 1999; Sandstrom and others, 2001); hh:mm, hour:minute; —, compound not detected. [-], no Chemical Abstract Service Number available for given compound]

Site number	Site name	Site identification number	p,p'-DDE [72-55-9] (34653)	Diazinon [333-41-5] (39572)	Diazoxon (E) [962-58-3] (61638)	Dichlorvos (E) [62-73-7] (38775)	Dimethoate [60-51-5] (82662)	EPTC [759-94-4] (82668)	Fonofos [944-22-9] (04095)
10	Westley rain gage at pump building near lateral 6 North	3733351211143001	—	0.038	na	—	—	0.032	—
10	Westley rain gage at pump building near lateral 6 North	3733351211143001	E0.041	—	na	E0.05	—	—	0.015
10	Westley rain gage at pump building near lateral 6 North	3733351211143001	0.025	0.110	na	—	—	0.034	—
10	Westley rain gage at pump building near lateral 6 North	3733351211143001	na	0.927	na	E0.08	—	na	—
10	Westley rain gage at pump building near lateral 6 North	3733351211143001	na	E0.056	—	—	—	na	—
10	Westley rain gage at pump building near lateral 6 North	3733351211143001	na	1.08	—	—	—	na	—
10	Westley rain gage at pump building near lateral 6 North	3733351211143001	na	0.352	—	—	—	na	—
10	Westley rain gage at pump building near lateral 6 North	3733351211143001	na	0.273	E0.08	—	—	na	—
10	Westley rain gage at pump building near lateral 6 North	3733351211143001	na	0.208	—	E0.09	—	na	—
10	Westley rain gage at pump building near lateral 6 North	3733351211143001	na	0.057	E0.05	—	—	na	—
10	Westley rain gage at pump building near lateral 6 North	3733351211143001	na	0.609	E0.23	E0.01	—	na	—
4	Modesto Irrigation District gage rooftop at Modesto	3738341210000601	—	2.78	na	—	—	—	—
4	Modesto Irrigation District gage rooftop at Modesto	3738341210000601	—	1.11	na	—	—	—	—
4	Modesto Irrigation District gage rooftop at Modesto	3738341210000601	—	0.348	na	—	—	—	—
4	Modesto Irrigation District gage rooftop at Modesto	3738341210000601	—	0.237	na	na	na	0.070	—
4	Modesto Irrigation District gage rooftop at Modesto	3738341210000601	—	0.074	na	—	E0.15	—	—
4	Modesto Irrigation District gage rooftop at Modesto	3738341210000601	0.030	0.330	na	—	—	—	—
4	Modesto Irrigation District gage rooftop at Modesto	3738341210000601	—	0.029	na	E0.06	—	na	—
4	Modesto Irrigation District gage rooftop at Modesto	3738341210000601	na	0.842	na	E0.07	—	na	—
4	Modesto Irrigation District gage rooftop at Modesto	3738341210000601	na	0.690	na	E0.12	—	na	—
4	Modesto Irrigation District gage rooftop at Modesto	3738341210000601	na	0.526	—	—	—	na	—
4	Modesto Irrigation District gage rooftop at Modesto	3738341210000601	na	0.399	na	—	—	na	—
4	Modesto Irrigation District gage rooftop at Modesto	3738341210000601	na	0.256	na	—	—	na	—
4	Modesto Irrigation District gage rooftop at Modesto	3738341210000601	na	0.138	—	—	—	na	—
4	Modesto Irrigation District gage rooftop at Modesto	3738341210000601	na	3.50	na	—	—	na	—
4	Modesto Irrigation District gage rooftop at Modesto	3738341210000601	na	12.0	E0.87	—	—	na	—
4	Modesto Irrigation District gage rooftop at Modesto	3738341210000601	na	12.7	E1.29	—	—	na	—
4	Modesto Irrigation District gage rooftop at Modesto	3738341210000601	na	1.40	E0.46	—	—	na	—
4	Modesto Irrigation District gage rooftop at Modesto	3738341210000601	na	0.269	E0.14	—	—	na	—
4	Modesto Irrigation District gage rooftop at Modesto	3738341210000601	na	0.202	—	—	—	na	—

Appendix 3. Calculated flux values presented as micrograms per square meter per day (μg/m^2/day) for wet-deposition samples with at least one detection collected during the 2002–04 study period at eight sites in the Central Valley, California.—Continued

[All compounds were analyzed by gas chromatography/mass spectrometry (GCMS) at the U.S. Geological Survey (USGS) National Water Quality Laboratory (NWQL). The Chemical Abstract Service Number is given below each compound name in parentheses. The five-digit number in parenthesis is a code used by the USGS to uniquely identify the given compound. Flux calculated as the analytical result in μg/L multiplied by the total sample volume collected in liters divided by the area of the respective sampler type (Funnel=0.0731 m^2, Autosampler= 0.0614 m^2) and sample composite time in days. Abbreviations: E, estimated reported laboratory value; na, not applicable; (E), estimated compound due to known poor performance by applied method (Childress and others, 1999; Sandstrom and others, 2001); hh:mm, hour:minute; —, compound not detected; [-], no Chemical Abstract Service Number available for given compound]

Site number	Site name	Site identification number	p,p'-DDE [72-55-9] (34653)	Diazinon [333-41-5] (39572)	Diazoxon (E) [962-58-3] (61638)	Dichlorvos (E) [62-73-7] (38775)	Dimethoate [60-51-5] (82662)	EPTC [759-94-4] (82668)	Fonofos [944-22-9] (04095)
4	Modesto Irrigation District gage rooftop at Modesto	373834121000601	na	0.828	—	—	—	na	—
4	Modesto Irrigation District gage rooftop at Modesto	373834121000601	na	0.645	E0.10	—	—	na	—
4	Modesto Irrigation District gage rooftop at Modesto	373834121000601	na	0.083	—	—	—	na	—
4	Modesto Irrigation District gage rooftop at Modesto	373834121000601	na	0.088	E0.03	—	—	na	—
4	Modesto Irrigation District gage rooftop at Modesto	373834121000601	na	0.115	—	—	—	na	—
4	Modesto Irrigation District gage rooftop at Modesto	373834121000601	na	0.216	E0.04	E0.07	—	na	—
4	Modesto Irrigation District gage rooftop at Modesto	373834121000601	na	0.077	E0.06	—	—	na	—
4	Modesto Irrigation District gage rooftop at Modesto	373834121000601	na	0.071	—	—	—	na	—
4	Modesto Irrigation District gage rooftop at Modesto	373834121000601	na	5.03	E0.64	E0.26	—	na	—
4	Modesto Irrigation District gage rooftop at Modesto	373834121000601	na	0.569	E0.08	—	—	na	—
4	Modesto Irrigation District gage rooftop at Modesto	373834121000601	na	0.188	E0.05	E0.03	—	na	—
4	Modesto Irrigation District gage rooftop at Modesto	373834121000601	na	5.38	E0.77	—	—	na	—
4	Modesto Irrigation District gage rooftop at Modesto	373834121000601	na	5.86	E0.92	E0.03	—	na	—
4	Modesto Irrigation District gage rooftop at Modesto	373834121000601	na	0.093	E0.05	—	—	na	—
4	Modesto Irrigation District gage rooftop at Modesto	373834121000601	na	0.349	E0.11	—	—	na	—
4	Modesto Irrigation District gage rooftop at Modesto	373834121000601	na	0.123	E0.12	E0.04	—	na	—
5	Modesto Irrigation District gage rooftop at Albers Road	373841120504801	—	2.90	na	—	—	—	—
5	Modesto Irrigation District gage rooftop at Albers Road	373841120504801	—	0.895	na	—	—	—	—
5	Modesto Irrigation District gage rooftop at Albers Road	373841120504801	—	1.41	na	—	—	0.038	—
5	Modesto Irrigation District gage rooftop at Albers Road	373841120504801	—	0.210	na	—	—	—	—
5	Modesto Irrigation District gage rooftop at Albers Road	373841120504801	—	0.153	na	—	—	—	—
5	Modesto Irrigation District gage rooftop at Albers Road	373841120504801	—	0.183	na	—	—	—	—
5	Modesto Irrigation District gage rooftop at Albers Road	373841120504801	na	0.372	na	—	—	na	—
5	Modesto Irrigation District gage rooftop at Albers Road	373841120504801	na	0.469	na	—	—	na	—
5	Modesto Irrigation District gage rooftop at Albers Road	373841120504801	na	0.120	—	—	—	na	—
5	Modesto Irrigation District gage rooftop at Albers Road	373841120504801	na	0.116	na	—	—	na	—
5	Modesto Irrigation District gage rooftop at Albers Road	373841120504801	na	0.104	—	—	—	na	—
5	Modesto Irrigation District gage rooftop at Albers Road	373841120504801	na	0.113	—	—	—	na	—
5	Modesto Irrigation District gage rooftop at Albers Road	373841120504801	na	0.378	na	—	—	na	—
5	Modesto Irrigation District gage rooftop at Albers Road	373841120504801	na	14.4	E0.99	—	—	na	—
5	Modesto Irrigation District gage rooftop at Albers Road	373841120504801	na	11.7	E0.56	—	—	na	—

Appendix 3. Calculated flux values presented as micrograms per square meter per day (µg/m²/day) for wet-deposition samples with at least one detection collected during the 2002–04 study period at eight sites in the Central Valley, California.—Continued

[All compounds were analyzed by gas chromatography/mass spectrometry (GCMS) at the U.S. Geological Survey (USGS) National Water Quality Laboratory (NWQL). The Chemical Abstract Service Number is given below each compound name in brackets. The five-digit number in parenthesis is a code used by the USGS to uniquely identify the given compound. Flux calculated as the analytical result in µg/L multiplied by the total sample volume collected in liters divided by the area of the respective sampler type (Funnel=0.0731 m², Autosampler= 0.0614 m²) and sample composite time in days. Abbreviations: E, estimated reported laboratory value; na, not applicable; (E), estimated compound due to known poor performance by applied method (Childress and others, 1999; Sandstrom and others, 2001); hh:mm, hour:minute; -, compound not detected. [-], no Chemical Abstract Service Number available for given compound]

Site number	Site name	Site identification number	p,p'-DDE [72-55-9] (34653)	Diazinon [333-41-5] (39572)	Diazoxon (E) [962-58-3] (61638)	Dichlorvos (E) [62-73-7] (38775)	Dimethoate [60-51-5] (82662)	EPTC [759-94-4] (82668)	Fonofos [944-22-9] (04095)
5	Modesto Irrigation District gage rooftop at Albers Road	37384112050480S01	na	0.149	-	-	-	na	-
5	Modesto Irrigation District gage rooftop at Albers Road	37384112050480S01	na	0.188	-	na	-	na	-
5	Modesto Irrigation District gage rooftop at Albers Road	37384112050480S01	na	0.152	-	na	-	na	-
5	Modesto Irrigation District gage rooftop at Albers Road	37384112050480S01	na	0.499	E0.07	-	-	na	-
5	Modesto Irrigation District gage rooftop at Albers Road	37384112050480S01	na	0.586	-	-	-	na	-
5	Modesto Irrigation District gage rooftop at Albers Road	37384112050480S01	na	0.073	E0.05	E0.01	-	na	-
5	Modesto Irrigation District gage rooftop at Albers Road	37384112050480S01	na	0.100	-	-	-	na	-
5	Modesto Irrigation District gage rooftop at Albers Road	37384112050480S01	na	0.052	-	-	-	na	-
5	Modesto Irrigation District gage rooftop at Albers Road	37384112050480S01	na	0.017	-	-	-	na	-
5	Modesto Irrigation District gage rooftop at Albers Road	37384112050480S01	na	0.031	-	-	-	na	-
5	Modesto Irrigation District gage rooftop at Albers Road	37384112050480S01	na	0.037	-	-	-	na	-
5	Modesto Irrigation District gage rooftop at Albers Road	37384112050480S01	na	0.312	E0.16	-	-	na	-
5	Modesto Irrigation District gage rooftop at Albers Road	37384112050480S01	na	0.322	-	-	-	na	-
5	Modesto Irrigation District gage rooftop at Albers Road	37384112050480S01	na	18.4	E0.97	E0.12	-	na	-
5	Modesto Irrigation District gage rooftop at Albers Road	37384112050480S01	na	0.795	E0.15	E0.02	-	na	-
5	Modesto Irrigation District gage rooftop at Albers Road	37384112050480S01	na	0.066	E0.01	E0.00	-	na	-
5	Modesto Irrigation District gage rooftop at Albers Road	37384112050480S01	na	2.12	E0.40	-	-	na	-
5	Modesto Irrigation District gage rooftop at Albers Road	37384112050480S01	na	1.04	E0.11	-	-	na	-
5	Modesto Irrigation District gage rooftop at Albers Road	37384112050480S01	na	0.521	E0.14	-	-	na	-
5	Modesto Irrigation District gage rooftop at Albers Road	37384112050480S01	na	0.418	E0.10	-	-	na	-
5	Modesto Irrigation District gage rooftop at Albers Road	37384112050480S01	na	0.246	E0.07	-	-	na	-
5	Modesto Irrigation District gage rooftop at Albers Road	37384112050480S01	na	0.191	E0.11	E0.03	-	na	-
16	Gridley High School precipitation gage at Gridley	39220512141020I	na	50.6	E1.64	-	-	na	-
16	Gridley High School precipitation gage at Gridley	39220512141020I	na	45.6	E6.52	-	-	na	-
16	Gridley High School precipitation gage at Gridley	39220512141020I	na	9.78	-	na	-	na	-
16	Gridley High School precipitation gage at Gridley	39220512141020I	na	2.99	-	na	-	na	-
16	Gridley High School precipitation gage at Gridley	39220512141020I	na	0.058	E0.03	-	-	na	-
16	Gridley High School precipitation gage at Gridley	39220512141020I	na	0.559	E0.08	-	-	na	-
16	Gridley High School precipitation gage at Gridley	39220512141020I	na	0.106	-	-	-	na	-
16	Gridley High School precipitation gage at Gridley	39220512141020I	na	0.160	-	-	-	na	-
16	Gridley High School precipitation gage at Gridley	39220512141020I	na	0.033	-	-	-	na	-

Appendix 3. Calculated flux values presented as micrograms per square meter per day (μg/m²/day) for wet-deposition samples with at least one detection collected during the 2002–04 study period at eight sites in the Central Valley, California.—Continued

[All compounds were analyzed by gas chromatography/mass spectrometry (GCMS) at the U.S. Geological Survey (USGS) National Water Quality Laboratory (NWQL). The Chemical Abstract Service Number is given below each compound name in brackets. The five-digit number in parenthesis is a code used by the USGS to uniquely identify the given compound. Flux calculated as the analytical result in μg/L multiplied by the total sample volume collected in liters divided by the area of the respective sampler type (Funnel=0.0731 m², Autosampler= 0.0614 m²) and sample composite time in days. Abbreviations: E, estimated reported laboratory value; na, not applicable; (E), estimated compound due to known poor performance by applied method (Childress and others, 1999; Sandstrom and others, 2001); hh:mm, hour:minute; –, compound not detected; [-], no Chemical Abstract Service Number available for given compound]

Site number	Site name	Site identification number	p,p'-DDE [72-55-9] (34653)	Diazinon [333-41-5] (39572)	Diazoxon (E) [962-58-3] (61638)	Dichlorvos (E) [62-73-7] (38775)	Dimethoate [60-51-5] (82662)	EPTC [759-94-4] (82668)	Fonofos [944-22-9] (04095)
16	Gridley High School precipitation gage at Gridley	392205121410201	na	0.100	–			na	–
16	Gridley High School precipitation gage at Gridley	392205121410201	na	0.056	–			na	–
16	Gridley High School precipitation gage at Gridley	392205121410201	na	0.048	–			na	–
16	Gridley High School precipitation gage at Gridley	392205121410201	na	0.035	–			na	–
16	Gridley High School precipitation gage at Gridley	392205121410201	na	0.053	E0.03			na	–
16	Gridley High School precipitation gage at Gridley	392205121410201	na	0.111	–			na	–
16	Gridley High School precipitation gage at Gridley	392205121410201	na	0.043	–			na	–
16	Gridley High School precipitation gage at Gridley	392205121410201	na	0.290	E0.08			na	–
16	Gridley High School precipitation gage at Gridley	392205121410201	na	0.094	E0.44			na	–
16	Gridley High School precipitation gage at Gridley	392205121410201	na	0.134	E0.65			na	–
16	Gridley High School precipitation gage at Gridley	392205121410201	na	6.19	E0.24			na	–
16	Gridley High School precipitation gage at Gridley	392205121410201	na	7.17	E0.52			na	–
16	Gridley High School precipitation gage at Gridley	392205121410201	na	2.33	E0.78			na	–
16	Gridley High School precipitation gage at Gridley	392205121410201	na	6.19	E0.21			na	–
16	Gridley High School precipitation gage at Gridley	392205121410201	na	4.82	E0.08			na	–
16	Gridley High School precipitation gage at Gridley	392205121410201	na	2.06				na	–
16	Gridley High School precipitation gage at Gridley	392205121410201	na	0.427		E0.02		na	–
15	Oroville Dam precipitation gage at spillway	393234121292701	na	E0.055	na			na	–
15	Oroville Dam precipitation gage at spillway	393234121292701	na	E0.044	na			na	–
15	Oroville Dam precipitation gage at spillway	393234121292701	na	0.702	E0.07			na	–
15	Oroville Dam precipitation gage at spillway	393234121292701	na	0.439	–	na		na	–
15	Oroville Dam precipitation gage at spillway	393234121292701	na	0.559	–			na	–
15	Oroville Dam precipitation gage at spillway	393234121292701	na	0.032	–			na	–
15	Oroville Dam precipitation gage at spillway	393234121292701	na	0.047	–			na	–
15	Oroville Dam precipitation gage at spillway	393234121292701	na	–	–			na	–
15	Oroville Dam precipitation gage at spillway	393234121292701	na	0.060	–			na	–
15	Oroville Dam precipitation gage at spillway	393234121292701	na	0.050	E0.04			na	–
15	Oroville Dam precipitation gage at spillway	393234121292701	na	0.278	E0.09			na	–
15	Oroville Dam precipitation gage at spillway	393234121292701	na	0.235	E0.09			na	–
15	Oroville Dam precipitation gage at spillway	393234121292701	na	E0.033	–	E0.03		na	–

Appendix 3. Calculated flux values presented as micrograms per square meter per day (μg/m²/day) for wet-deposition samples with at least one detection collected during the 2002–04 study period at eight sites in the Central Valley, California.—Continued

[All compounds were analyzed by gas chromatography mass spectrometry (GCMS) at the U.S. Geological Survey (USGS) National Water Quality Laboratory (NWQL). The Chemical Abstract Service Number is given below each compound name in brackets. The five-digit number in parenthesis is a code used by the USGS to uniquely identify the given compound. Flux calculated as the analytical result in μg/L multiplied by the total sample volume collected in liters divided by the area of the respective sampler type (Funnel=0.0731 m², Autosampler= 0.0614 m²) and sample composite time in days. Abbreviations: E, estimated reported laboratory value; na, not applicable; (E), estimated compound due to known poor performance by applied method (Childress and others, 1999; Sandstrom and others, 2001); hh:mm, hour:minute; -, compound not detected. [-], no Chemical Abstract Service Number available for given compound]

Site number	Site name	Site identification number	Fonofos oxon [-] (61649)	Iprodione (E) [36734-19-7] (61593)	Cyhalothrin-lambda [91465-08-6] (61595)	Malathion [121-75-5] (39532)	Malaoxon (E) [1634-78-2] (61652)	Metalaxyl [57837-19-1] (61596)	Methidathion [950-37-8] (61598)
11	Newman rain gage at wasteway levee near Draper Road	371735121031201	—	—	—	—	—	—	—
11	Newman rain gage at wasteway levee near Draper Road	371735121031201	—	—	—	—	—	—	—
11	Newman rain gage at wasteway levee near Draper Road	371735121031201	—	E2.66	—	1.490	E0.243	—	—
11	Newman rain gage at wasteway levee near Draper Road	371735121031201	—	E1.12	E0.033	E0.033	E0.049	—	—
11	Newman rain gage at wasteway levee near Draper Road	371735121031201	—	E1.76	E0.082	0.203	E0.221	—	—
11	Newman rain gage at wasteway levee near Draper Road	371735121031201	—	E0.217	na	E0.434	E0.337	—	—
11	Newman rain gage at wasteway levee near Draper Road	371735121031201	—	E0.110	na	—	—	—	—
11	Newman rain gage at wasteway levee near Draper Road	371735121031201	—	—	na	—	—	—	E0.060
11	Newman rain gage at wasteway levee near Draper Road	371735121031201	—	E2.85	na	E0.076	E0.108	—	—
11	Newman rain gage at wasteway levee near Draper Road	371735121031201	—	E0.495	na	0.092	E0.186	—	—
11	Newman rain gage at wasteway levee near Draper Road	371735121031201	—	—	na	0.052	E0.089	—	—
11	Newman rain gage at wasteway levee near Draper Road	371735121031201	—	—	na	—	—	—	0.370
11	Newman rain gage at wasteway levee near Draper Road	371735121031201	—	E0.127	na	—	—	—	0.280
11	Newman rain gage at wasteway levee near Draper Road	371735121031201	—	E2.73	na	—	—	—	0.032
11	Newman rain gage at wasteway levee near Draper Road	371735121031201	—	E3.47	na	—	—	—	0.090
12	Turlock rain gage near Idaho Road	372713120534901	—	—	—	E0.043	—	—	0.094
12	Turlock rain gage near Idaho Road	372713120534901	—	—	—	—	—	—	—
12	Turlock rain gage near Idaho Road	372713120534901	—	—	—	—	—	—	—
12	Turlock rain gage near Idaho Road	372713120534901	—	E4.75	—	0.456	E0.119	—	0.016
12	Turlock rain gage near Idaho Road	372713120534901	—	—	—	E0.086	—	0.034	—
12	Turlock rain gage near Idaho Road	372713120534901	—	E1.14	—	E0.053	—	—	—
12	Turlock rain gage near Idaho Road	372713120534901	—	E2.52	—	2.49	—	—	—
12	Turlock rain gage near Idaho Road	372713120534901	—	—	na	E0.285	E0.223	—	—
12	Turlock rain gage near Idaho Road	372713120534901	—	—	na	E0.049	E0.285	—	—
12	Turlock rain gage near Idaho Road	372713120534901	—	—	na	—	—	—	—
12	Turlock rain gage near Idaho Road	372713120534901	na	E0.755	na	E0.09	—	—	0.199
12	Turlock rain gage near Idaho Road	372713120534901	—	E4.17	na	E0.131	E0.102	—	—
12	Turlock rain gage near Idaho Road	372713120534901	—	E0.586	na	E0.071	E0.112	—	—
12	Turlock rain gage near Idaho Road	372713120534901	—	E0.042	na	E0.022	—	—	—

Appendix 3. Calculated flux values presented as micrograms per square meter per day (µg/m²/day) for wet-deposition samples with at least one detection collected during the 2002–04 study period at eight sites in the Central Valley, California.—Continued

[All compounds were analyzed by gas chromatography/mass spectrometry (GCMS) at the U.S. Geological Survey (USGS) National Water Quality Laboratory (NWQL). The Chemical Abstract Service Number is given below each compound name in parentheses. The five-digit number in parenthesis is a code used by the USGS to uniquely identify the given compound. Flux calculated as the analytical result in µg/L multiplied by the total sample volume collected in liters divided by the area of the respective sampler type (Funnel=0.0731 m², Autosampler= 0.0614 m²) and sample composite time in days. Abbreviations: E, estimated reported laboratory value; na, not applicable; (E), estimated compound due to known poor performance by applied method (Childress and others, 1999; Sandstrom and others, 2001); hh:mm, hour:minute; —, compound not detected; [-], no Chemical Abstract Service Number available for given compound]

Site number	Site name	Site identification number	Fonofos oxon [-] (61649)	Iprodione (E) [36734-19-7] (61593)	Cyhalothrin-lambda [91465-08-6] (61595)	Malathion [121-75-5] (39532)	Malaoxon (E) [1634-78-2] (61652)	Metalaxyl [57837-19-1] (61596)	Methidathion [950-37-8] (61598)
12	Turlock rain gage near Idaho Road	37271312053434901	—	—	na	E0.032	E0.056	—	0.110
12	Turlock rain gage near Idaho Road	37271312053434901	—	—	na	—	—	—	E0.027
12	Turlock rain gage near Idaho Road	37271312053434901	—	E0.252	na	—	—	—	0.020
12	Turlock rain gage near Idaho Road	37271312053434901	—	E4.23	na	—	—	—	—
12	Turlock rain gage near Idaho Road	37271312053434901	—	E3.94	na	—	—	—	—
12	Turlock rain gage near Idaho Road	37271312053434901	—	E3.34	na	0.352	E0.190	—	—
13	Turlock Airport rain gage	372857120414001	—	—	—	E0.08	—	—	—
13	Turlock Airport rain gage	372857120414001	—	—	—	—	—	—	—
13	Turlock Airport rain gage	372857120414001	—	—	—	—	—	—	—
13	Turlock Airport rain gage	372857120414001	—	E9.27	na	1.46	E0.177	—	E0.024
13	Turlock Airport rain gage	372857120414001	—	E3.65	na	—	—	—	—
13	Turlock Airport rain gage	372857120414001	—	E1.013	na	E0.057	E0.047	—	—
13	Turlock Airport rain gage	372857120414001	—	E3.34	na	0.710	E0.167	—	—
13	Turlock Airport rain gage	372857120414001	—	E0.507	na	E0.479	E0.620	—	0.197
13	Turlock Airport rain gage	372857120414001	—	—	na	—	—	—	5.206
13	Turlock Airport rain gage	372857120414001	—	E6.83	na	E0.016	—	—	0.054
13	Turlock Airport rain gage	372857120414001	—	E14.1	na	E0.019	—	—	0.031
13	Turlock Airport rain gage	372857120414001	—	E5.81	na	E0.196	E0.116	—	—
13	Turlock Airport rain gage	372857120414001	—	E0.374	na	—	—	—	—
13	Turlock Airport rain gage	372857120414001	—	E0.191	na	—	—	—	—
13	Turlock Airport rain gage	372857120414001	—	E0.091	na	0.284	E0.061	—	0.457
13	Turlock Airport rain gage	372857120414001	—	E11.9	na	—	—	—	0.028
13	Turlock Airport rain gage	372857120414001	—	E4.90	na	—	—	—	—
13	Turlock Airport rain gage	372857120414001	—	E10.5	na	E0.314	E0.279	—	0.093
10	Westley rain gage at pump building near lateral 6 North	373335121143001	—	—	—	E0.036	—	—	0.171
10	Westley rain gage at pump building near lateral 6 North	373335121143001	—	—	—	—	—	—	0.047
10	Westley rain gage at pump building near lateral 6 North	373335121143001	—	E8.55	—	0.265	E0.111	—	0.024

Appendix 3. Calculated flux values presented as micrograms per square meter per day (µg/m²/day) for wet-deposition samples with at least one detection collected during the 2002–04 study period at eight sites in the Central Valley, California.—Continued

[All compounds were analyzed by gas chromatography/mass spectrometry (GCMS) at the U.S. Geological Survey (USGS) National Water Quality Laboratory (NWQL). The Chemical Abstract Service Number is given below each compound name in parentheses. The five-digit number in parenthesis is a code used by the USGS to uniquely identify the given compound. Flux calculated as the analytical result in µg/L multiplied by the total sample volume collected in liters divided by the area of the respective sampler type (Funnel=0.0731 m², Autosampler= 0.0614 m²) and sample composite time in days. Abbreviations: E, estimated reported laboratory value; na, not applicable; (E), estimated compound due to known poor performance by applied method (Childress and others, 1999; Sandstrom and others, 2001); hh:mm, hour:minute; -, compound not detected; [-], no Chemical Abstract Service Number available for given compound]

Site number	Site name	Site identification number	Fonofos oxon [-] (61649)	Iprodione (E) [36734-19-7] (61593)	Cyhalothrin-lambda [91465-08-6] (61595)	Malathion [121-75-5] (39532)	Malaoxon (E) [1634-78-2] (61652)	Metalaxyl [57837-19-1] (61596)	Methidathion [950-37-8] (61598)
10	Westley rain gage at pump building near lateral 6 North	373335121143001	-	E3.02	-	0.370	E0.059	-	0.018
10	Westley rain gage at pump building near lateral 6 North	373335121143001	E0.036	E1.72	-	E0.046	E0.056	-	-
10	Westley rain gage at pump building near lateral 6 North	373335121143001	-	E3.69	-	0.135	E0.16	-	-
10	Westley rain gage at pump building near lateral 6 North	373335121143001	-	E0.569	na	E0.422	E0.759	-	-
10	Westley rain gage at pump building near lateral 6 North	373335121143001	-	E0.113	na	-	-	-	-
10	Westley rain gage at pump building near lateral 6 North	373335121143001	-	-	na	-	-	-	0.084
10	Westley rain gage at pump building near lateral 6 North	373335121143001	-	E77.6	na	E0.074	-	-	-
10	Westley rain gage at pump building near lateral 6 North	373335121143001	-	E10.3	na	E0.029	-	-	-
10	Westley rain gage at pump building near lateral 6 North	373335121143001	-	E0.121	na	E0.112	E0.088	0.078	-
10	Westley rain gage at pump building near lateral 6 North	373335121143001	-	E3.64	na	E0.139	E0.19	-	-
10	Westley rain gage at pump building near lateral 6 North	373335121143001	-	E3.48	na	-	-	-	-
4	Modesto Irrigation District gage rooftop at Modesto	373834121000601	-	-	-	E0.032	-	-	-
4	Modesto Irrigation District gage rooftop at Modesto	373834121000601	-	-	-	-	-	-	-
4	Modesto Irrigation District gage rooftop at Modesto	373834121000601	-	E1.36	-	0.084	E0.051	-	-
4	Modesto Irrigation District gage rooftop at Modesto	373834121000601	na	na	na	E0.051	na	na	na
4	Modesto Irrigation District gage rooftop at Modesto	373834121000601	-	E3.37	-	E0.105	E0.051	-	0.031
4	Modesto Irrigation District gage rooftop at Modesto	373834121000601	-	-	-	1.830	E0.41	-	-
4	Modesto Irrigation District gage rooftop at Modesto	373834121000601	-	-	-	0.030	E0.056	0.013	-
4	Modesto Irrigation District gage rooftop at Modesto	373834121000601	-	-	na	E0.477	E1.27	-	-
4	Modesto Irrigation District gage rooftop at Modesto	373834121000601	-	-	na	E0.469	-	-	-
4	Modesto Irrigation District gage rooftop at Modesto	373834121000601	-	-	na	E0.088	-	-	-
4	Modesto Irrigation District gage rooftop at Modesto	373834121000601	-	-	na	E0.059	-	-	-
4	Modesto Irrigation District gage rooftop at Modesto	373834121000601	-	-	na	E0.022	-	-	E0.017
4	Modesto Irrigation District gage rooftop at Modesto	373834121000601	-	-	na	-	-	-	0.067
4	Modesto Irrigation District gage rooftop at Modesto	373834121000601	-	-	na	E0.038	E0.068	-	0.350
4	Modesto Irrigation District gage rooftop at Modesto	373834121000601	-	-	na	E0.112	-	-	0.436
4	Modesto Irrigation District gage rooftop at Modesto	373834121000601	-	-	na	E0.15	-	-	0.462
4	Modesto Irrigation District gage rooftop at Modesto	373834121000601	-	E0.341	na	E0.021	-	-	0.242
4	Modesto Irrigation District gage rooftop at Modesto	373834121000601	-	E4.33	na	E0.034	-	-	0.034
4	Modesto Irrigation District gage rooftop at Modesto	373834121000601	-	E0.478	na	-	-	-	0.021

Appendix 3. Calculated flux values presented as micrograms per square meter per day (µg/m²/day) for wet-deposition samples with at least one detection collected during the 2002–04 study period at eight sites in the Central Valley, California.—Continued

[All compounds were analyzed by gas chromatography/mass spectrometry (GCMS) at the U.S. Geological Survey (USGS) National Water Quality Laboratory (NWQL). The Chemical Abstract Service Number is given below each compound name in brackets. The five-digit number in parenthesis is a code used by the USGS to uniquely identify the given compound. Flux calculated as the analytical result in µg/L multiplied by the total sample volume collected in liters divided by the area of the respective sampler type (Funnel=0.0731 m², Autosampler= 0.0614 m²) and sample composite time in days. Abbreviations: E, estimated compound value, na, not applicable; (E), estimated compound due to known poor performance by applied method (Childress and others, 1999; Sandstrom and others, 2001); hh:mm, hour:minute; –, compound not detected; [–], no Chemical Abstract Service Number available for given compound]

Site number	Site name	Site identification number	Fonofos oxon [–] (61649)	Iprodione (E) [36734-19-7] (61593)	Cyhalothrin-lambda [91465-08-6] (61595)	Malathion [121-75-5] (39532)	Malaoxon (E) [1634-78-2] (61652)	Metalaxyl [57837-19-1] (61596)	Methidathion [950-37-8] (61598)
4	Modesto Irrigation District gage rooftop at Modesto	3738341210000601	–	E1.21	na	E0.194	E0.153	–	–
4	Modesto Irrigation District gage rooftop at Modesto	3738341210000601	–	E0.195	na	E0.098	–	–	–
4	Modesto Irrigation District gage rooftop at Modesto	3738341210000601	–	E0.105	na	E0.048	–	–	–
4	Modesto Irrigation District gage rooftop at Modesto	3738341210000601	–	E0.057	na	0.353	E0.108	–	–
4	Modesto Irrigation District gage rooftop at Modesto	3738341210000601	–	E0.054	na	E0.049	–	–	–
4	Modesto Irrigation District gage rooftop at Modesto	3738341210000601	–	–	na	0.114	E0.143	–	0.043
4	Modesto Irrigation District gage rooftop at Modesto	3738341210000601	–	E0.047	na	–	–	–	–
4	Modesto Irrigation District gage rooftop at Modesto	3738341210000601	–	–	na	–	–	–	0.269
4	Modesto Irrigation District gage rooftop at Modesto	3738341210000601	–	E0.295	na	–	–	0.449	E0.038
4	Modesto Irrigation District gage rooftop at Modesto	3738341210000601	–	–	na	–	–	–	–
4	Modesto Irrigation District gage rooftop at Modesto	3738341210000601	–	–	na	–	–	–	0.872
4	Modesto Irrigation District gage rooftop at Modesto	3738341210000601	–	–	na	–	–	–	0.523
4	Modesto Irrigation District gage rooftop at Modesto	3738341210000601	–	E5.12	na	–	–	–	0.022
4	Modesto Irrigation District gage rooftop at Modesto	3738341210000601	–	E4.13	na	–	–	–	–
4	Modesto Irrigation District gage rooftop at Modesto	3738341210000601	–	E2.04	na	0.312	E0.131	–	–
5	Modesto Irrigation District gage rooftop at Albers Road	3738411205048 01	–	–	–	–	–	–	0.035
5	Modesto Irrigation District gage rooftop at Albers Road	3738411205048 01	–	–	–	–	–	–	0.051
5	Modesto Irrigation District gage rooftop at Albers Road	3738411205048 01	–	–	–	E0.025	–	–	0.054
5	Modesto Irrigation District gage rooftop at Albers Road	3738411205048 01	–	E9.55	–	0.768	E0.167	–	–
5	Modesto Irrigation District gage rooftop at Albers Road	3738411205048 01	–	E2.59	–	0.102	–	–	–
5	Modesto Irrigation District gage rooftop at Albers Road	3738411205048 01	–	E3.37	–	0.800	E0.299	–	E0.143
5	Modesto Irrigation District gage rooftop at Albers Road	3738411205048 01	–	E0.486	na	E0.257	–	–	–
5	Modesto Irrigation District gage rooftop at Albers Road	3738411205048 01	–	E0.44	na	E0.293	E0.322	–	–
5	Modesto Irrigation District gage rooftop at Albers Road	3738411205048 01	–	–	na	–	–	–	–
5	Modesto Irrigation District gage rooftop at Albers Road	3738411205048 01	–	E0.095	na	–	–	–	–
5	Modesto Irrigation District gage rooftop at Albers Road	3738411205048 01	–	–	na	–	–	–	–
5	Modesto Irrigation District gage rooftop at Albers Road	3738411205048 01	–	–	na	–	–	–	0.032
5	Modesto Irrigation District gage rooftop at Albers Road	3738411205048 01	–	–	na	–	–	–	0.706
5	Modesto Irrigation District gage rooftop at Albers Road	3738411205048 01	–	–	na	–	–	–	0.406

Appendix 3. Calculated flux values presented as micrograms per square meter per day (µg/m²/day) for wet-deposition samples with at least one detection collected during the 2002–04 study period at eight sites in the Central Valley, California.—Continued

[All compounds were analyzed by gas chromatography/mass spectrometry (GCMS) at the U.S. Geological Survey (USGS) National Water Quality Laboratory (NWQL). The Chemical Abstract Service Number is given below each compound name in brackets. The five-digit number in parenthesis is a code used by the USGS to uniquely identify the given compound. Flux calculated as the analytical result in µg/L multiplied by the total sample volume collected in liters divided by the area of the respective sampler type (Funnel=0.0731 m², Autosampler= 0.0614 m²) and sample composite time in days. Abbreviations: E, estimated reported laboratory value; na, not applicable; (E), estimated compound due to known poor performance by applied method (Childress and others, 1999; Sandstrom and others, 2001); hh:mm, hour:minute; —, compound not detected; [-], no Chemical Abstract Service Number available for given compound]

Site number	Site name	Site identification number	Fonofos oxon [-] (61649)	Iprodione (E) [36734-19-7] (61593)	Cyhalothrin-lambda [91465-08-6] (61595)	Malathion [121-75-5] (39532)	Malaoxon (E) [1634-78-2] (61652)	Metalaxyl [57837-19-1] (61596)	Methidathion [950-37-8] (61598)
5	Modesto Irrigation District gage rooftop at Albers Road	373841120504801	—	E0.132	na	—	—	—	0.010
5	Modesto Irrigation District gage rooftop at Albers Road	373841120504801	na	E7.43	na	—	—	—	—
5	Modesto Irrigation District gage rooftop at Albers Road	373841120504801	na	E0.149	na	—	—	—	—
5	Modesto Irrigation District gage rooftop at Albers Road	373841120504801	—	E4.36	na	E0.066	—	—	—
5	Modesto Irrigation District gage rooftop at Albers Road	373841120504801	—	E0.100	na	—	E0.088	—	—
5	Modesto Irrigation District gage rooftop at Albers Road	373841120504801	—	E0.178	na	0.771	E0.100	—	—
5	Modesto Irrigation District gage rooftop at Albers Road	373841120504801	—	E0.047	na	E0.147	E0.114	—	—
5	Modesto Irrigation District gage rooftop at Albers Road	373841120504801	—	E0.065	na	—	—	—	
5	Modesto Irrigation District gage rooftop at Albers Road	373841120504801	—		na	E0.024	E0.044	—	0.458
5	Modesto Irrigation District gage rooftop at Albers Road	373841120504801	—		na	E0.040	E0.050	—	—
5	Modesto Irrigation District gage rooftop at Albers Road	373841120504801	—		na	—	—	—	0.914
5	Modesto Irrigation District gage rooftop at Albers Road	373841120504801	—		na	—	—	—	0.093
5	Modesto Irrigation District gage rooftop at Albers Road	373841120504801	—		na	—	—	—	0.071
5	Modesto Irrigation District gage rooftop at Albers Road	373841120504801	—		na	—	—	—	1.06
5	Modesto Irrigation District gage rooftop at Albers Road	373841120504801	—		na	—	—	—	0.180
5	Modesto Irrigation District gage rooftop at Albers Road	373841120504801	—	E0.368	na	E0.084	E0.074	—	0.293
5	Modesto Irrigation District gage rooftop at Albers Road	373841120504801	—	E22.2	na	—	—	—	0.094
5	Modesto Irrigation District gage rooftop at Albers Road	373841120504801	—	E6.60	na	—	—	—	0.075
5	Modesto Irrigation District gage rooftop at Albers Road	373841120504801	—	E3.70	na	0.37	E0.191	—	0.123
16	Gridley High School precipitation gage at Gridley	392205121410201	—	E3.23	na	—	—	—	0.246
16	Gridley High School precipitation gage at Gridley	392205121410201	—	E5.44	na	—	—	—	1.66
16	Gridley High School precipitation gage at Gridley	392205121410201	—	E1.24	na	—	—	—	—
16	Gridley High School precipitation gage at Gridley	392205121410201	—		na	—	—	—	—
16	Gridley High School precipitation gage at Gridley	392205121410201	—	E0.102	na	E0.039	—	—	—
16	Gridley High School precipitation gage at Gridley	392205121410201	—	E0.284	na	E0.057	—	—	—
16	Gridley High School precipitation gage at Gridley	392205121410201	—		na	E0.037	—	—	—
16	Gridley High School precipitation gage at Gridley	392205121410201	—		na	—	—	—	—
16	Gridley High School precipitation gage at Gridley	392205121410201	—		na	—	—	—	—

Appendix 3. Calculated flux values presented as micrograms per square meter per day (μg/m²/day) for wet-deposition samples with at least one detection collected during the 2002–04 study period at eight sites in the Central Valley, California.—Continued

[All compounds were analyzed by gas chromatography/mass spectrometry (GCMS) at the U.S. Geological Survey (USGS) National Water Quality Laboratory (NWQL). The Chemical Abstract Service Number is given below each compound name in parentheses. The five-digit number in parenthesis is a code used by the USGS to uniquely identify the given compound. Flux calculated as the analytical result in μg/L multiplied by the total sample volume collected in liters divided by the area of the respective sampler type (Funnel=0.0731 m², Autosampler= 0.0614 m²) and sample composite time in days. Abbreviations: E, estimated reported laboratory value; na, not applicable; (E), estimated compound due to known poor performance by applied method (Childress and others, 1999; Sandstrom and others, 2001); hh:mm, hour:minute; —, compound not detected; [-], no Chemical Abstract Service Number available for given compound]

Site number	Site name	Site identification number	Fonofos oxon [-] (61649)	Iprodione (E) [36734-19-7] (61593)	Cyhalothrin-lambda [91465-08-6] (61595)	Malathion [121-75-5] (39532)	Malaoxon (E) [1634-78-2] (61652)	Metalaxyl [57837-19-1] (61596)	Methidathion [950-37-8] (61598)
16	Gridley High School precipitation gage at Gridley	392205121410201	—	—	na	E0.067	E0.111	—	—
16	Gridley High School precipitation gage at Gridley	392205121410201	—	—	na	E0.056	—	—	—
16	Gridley High School precipitation gage at Gridley	392205121410201	—	—	na	—	—	—	—
16	Gridley High School precipitation gage at Gridley	392205121410201	—	—	na	E0.040	E0.059	—	—
16	Gridley High School precipitation gage at Gridley	392205121410201	—	—	na	E0.043	—	—	—
16	Gridley High School precipitation gage at Gridley	392205121410201	—	—	na	—	—	—	—
16	Gridley High School precipitation gage at Gridley	392205121410201	—	—	na	—	—	—	E0.039
16	Gridley High School precipitation gage at Gridley	392205121410201	—	—	na	—	—	—	—
16	Gridley High School precipitation gage at Gridley	392205121410201	—	—	na	E0.031	—	—	0.055
16	Gridley High School precipitation gage at Gridley	392205121410201	—	—	na	E0.034	—	—	0.058
16	Gridley High School precipitation gage at Gridley	392205121410201	—	—	na	—	—	—	0.091
16	Gridley High School precipitation gage at Gridley	392205121410201	—	—	na	E0.035	—	—	0.044
16	Gridley High School precipitation gage at Gridley	392205121410201	—	—	na	E0.093	—	—	0.064
16	Gridley High School precipitation gage at Gridley	392205121410201	—	—	na	—	—	—	0.171
16	Gridley High School precipitation gage at Gridley	392205121410201	—	E0.276	na	—	—	—	—
16	Gridley High School precipitation gage at Gridley	392205121410201	—	E1.57	na	—	—	—	0.075
15	Oroville Dam precipitation gage at spillway	393234121292701	—	E0.301	na	—	—	—	—
15	Oroville Dam precipitation gage at spillway	393234121292701	—	—	na	—	—	—	—
15	Oroville Dam precipitation gage at spillway	393234121292701	—	—	na	—	—	—	—
15	Oroville Dam precipitation gage at spillway	393234121292701	—	E0.853	na	—	—	—	—
15	Oroville Dam precipitation gage at spillway	393234121292701	—	E0.462	na	E0.052	—	—	0.090
15	Oroville Dam precipitation gage at spillway	393234121292701	—	—	na	—	—	—	—
15	Oroville Dam precipitation gage at spillway	393234121292701	—	—	na	E0.023	—	—	—
15	Oroville Dam precipitation gage at spillway	393234121292701	—	—	na	—	—	—	—
15	Oroville Dam precipitation gage at spillway	393234121292701	—	—	na	—	—	—	0.033
15	Oroville Dam precipitation gage at spillway	393234121292701	—	—	na	—	—	—	—
15	Oroville Dam precipitation gage at spillway	393234121292701	—	—	na	—	—	—	—
15	Oroville Dam precipitation gage at spillway	393234121292701	—	E0.235	na	—	—	—	—
15	Oroville Dam precipitation gage at spillway	393234121292701	—	E4.34	na	—	—	—	—

Appendix 3. Calculated flux values presented as micrograms per square meter per day (μg/m²/day) for wet-deposition samples with at least one detection collected during the 2002–04 study period at eight sites in the Central Valley, California.—Continued

[All compounds were analyzed by gas chromatography/mass spectrometry (GCMS) at the U.S. Geological Survey (USGS) National Water Quality Laboratory (NWQL). The Chemical Abstract Service Number is given below each compound name in brackets. The five-digit number in parenthesis is a code used by the USGS to uniquely identify the given compound. Flux calculated as the analytical result in μg/L multiplied by the total sample volume collected in liters divided by the area of the respective sampler type (Funnel=0.0731 m², Autosampler= 0.0614 m²) and sample composite time in days. Abbreviations: E, estimated reported laboratory value; na, not applicable; (E), estimated compound due to known poor performance by applied method (Childress and others, 1999; Sandstrom and others, 2001); hh:mm, hour:minute; –, compound not detected; [-], no Chemical Abstract Service Number available for given compound]

Site number	Site name	Site identification number	Metolachlor [51218-45-2] (39415)	Metribuzin [21087-64-9] (82630)	Molinate [2212-67-1] (82671)	Myclobutanil (E) [88671-89-0] (61599)	Napropamide [15299-99-7] (82684)	Oxyfluorfen [42874-03-3] (61600)	Parathion-methyl [298-00-0] (61664)
11	Newman rain gage at wasteway levee near Draper Road	371735121031201	–	–	–	–	–	–	–
11	Newman rain gage at wasteway levee near Draper Road	371735121031201	–	–	–	–	–	0.053	–
11	Newman rain gage at wasteway levee near Draper Road	371735121031201	–	–	–	E0.095	–	–	–
11	Newman rain gage at wasteway levee near Draper Road	371735121031201	–	–	–	E0.114	–	–	–
11	Newman rain gage at wasteway levee near Draper Road	371735121031201	–	–	–	E0.099	–	–	–
11	Newman rain gage at wasteway levee near Draper Road	371735121031201	E0.265	–	na	E0.289	na	na	0.626
11	Newman rain gage at wasteway levee near Draper Road	371735121031201	E0.049	0.049	na	E0.049	na	na	–
11	Newman rain gage at wasteway levee near Draper Road	371735121031201	E0.034	–	na	–	na	na	–
11	Newman rain gage at wasteway levee near Draper Road	371735121031201	E0.083	–	na	–	na	na	–
11	Newman rain gage at wasteway levee near Draper Road	371735121031201	E0.070	–	na	E0.120	na	na	–
11	Newman rain gage at wasteway levee near Draper Road	371735121031201	E0.026	–	na	E0.074	na	na	–
11	Newman rain gage at wasteway levee near Draper Road	371735121031201	0.085	0.055	na	E0.053	na	na	0.080
11	Newman rain gage at wasteway levee near Draper Road	371735121031201	E0.021	–	na	–	na	na	–
11	Newman rain gage at wasteway levee near Draper Road	371735121031201	E0.059	–	na	E0.170	na	na	–
11	Newman rain gage at wasteway levee near Draper Road	371735121031201	0.023	–	na	E0.141	na	na	–
11	Newman rain gage at wasteway levee near Draper Road	371735121031201	E0.030	–	na	E0.221	na	na	–
12	Turlock rain gage near Idaho Road	372713120534901	E0.031	–	–	–	0.070	–	–
12	Turlock rain gage near Idaho Road	372713120534901	0.038	–	–	–	–	–	–
12	Turlock rain gage near Idaho Road	372713120534901	E0.017	–	–	–	0.044	0.078	–
12	Turlock rain gage near Idaho Road	372713120534901	E0.019	–	–	E0.639	0.043	–	–
12	Turlock rain gage near Idaho Road	372713120534901	–	–	–	E0.075	–	–	–
12	Turlock rain gage near Idaho Road	372713120534901	E0.025	–	–	–	–	–	–
12	Turlock rain gage near Idaho Road	372713120534901	E0.114	–	na	E0.153	na	na	–
12	Turlock rain gage near Idaho Road	372713120534901	E0.039	–	na	E0.114	na	na	–
12	Turlock rain gage near Idaho Road	372713120534901	E0.016	–	na	–	na	na	–
12	Turlock rain gage near Idaho Road	372713120534901	E0.054	–	na	–	na	na	–
12	Turlock rain gage near Idaho Road	372713120534901	E0.008	–	na	na	na	na	–
12	Turlock rain gage near Idaho Road	372713120534901	E0.066	–	na	E0.204	na	na	–
12	Turlock rain gage near Idaho Road	372713120534901	E0.022	–	na	E0.093	na	na	–
12	Turlock rain gage near Idaho Road	372713120534901	0.159	–	na	E0.198	na	na	–

Appendix 3.   Calculated flux values presented as micrograms per square meter per day ($\mu g/m^2/day$) for wet-deposition samples with at least one detection collected during the 2002–04 study period at eight sites in the Central Valley, California.—Continued

[All compounds were analyzed by gas chromatography/mass spectrometry (GCMS) at the U.S. Geological Survey (USGS) National Water Quality Laboratory (NWQL). The Chemical Abstract Service Number is given below each compound name in brackets. The five-digit number in parenthesis is a code used by the USGS to uniquely identify the given compound. Flux calculated as the analytical result in $\mu g/L$ multiplied by the total sample volume collected in liters divided by the area of the respective sampler type (Funnel=$0.0731\ m^2$, Autosampler= $0.0614\ m^2$) and sample composite time in days. Abbreviations: E, estimated reported laboratory value; na, not applicable; (E), estimated compound due to known poor performance by applied method (Childress and others, 1999; Sandstrom and others, 2001); hh:mm, hour:minute; –, compound not detected; [–], no Chemical Abstract Service Number available for given compound]

Site number	Site name	Site identification number	Metolachlor [51218-45-2] (39415)	Metribuzin [21087-64-9] (82630)	Molinate [2212-67-1] (82671)	Myclobutanil (E) [88671-89-0] (61599)	Napropamide [15299-99-7] (82684)	Oxyfluorfen [42874-03-3] (61600)	Parathion-methyl [298-00-0] (61664)
12	Turlock rain gage near Idaho Road	372713120534901	E0.023	–	na	E0.027	na	na	0.038
12	Turlock rain gage near Idaho Road	372713120534901	–	–	na	–	na	na	–
12	Turlock rain gage near Idaho Road	372713120534901	E0.027	–	na	E0.071	na	na	–
12	Turlock rain gage near Idaho Road	372713120534901	0.017	–	na	E0.425	na	na	–
12	Turlock rain gage near Idaho Road	372713120534901	E0.023	–	na	E0.333	na	na	–
12	Turlock rain gage near Idaho Road	372713120534901	0.060	–	na	E0.111	na	na	–
13	Turlock Airport rain gage	372857120414001	–	–	–	–	0.125	0.143	–
13	Turlock Airport rain gage	372857120414001	–	–	–	–	0.053	–	–
13	Turlock Airport rain gage	372857120414001	–	–	–	–	0.047	0.071	–
13	Turlock Airport rain gage	372857120414001	–	–	–	E0.750	–	–	–
13	Turlock Airport rain gage	372857120414001	–	–	–	E0.092	–	–	–
13	Turlock Airport rain gage	372857120414001	–	–	–	E0.081	–	–	–
13	Turlock Airport rain gage	372857120414001	–	–	na	–	na	na	0.282
13	Turlock Airport rain gage	372857120414001	E0.032	–	na	–	na	na	–
13	Turlock Airport rain gage	372857120414001	–	–	na	–	na	na	–
13	Turlock Airport rain gage	372857120414001	E0.006	–	na	E0.224	na	na	–
13	Turlock Airport rain gage	372857120414001	E0.013	–	na	E0.471	na	na	–
13	Turlock Airport rain gage	372857120414001	E0.036	–	na	E0.247	na	na	–
13	Turlock Airport rain gage	372857120414001	E0.032	–	na	E0.470	na	na	–
13	Turlock Airport rain gage	372857120414001	0.064	–	na	E0.304	na	na	–
13	Turlock Airport rain gage	372857120414001	E0.020	–	na	E0.046	na	na	0.049
13	Turlock Airport rain gage	372857120414001	–	–	na	–	na	na	–
13	Turlock Airport rain gage	372857120414001	–	–	na	E0.861	na	na	–
13	Turlock Airport rain gage	372857120414001	–	–	na	E0.354	na	na	–
13	Turlock Airport rain gage	372857120414001	E0.081	–	na	E0.314	na	na	–
10	Westley rain gage at pump building near lateral 6 North	373335121143001	0.027	–	–	–	0.071	0.088	–
10	Westley rain gage at pump building near lateral 6 North	373335121143001	E0.026	–	–	–	–	0.467	–
10	Westley rain gage at pump building near lateral 6 North	373335121143001	0.016	–	–	E1.14	–	0.062	–

Appendix 3. Calculated flux values presented as micrograms per square meter per day ($\mu g/m^2/day$) for wet-deposition samples with at least one detection collected during the 2002–04 study period at eight sites in the Central Valley, California.—Continued

[All compounds were analyzed by gas chromatography/mass spectrometry (GCMS) at the U.S. Geological Survey (USGS) National Water Quality Laboratory (NWQL). The Chemical Abstract Service Number is given below each compound name in brackets. The five-digit number in parenthesis is a code used by the USGS to uniquely identify the given compound. Flux calculated as the analytical result in $\mu g/L$ multiplied by the total sample volume collected in liters divided by the area of the respective sampler type (Funnel=0.0731 m^2, Autosampler= 0.0614 m^2) and sample composite time in days. Abbreviations: E, estimated reported laboratory value; na, not applicable; (E), estimated compound due to known poor performance by applied method (Childress and others, 1999; Sandstrom and others, 2001); hh:mm, hour:minute; –, compound not detected. [-], no Chemical Abstract Service Number available for given compound]

Site number	Site name	Site identification number	Metolachlor [51218-45-2] (39415)	Metribuzin [21087-64-9] (82630)	Molinate [2212-67-1] (82671)	Myclobutanil (E) [88671-89-0] (61599)	Napropamide [15299-99-7] (82684)	Oxyfluorfen [42874-03-3] (61600)	Parathion-methyl [298-00-0] (61664)
10	Westley rain gage at pump building near lateral 6 North	373335121143001	–	–	–	E0.819	0.034	0.052	–
10	Westley rain gage at pump building near lateral 6 North	373335121143001	–	–	–	E0.067	–	–	–
10	Westley rain gage at pump building near lateral 6 North	373335121143001	0.067	–	–	E0.523	–	0.388	–
10	Westley rain gage at pump building near lateral 6 North	373335121143001	0.400	1.12	na	E0.316	na	na	1.391
10	Westley rain gage at pump building near lateral 6 North	373335121143001	E0.072	0.233	na	E0.048	na	na	0.056
10	Westley rain gage at pump building near lateral 6 North	373335121143001	E0.025	0.029	na	–	na	na	–
10	Westley rain gage at pump building near lateral 6 North	373335121143001	0.177	0.121	na	–	na	na	–
10	Westley rain gage at pump building near lateral 6 North	373335121143001	E0.036	0.025	na	E0.754	na	na	–
10	Westley rain gage at pump building near lateral 6 North	373335121143001	E0.072	–	na	E0.128	na	na	0.096
10	Westley rain gage at pump building near lateral 6 North	373335121143001	0.130	6.84	na	E0.061	na	na	0.277
10	Westley rain gage at pump building near lateral 6 North	373335121143001	0.028	–	na	0.089	na	na	0.019
10	Westley rain gage at pump building near lateral 6 North	373335121143001	E0.034	–	na	0.149	na	na	–
4	Modesto Irrigation District gage rooftop at Modesto	373834121000601	–	–	–	–	0.056	0.062	–
4	Modesto Irrigation District gage rooftop at Modesto	373834121000601	0.043	–	–	–	0.568	0.074	–
4	Modesto Irrigation District gage rooftop at Modesto	373834121000601	–	–	–	E0.210	–	na	–
4	Modesto Irrigation District gage rooftop at Modesto	373834121000601	–	–	–	na	na	na	–
4	Modesto Irrigation District gage rooftop at Modesto	373834121000601	–	–	–	E0.051	na	na	–
4	Modesto Irrigation District gage rooftop at Modesto	373834121000601	–	–	–	E0.180	na	na	–
4	Modesto Irrigation District gage rooftop at Modesto	373834121000601	0.190	–	0.095	E0.029	–	–	0.309
4	Modesto Irrigation District gage rooftop at Modesto	373834121000601	E0.168	–	na	–	na	na	0.414
4	Modesto Irrigation District gage rooftop at Modesto	373834121000601	E0.138	–	na	E0.221	na	na	–
4	Modesto Irrigation District gage rooftop at Modesto	373834121000601	E0.044	–	na	–	na	na	–
4	Modesto Irrigation District gage rooftop at Modesto	373834121000601	E0.035	–	na	–	na	na	–
4	Modesto Irrigation District gage rooftop at Modesto	373834121000601	E0.017	–	na	–	na	na	–
4	Modesto Irrigation District gage rooftop at Modesto	373834121000601	E0.022	–	na	–	na	na	–
4	Modesto Irrigation District gage rooftop at Modesto	373834121000601	E0.016	–	na	–	na	na	–
4	Modesto Irrigation District gage rooftop at Modesto	373834121000601	E0.050	–	na	–	na	na	–
4	Modesto Irrigation District gage rooftop at Modesto	373834121000601	E0.064	–	na	–	na	na	–
4	Modesto Irrigation District gage rooftop at Modesto	373834121000601	E0.011	–	na	E0.058	na	na	–
4	Modesto Irrigation District gage rooftop at Modesto	373834121000601	E0.017	–	na	E0.614	na	na	–
4	Modesto Irrigation District gage rooftop at Modesto	373834121000601	E0.015	–	na	E0.141	na	na	–

Appendix 3. Calculated flux values presented as micrograms per square meter per day (μg/m²/day) for wet-deposition samples with at least one detection collected during the 2002–04 study period at eight sites in the Central Valley, California.—Continued

[All compounds were analyzed by gas chromatography/mass spectrometry (GCMS) at the U.S. Geological Survey (USGS) National Water Quality Laboratory (NWQL). The Chemical Abstract Service Number is given below each compound name in brackets. The five-digit number in parenthesis is a code used by the USGS to uniquely identify the given compound. Flux calculated as the analytical result in μg/L multiplied by the total sample volume collected in liters divided by the area of the respective sampler type (Funnel=0.0731 m², Autosampler= 0.0614 m²) and sample composite time in days. Abbreviations: E, estimated reported laboratory value; na, not applicable; (E), estimated compound due to known poor performance by applied method (Childress and others, 1999; Sandstrom and others, 2001); hh:mm, hour:minute; -, compound not detected; [-], no Chemical Abstract Service Number available for given compound]

Site number	Site name	Site identification number	Metolachlor [51218-45-2] (39415)	Metribuzin [21087-64-9] (82630)	Molinate [2212-67-1] (82671)	Myclobutanil (E) [88671-89-0] (61599)	Napropamide [15299-99-7] (82684)	Oxyfluorfen [42874-03-3] (61600)	Parathion-methyl [298-00-0] (61664)
4	Modesto Irrigation District gage rooftop at Modesto	373834121000601	E0.061	-	na	E0.102	na	na	-
4	Modesto Irrigation District gage rooftop at Modesto	373834121000601	E0.049	-	na	-	na	na	-
4	Modesto Irrigation District gage rooftop at Modesto	373834121000601	-	-	na	E0.037	na	na	-
4	Modesto Irrigation District gage rooftop at Modesto	373834121000601	E0.020	-	na	E0.040	na	na	-
4	Modesto Irrigation District gage rooftop at Modesto	373834121000601	0.161	-	na	E0.281	na	na	0.158
4	Modesto Irrigation District gage rooftop at Modesto	373834121000601	E0.04	-	na		na	na	-
4	Modesto Irrigation District gage rooftop at Modesto	373834121000601	E0.013	-	na		na	na	-
4	Modesto Irrigation District gage rooftop at Modesto	373834121000601	E0.013	-	na		na	na	-
4	Modesto Irrigation District gage rooftop at Modesto	373834121000601	-	-	na		na	na	-
4	Modesto Irrigation District gage rooftop at Modesto	373834121000601	-	E0.038	na		na	na	-
4	Modesto Irrigation District gage rooftop at Modesto	373834121000601	-	-	na		na	na	-
4	Modesto Irrigation District gage rooftop at Modesto	373834121000601	E0.016	-	na	E1.10	na	na	-
4	Modesto Irrigation District gage rooftop at Modesto	373834121000601	-	-	na	E0.491	na	na	-
4	Modesto Irrigation District gage rooftop at Modesto	373834121000601	0.086	-	na	E0.152	na	na	-
5	Modesto Irrigation District gage rooftop at Albers Road	373841120504801	-	-	-	-	0.064	-	-
5	Modesto Irrigation District gage rooftop at Albers Road	373841120504801	-	-	-	-	-	0.086	-
5	Modesto Irrigation District gage rooftop at Albers Road	373841120504801	E0.013	-	-	-	-	0.191	-
5	Modesto Irrigation District gage rooftop at Albers Road	373841120504801	0.022	-	-	E0.685	-	-	-
5	Modesto Irrigation District gage rooftop at Albers Road	373841120504801	-	-	-	E0.413	-	-	-
5	Modesto Irrigation District gage rooftop at Albers Road	373841120504801	E0.031	-	-	E0.305	na	-	-
5	Modesto Irrigation District gage rooftop at Albers Road	373841120504801	E0.114	-	-	E0.172	na	na	0.858
5	Modesto Irrigation District gage rooftop at Albers Road	373841120504801	E0.117	-	-	E0.176	na	na	0.821
5	Modesto Irrigation District gage rooftop at Albers Road	373841120504801	E0.033	-	na		na	na	-
5	Modesto Irrigation District gage rooftop at Albers Road	373841120504801	E0.021	-	na	E0.032	na	na	-
5	Modesto Irrigation District gage rooftop at Albers Road	373841120504801	-	-	na		na	na	-
5	Modesto Irrigation District gage rooftop at Albers Road	373841120504801	E0.008	-	na	E0.011	na	na	-
5	Modesto Irrigation District gage rooftop at Albers Road	373841120504801	-	-	na		na	na	-
5	Modesto Irrigation District gage rooftop at Albers Road	373841120504801	-	-	na		na	na	-

Appendix 3. Calculated flux values presented as micrograms per square meter per day (μg/m^2/day) for wet-deposition samples with at least one detection collected during the 2002–04 study period at eight sites in the Central Valley, California.—Continued

[All compounds were analyzed by gas chromatography/mass spectrometry (GCMS) at the U.S. Geological Survey (USGS) National Water Quality Laboratory (NWQL). The Chemical Abstract Service Number is given below each compound name in parentheses. The five-digit number in parenthesis is a code used by the USGS to uniquely identify the given compound. Flux calculated as the analytical result in μg/L multiplied by the total sample volume collected in liters divided by the area of the respective sampler type (Funnel=0.0731 m^2, Autosampler= 0.0614 m^2) and sample composite time in days. Abbreviations: E, estimated reported laboratory value; na, not applicable; (E), estimated compound due to known poor performance by applied method (Childress and others, 1999; Sandstrom and others, 2001); hh:mm, hour:minute; –, compound not detected; [-], no Chemical Abstract Service Number available for given compound]

Site number	Site name	Site identification number	Metolachlor [51218-45-2] (39415)	Metribuzin [21087-64-9] (82630)	Molinate [2212-67-1] (82671)	Myclobutanil (E) [88671-89-0] (61599)	Napropamide [15299-99-7] (82684)	Oxyfluorfen [42874-03-3] (61600)	Parathion-methyl [298-00-0] (61664)
5	Modesto Irrigation District gage rooftop at Albers Road	3738411205048O1	E0.005	–	na	E0.020	na	na	–
5	Modesto Irrigation District gage rooftop at Albers Road	3738411205048O1	E0.008	–	na	E0.538	na	na	–
5	Modesto Irrigation District gage rooftop at Albers Road	3738411205048O1	–	–	na	E0.069	na	na	–
5	Modesto Irrigation District gage rooftop at Albers Road	3738411205048O1	E0.022	–	na	E0.397	na	na	–
5	Modesto Irrigation District gage rooftop at Albers Road	3738411205048O1	E0.023	–	na	E0.108	na	na	–
5	Modesto Irrigation District gage rooftop at Albers Road	3738411205048O1	E0.018	–	na	E0.059	na	na	–
5	Modesto Irrigation District gage rooftop at Albers Road	3738411205048O1	E0.047	–	na	E0.060	na	na	–
5	Modesto Irrigation District gage rooftop at Albers Road	3738411205048O1	0.055	–	na	E0.292	na	na	–
5	Modesto Irrigation District gage rooftop at Albers Road	3738411205048O1	E0.023	–	na	E0.033	na	na	0.120
5	Modesto Irrigation District gage rooftop at Albers Road	3738411205048O1	E0.011	–	na	E0.013	na	na	0.107
5	Modesto Irrigation District gage rooftop at Albers Road	3738411205048O1	–	–	na		na	na	–
5	Modesto Irrigation District gage rooftop at Albers Road	3738411205048O1	–	–	na		na	na	–
5	Modesto Irrigation District gage rooftop at Albers Road	3738411205048O1	–	–	na		na	na	–
5	Modesto Irrigation District gage rooftop at Albers Road	3738411205048O1	E0.004	–	na		na	na	–
5	Modesto Irrigation District gage rooftop at Albers Road	3738411205048O1	–	–	na		na	na	–
5	Modesto Irrigation District gage rooftop at Albers Road	3738411205048O1	–	–	na		na	na	–
5	Modesto Irrigation District gage rooftop at Albers Road	3738411205048O1	E0.019	–	na	E4.18	na	na	–
5	Modesto Irrigation District gage rooftop at Albers Road	3738411205048O1	E0.017	–	na	E1.09	na	na	–
5	Modesto Irrigation District gage rooftop at Albers Road	3738411205048O1	–	–	na	E0.684	na	na	–
5	Modesto Irrigation District gage rooftop at Albers Road	3738411205048O1	E0.067	–	na	E0.202	na	na	–
16	Gridley High School precipitation gage at Gridley	392205121410201	–	–	na	E0.328	na	na	–
16	Gridley High School precipitation gage at Gridley	392205121410201	–	–	na	E1.99	na	na	–
16	Gridley High School precipitation gage at Gridley	392205121410201	–	–	na	na	na	na	–
16	Gridley High School precipitation gage at Gridley	392205121410201	–	–	na	na	na	na	–
16	Gridley High School precipitation gage at Gridley	392205121410201	E0.017	–	na	–	na	na	–
16	Gridley High School precipitation gage at Gridley	392205121410201	E0.032	–	na	–	na	na	–
16	Gridley High School precipitation gage at Gridley	392205121410201	E0.032	–	na	–	na	na	–
16	Gridley High School precipitation gage at Gridley	392205121410201	E0.068	–	na	–	na	na	–
16	Gridley High School precipitation gage at Gridley	392205121410201	E0.015	–	na	–	na	na	–

Appendix 3. Calculated flux values presented as micrograms per square meter per day (µg/m²/day) for wet-deposition samples with at least one detection collected during the 2002–04 study period at eight sites in the Central Valley, California.—Continued

[All compounds were analyzed by gas chromatography/mass spectrometry (GCMS) at the U.S. Geological Survey (USGS) National Water Quality Laboratory (NWQL). The Chemical Abstract Service Number is given below each compound name in brackets. The five-digit number in parenthesis is a code used by the USGS to uniquely identify the given compound. Flux calculated as the analytical result in µg/L multiplied by the total sample volume collected in liters divided by the area of the respective sampler type (Funnel=0.0731 m², Autosampler= 0.0614 m²) and sample composite time in days. Abbreviations: E, estimated reported laboratory value; na, not applicable; (E), estimated compound due to known poor performance by applied method (Childress and others, 1999; Sandstrom and others, 2001); hh:mm, hour:minute; –, compound not detected; [-], no Chemical Abstract Service Number available for given compound]

Site number	Site name	Metolachlor [51218-45-2] (39415)	Metribuzin [21087-64-9] (82630)	Molinate [2212-67-1] (82671)	Myclobutanil (E) [88671-89-0] (61599)	Napropamide [15299-99-7] (82684)	Oxyfluorfen [42874-03-3] (61600)	Parathion-methyl [298-00-0] (61664)
16	Gridley High School precipitation gage at Gridley	–	–	na	–	na	na	0.657
16	Gridley High School precipitation gage at Gridley	E0.028	–	na	–	na	na	0.193
16	Gridley High School precipitation gage at Gridley	–	–	na	–	na	na	0.204
16	Gridley High School precipitation gage at Gridley	–	–	na	–	na	na	–
16	Gridley High School precipitation gage at Gridley	–	–	na	–	na	na	0.084
16	Gridley High School precipitation gage at Gridley	E0.022	–	na	–	na	na	0.050
16	Gridley High School precipitation gage at Gridley	–	–	na	–	na	na	0.043
16	Gridley High School precipitation gage at Gridley	–	–	na	–	na	na	–
16	Gridley High School precipitation gage at Gridley	–	–	na	–	na	na	–
16	Gridley High School precipitation gage at Gridley	–	–	na	–	na	na	–
16	Gridley High School precipitation gage at Gridley	–	–	na	–	na	na	–
16	Gridley High School precipitation gage at Gridley	–	–	na	–	na	na	–
16	Gridley High School precipitation gage at Gridley	–	–	na	–	na	na	–
16	Gridley High School precipitation gage at Gridley	–	–	na	E0.127	na	na	–
16	Gridley High School precipitation gage at Gridley	–	–	na	E0.184	na	na	0.062
16	Gridley High School precipitation gage at Gridley	–	–	na	–	na	na	–
15	Oroville Dam precipitation gage at spillway	–	–	na	–	na	na	na
15	Oroville Dam precipitation gage at spillway	–	–	na	–	na	na	na
15	Oroville Dam precipitation gage at spillway	–	–	na	–	na	na	na
15	Oroville Dam precipitation gage at spillway	–	–	na	na	na	na	na
15	Oroville Dam precipitation gage at spillway	E0.022	–	na	–	na	na	na
15	Oroville Dam precipitation gage at spillway	M	–	na	–	na	na	na
15	Oroville Dam precipitation gage at spillway	E0.039	–	na	–	na	na	na
15	Oroville Dam precipitation gage at spillway	–	–	na	–	na	na	na
15	Oroville Dam precipitation gage at spillway	–	–	na	–	na	na	na
15	Oroville Dam precipitation gage at spillway	–	–	na	–	na	na	na
15	Oroville Dam precipitation gage at spillway	–	–	na	–	na	na	na
15	Oroville Dam precipitation gage at spillway	E0.043	–	na	E0.107	na	na	na
15	Oroville Dam precipitation gage at spillway	–	E0.033	na	E0.322	na	na	na

Appendix 3.   Calculated flux values presented as micrograms per square meter per day (μg/m²/day) for wet-deposition samples with at least one detection collected during the 2002–04 study period at eight sites in the Central Valley, California.—Continued

[All compounds were analyzed by gas chromatography mass spectrometry (GCMS) at the U.S. Geological Survey (USGS) National Water Quality Laboratory (NWQL). The Chemical Abstract Service Number is given below each compound name in brackets. The five-digit number in parenthesis is a code used by the USGS to uniquely identify the given compound. Flux calculated as the analytical result in μg/L multiplied by the total sample volume collected in liters divided by the area of the respective sampler type (Funnel=0.0731 m², Autosampler= 0.0614 m²) and sample composite time in days. Abbreviations: E, estimated reported laboratory value; na, not applicable; (E), estimated compound due to known poor performance by applied method (Childress and others, 1999; Sandstrom and others, 2001); hh:mm, hour:minute; —, compound not detected; [-], no Chemical Abstract Service Number available for given compound]

Site number	Site name	Site identification number	Paraoxon-methyl (E) [950-35-6] (61664)	Pendimethalin [40487-42-1] (82683)	cis-Permethrin [54774-45-7] (82687)	cis-Propi-conazole [-] (79846)	trans-Propiconazole [-] (79847)	Phosmet [732-11-6] (61601)
11	Newman rain gage at wasteway levee near Draper Road	37173512103201	—	—	—	—	—	—
11	Newman rain gage at wasteway levee near Draper Road	37173512103201	—	0.079	—	—	—	—
11	Newman rain gage at wasteway levee near Draper Road	37173512103201	—	0.278	—	E0.010	0.02	—
11	Newman rain gage at wasteway levee near Draper Road	37173512103201	—	—	—	—	—	—
11	Newman rain gage at wasteway levee near Draper Road	37173512103201	—	0.177	—	0.043	E0.04	E0.241
11	Newman rain gage at wasteway levee near Draper Road	37173512103201	—	2.67	—	na	na	—
11	Newman rain gage at wasteway levee near Draper Road	37173512103201	—	0.621	—	na	na	—
11	Newman rain gage at wasteway levee near Draper Road	37173512103201	—	—	—	na	na	—
11	Newman rain gage at wasteway levee near Draper Road	37173512103201	—	0.774	—	na	na	E0.133
11	Newman rain gage at wasteway levee near Draper Road	37173512103201	—	0.380	—	na	na	—
11	Newman rain gage at wasteway levee near Draper Road	37173512103201	—	—	—	na	na	E0.025
11	Newman rain gage at wasteway levee near Draper Road	37173512103201	—	0.091	—	na	na	—
11	Newman rain gage at wasteway levee near Draper Road	37173512103201	—	E0.308	—	na	na	—
11	Newman rain gage at wasteway levee near Draper Road	37173512103201	—	E0.560	—	na	na	—
11	Newman rain gage at wasteway levee near Draper Road	37173512103201	—	0.069	—	na	na	—
11	Newman rain gage at wasteway levee near Draper Road	37173512103201	—	E0.361	—	na	na	—
12	Turlock rain gage near Idaho Road	37271312053490l	—	—	—	—	—	—
12	Turlock rain gage near Idaho Road	37271312053490l	—	0.160	E0.02	—	—	—
12	Turlock rain gage near Idaho Road	37271312053490l	—	0.110	—	—	—	—
12	Turlock rain gage near Idaho Road	37271312053490l	—	0.422	—	—	—	—
12	Turlock rain gage near Idaho Road	37271312053490l	—	—	—	0.040	0.11	—
12	Turlock rain gage near Idaho Road	37271312053490l	—	—	—	—	—	—
12	Turlock rain gage near Idaho Road	37271312053490l	—	0.172	—	0.057	0.06	E0.204
12	Turlock rain gage near Idaho Road	37271312053490l	—	0.628	—	na	na	E0.171
12	Turlock rain gage near Idaho Road	37271312053490l	—	0.483	—	na	na	—
12	Turlock rain gage near Idaho Road	37271312053490l	—	0.094	—	na	na	—
12	Turlock rain gage near Idaho Road	37271312053490l	—	0.506	—	na	na	—
12	Turlock rain gage near Idaho Road	37271312053490l	na	E0.086	—	na	na	—
12	Turlock rain gage near Idaho Road	37271312053490l	—	0.438	—	na	na	E0.073
12	Turlock rain gage near Idaho Road	37271312053490l	—	—	—	na	na	—
12	Turlock rain gage near Idaho Road	37271312053490l	—	E0.056	—	na	na	E0.042

Appendix 3. Calculated flux values presented as micrograms per square meter per day ($\mu g/m^2/day$) for wet-deposition samples with at least one detection collected during the 2002–04 study period at eight sites in the Central Valley, California.—Continued

[All compounds were analyzed by gas chromatography/mass spectrometry (GCMS) at the U.S. Geological Survey (USGS) National Water Quality Laboratory (NWQL). The Chemical Abstract Service Number is given below each compound name in brackets. The five-digit number in parenthesis is a code used by the USGS to uniquely identify the given compound. Flux calculated as the analytical result in $\mu g/L$ multiplied by the total sample volume collected in liters divided by the area of the respective sampler type (Funnel=$0.0731\ m^2$, Autosampler= $0.0614\ m^2$) and sample composite time in days. Abbreviations: E, estimated reported laboratory value; na, not applicable; (E), estimated compound due to known poor performance by applied method (Childress and others, 1999; Sandstrom and others, 2001); hh:mm, hour:minute; —, compound not detected; [-], no Chemical Abstract Service Number available for given compound]

Site number	Site name	Site identification number	Paraoxon-methyl (E) [950-35-6] (61664)	Pendimethalin [40487-42-1] (82683)	cis-Permethrin [54774-45-7] (82687)	cis-Propi-conazole [-] (79846)	trans-Propiconazole [-] (79847)	Phosmet [732-11-6] (61601)
12	Turlock rain gage near Idaho Road	372713120534901	—	E0.038	—	na	na	E0.027
12	Turlock rain gage near Idaho Road	372713120534901	—	E0.394	E0.04	na	na	—
12	Turlock rain gage near Idaho Road	372713120534901	—	E0.153	—	na	na	—
12	Turlock rain gage near Idaho Road	372713120534901	—	0.036	—	na	na	—
12	Turlock rain gage near Idaho Road	372713120534901	—	E0.187	—	na	na	—
12	Turlock rain gage near Idaho Road	372713120534901	—	0.072	—	na	na	na
13	Turlock Airport rain gage	372857120414001	—	E0.152	E0.04	—	—	—
13	Turlock Airport rain gage	372857120414001	—	—	—	—	—	—
13	Turlock Airport rain gage	372857120414001	—	0.137	—	—	—	—
13	Turlock Airport rain gage	372857120414001	—	0.215	—	0.096	0.14	E0.337
13	Turlock Airport rain gage	372857120414001	—	—	—	0.019	0.04	—
13	Turlock Airport rain gage	372857120414001	—	0.157	—	0.047	0.07	E0.338
13	Turlock Airport rain gage	372857120414001	—	0.818	—	0.062	0.07	—
13	Turlock Airport rain gage	372857120414001	—	0.340	—	na	na	—
13	Turlock Airport rain gage	372857120414001	—	E0.065	—	na	na	—
13	Turlock Airport rain gage	372857120414001	—	E0.279	—	na	na	—
13	Turlock Airport rain gage	372857120414001	—	0.113	—	na	na	—
13	Turlock Airport rain gage	372857120414001	—	0.086	—	na	na	—
13	Turlock Airport rain gage	372857120414001	—	0.276	—	na	na	E0.087
13	Turlock Airport rain gage	372857120414001	—	E0.107	—	na	na	—
13	Turlock Airport rain gage	372857120414001	—	E0.078	—	na	na	E0.093
13	Turlock Airport rain gage	372857120414001	—	E0.037	—	na	na	E0.039
13	Turlock Airport rain gage	372857120414001	—	E0.561	—	na	na	—
13	Turlock Airport rain gage	372857120414001	—	0.182	—	na	na	—
13	Turlock Airport rain gage	372857120414001	—	E0.122	—	na	na	—
13	Turlock Airport rain gage	372857120414001	—	0.349	—	na	na	na
10	Westley rain gage at pump building near lateral 6 North	373335121143001	—	0.262	—	—	—	—
10	Westley rain gage at pump building near lateral 6 North	373335121143001	—	1.13	—	—	—	—
10	Westley rain gage at pump building near lateral 6 North	373335121143001	—	0.200	—	0.087	0.11	—

Appendix 3. Calculated flux values presented as micrograms per square meter per day ($\mu g/m^2/day$) for wet-deposition samples with at least one detection collected during the 2002–04 study period at eight sites in the Central Valley, California.—Continued

[All compounds were analyzed by gas chromatography/mass spectrometry (GCMS) at the U.S. Geological Survey (USGS) National Water Quality Laboratory (NWQL). The Chemical Abstract Service Number is given below each compound name in parenthesis. The five-digit number in parenthesis is a code used by the USGS to uniquely identify the given compound. Flux calculated as the analytical result in $\mu g/L$ multiplied by the total sample volume collected in liters divided by the area of the respective sampler type (Funnel=0.0731 m^2, Autosampler= 0.0614 m^2) and sample composite time in days. Abbreviations: E, estimated reported laboratory value; na, not applicable; (E), estimated compound due to known poor performance by applied method (Childress and others, 1999; Sandstrom and others, 2001); hh:mm, hour:minute; –, compound not detected; [-], no Chemical Abstract Service Number available for given compound]

Site number	Site name	Site identification number	Paraoxon-methyl (E) [950-35-6] (61664)	Pendimethalin [40487-42-1] (82683)	cis-Permethrin [54774-45-7] (82687)	cis-Propiconazole [-] (79846)	trans-Propiconazole [-] (79847)	Phosmet [732-11-6] (61601)
10	Westley rain gage at pump building near lateral 6 North	37333512114300I	–	0.146	–	0.083	0.11	–
10	Westley rain gage at pump building near lateral 6 North	37333512114300I	–	0.221	–	E0.021	E0.05	–
10	Westley rain gage at pump building near lateral 6 North	37333512114300I	–	0.750	–	E0.059	E0.08	–
10	Westley rain gage at pump building near lateral 6 North	37333512114300I	–	1.79	–	na	na	E0.316
10	Westley rain gage at pump building near lateral 6 North	37333512114300I	–	1.09	–	na	na	–
10	Westley rain gage at pump building near lateral 6 North	37333512114300I	–	0.158	–	na	na	–
10	Westley rain gage at pump building near lateral 6 North	37333512114300I	–	0.791	–	na	na	–
10	Westley rain gage at pump building near lateral 6 North	37333512114300I	–	0.352	–	na	na	–
10	Westley rain gage at pump building near lateral 6 North	37333512114300I	–	0.795	–	na	na	E0.128
10	Westley rain gage at pump building near lateral 6 North	37333512114300I	E0.087	0.329	E0.05	na	na	E0.087
10	Westley rain gage at pump building near lateral 6 North	37333512114300I	–	0.227	–	na	na	–
10	Westley rain gage at pump building near lateral 6 North	37333512114300I	–	E2.70	–	na	na	–
4	Modesto Irrigation District gage rooftop at Modesto	37383412100060I	–	0.165	–	–	–	–
4	Modesto Irrigation District gage rooftop at Modesto	37383412100060I	–	0.326	–	–	–	–
4	Modesto Irrigation District gage rooftop at Modesto	37383412100060I	–	0.367	–	–	–	–
4	Modesto Irrigation District gage rooftop at Modesto	37383412100060I	na	0.237	–	na	na	na
4	Modesto Irrigation District gage rooftop at Modesto	37383412100060I	–	0.191	–	na	na	–
4	Modesto Irrigation District gage rooftop at Modesto	37383412100060I	–	0.610	–	E0.07	E0.10	–
4	Modesto Irrigation District gage rooftop at Modesto	37383412100060I	–	0.028	–	–	–	–
4	Modesto Irrigation District gage rooftop at Modesto	37383412100060I	–	0.926	–	na	na	–
4	Modesto Irrigation District gage rooftop at Modesto	37383412100060I	E0.552	0.911	–	na	na	–
4	Modesto Irrigation District gage rooftop at Modesto	37383412100060I	–	0.580	–	na	na	–
4	Modesto Irrigation District gage rooftop at Modesto	37383412100060I	–	0.375	–	na	na	–
4	Modesto Irrigation District gage rooftop at Modesto	37383412100060I	–	0.299	–	na	na	–
4	Modesto Irrigation District gage rooftop at Modesto	37383412100060I	–	E0.085	–	na	na	–
4	Modesto Irrigation District gage rooftop at Modesto	37383412100060I	–	0.069	–	na	na	–
4	Modesto Irrigation District gage rooftop at Modesto	37383412100060I	–	0.598	–	na	na	–
4	Modesto Irrigation District gage rooftop at Modesto	37383412100060I	–	0.666	–	na	na	–
4	Modesto Irrigation District gage rooftop at Modesto	37383412100060I	–	0.120	–	na	na	–
4	Modesto Irrigation District gage rooftop at Modesto	37383412100060I	–	0.187	–	na	na	–
4	Modesto Irrigation District gage rooftop at Modesto	37383412100060I	–	0.132	–	na	na	–

Appendix 3. Calculated flux values presented as micrograms per square meter per day (µg/m²/day) for wet-deposition samples with at least one detection collected during the 2002–04 study period at eight sites in the Central Valley, California.—Continued

[All compounds were analyzed by gas chromatography/mass spectrometry (GCMS) at the U.S. Geological Survey (USGS) National Water Quality Laboratory (NWQL). The Chemical Abstract Service Number is given below each compound name in brackets. The five-digit number in parenthesis is a code used by the USGS to uniquely identify the given compound. Flux calculated as the analytical result in µg/L multiplied by the total sample volume collected in liters divided by the area of the respective sampler type (Funnel=0.0731 m², Autosampler= 0.0614 m²) and sample composite time in days. Abbreviations: E, estimated reported laboratory value; na, not applicable; (E), estimated compound due to known poor performance by applied method (Childress and others, 1999; Sandstrom and others, 2001); hh:mm, hour:minute; –, compound not detected; [-], no Chemical Abstract Service Number available for given compound]

Site number	Site name	Site identification number	Paraoxon-methyl (E) [950-35-6] (61664)	Pendimethalin [40487-42-1] (82683)	cis-Permethrin [54774-45-7] (82687)	cis-Propiconazole [-] (79846)	trans-Propiconazole [-] (79847)	Phosmet [732-11-6] (61601)
4	Modesto Irrigation District gage rooftop at Modesto	37383412100601	–	0.961	–	na	na	–
4	Modesto Irrigation District gage rooftop at Modesto	37383412100601	–	0.450	–	na	na	–
4	Modesto Irrigation District gage rooftop at Modesto	37383412100601	–	0.071	–	na	na	–
4	Modesto Irrigation District gage rooftop at Modesto	37383412100601	–	0.094	–	na	na	–
4	Modesto Irrigation District gage rooftop at Modesto	37383412100601	–	0.102	–	na	na	–
4	Modesto Irrigation District gage rooftop at Modesto	37383412100601	E0.073	E0.059	–	na	na	E0.073
4	Modesto Irrigation District gage rooftop at Modesto	37383412100601	–	–	–	na	na	–
4	Modesto Irrigation District gage rooftop at Modesto	37383412100601	–	0.055	–	na	na	–
4	Modesto Irrigation District gage rooftop at Modesto	37383412100601	–	1.01	–	na	na	–
4	Modesto Irrigation District gage rooftop at Modesto	37383412100601	–	0.331	–	na	na	–
4	Modesto Irrigation District gage rooftop at Modesto	37383412100601	–	E0.107	–	na	na	–
4	Modesto Irrigation District gage rooftop at Modesto	37383412100601	–	E1.11	–	na	na	–
4	Modesto Irrigation District gage rooftop at Modesto	37383412100601	–	E0.8	–	na	na	–
4	Modesto Irrigation District gage rooftop at Modesto	37383412100601	–	0.073	–	na	na	–
4	Modesto Irrigation District gage rooftop at Modesto	37383412100601	–	E0.436	–	na	na	–
4	Modesto Irrigation District gage rooftop at Modesto	37383412100601	–	0.094	–	na	na	na
5	Modesto Irrigation District gage rooftop at Albers Road	37384112050801	–	0.199	–	–	–	–
5	Modesto Irrigation District gage rooftop at Albers Road	37384112050801	–	0.145	–	–	–	–
5	Modesto Irrigation District gage rooftop at Albers Road	37384112050801	–	0.897	–	–	–	–
5	Modesto Irrigation District gage rooftop at Albers Road	37384112050801	–	–	–	0.193	0.18	–
5	Modesto Irrigation District gage rooftop at Albers Road	37384112050801	–	–	–	E0.062	E0.11	–
5	Modesto Irrigation District gage rooftop at Albers Road	37384112050801	–	0.171	–	0.140	2.20	–
5	Modesto Irrigation District gage rooftop at Albers Road	37384112050801	–	E0.429	–	na	na	–
5	Modesto Irrigation District gage rooftop at Albers Road	37384112050801	–	E0.469	–	na	na	E0.235
5	Modesto Irrigation District gage rooftop at Albers Road	37384112050801	–	0.328	–	na	na	–
5	Modesto Irrigation District gage rooftop at Albers Road	37384112050801	–	0.392	–	na	na	–
5	Modesto Irrigation District gage rooftop at Albers Road	37384112050801	–	0.121	–	na	na	–
5	Modesto Irrigation District gage rooftop at Albers Road	37384112050801	–	0.106	–	na	na	–
5	Modesto Irrigation District gage rooftop at Albers Road	37384112050801	–	0.061	–	na	na	–
5	Modesto Irrigation District gage rooftop at Albers Road	37384112050801	–	0.575	–	na	na	–
5	Modesto Irrigation District gage rooftop at Albers Road	37384112050801	–	0.378	–	na	na	–

Appendix 3.   Calculated flux values presented as micrograms per square meter per day (µg/m²/day) for wet-deposition samples with at least one detection collected during the 2002–04 study period at eight sites in the Central Valley, California.—Continued

[All compounds were analyzed by gas chromatography/mass spectrometry (GCMS) at the U.S. Geological Survey (USGS) National Water Quality Laboratory (NWQL). The Chemical Abstract Service Number is given below each compound name in brackets. The five-digit number in parenthesis is a code used by the USGS to uniquely identify the given compound. Flux calculated as the analytical result in µg/L multiplied by the total sample volume collected in liters divided by the area of the respective sampler type (Funnel=0.0731 m², Autosampler= 0.0614 m²) and sample composite time in days. Abbreviations: E, estimated reported laboratory value; na, not applicable; (E), estimated compound due to known poor performance by applied method (Childress and others, 1999; Sandstrom and others, 2001); hh:mm, hour:minute; —, compound not detected; [-], no Chemical Abstract Service Number available for given compound]

Site number	Site name	Site identification number	Paraoxon-methyl (E) [950-35-6] (61664)	Pendimethalin [40487-42-1] (82683)	cis-Permethrin [54774-45-7] (82687)	cis-Propi-conazole [-] (79846)	trans-Propiconazole [-] (79847)	Phosmet [732-11-6] (61601)
5	Modesto Irrigation District gage rooftop at Albers Road	37384112050480I	—	0.058	—	na	na	—
5	Modesto Irrigation District gage rooftop at Albers Road	37384112050480I	na	E0.057	—	na	na	—
5	Modesto Irrigation District gage rooftop at Albers Road	37384112050480I	na	E0.049	—	na	na	—
5	Modesto Irrigation District gage rooftop at Albers Road	37384112050480I	—	0.639	—	na	na	—
5	Modesto Irrigation District gage rooftop at Albers Road	37384112050480I	—	0.971	—	na	na	—
5	Modesto Irrigation District gage rooftop at Albers Road	37384112050480I	—	0.210	—	na	na	—
5	Modesto Irrigation District gage rooftop at Albers Road	37384112050480I	—	E0.087	—	na	na	E0.053
5	Modesto Irrigation District gage rooftop at Albers Road	37384112050480I	—	0.104	—	na	na	E0.049
5	Modesto Irrigation District gage rooftop at Albers Road	37384112050480I	—	—	—	na	na	—
5	Modesto Irrigation District gage rooftop at Albers Road	37384112050480I	—	E0.031	—	na	na	E0.053
5	Modesto Irrigation District gage rooftop at Albers Road	37384112050480I	—	—	—	na	na	E0.033
5	Modesto Irrigation District gage rooftop at Albers Road	37384112050480I	—	0.182	—	na	na	—
5	Modesto Irrigation District gage rooftop at Albers Road	37384112050480I	—	E0.104	—	na	na	—
5	Modesto Irrigation District gage rooftop at Albers Road	37384112050480I	—	0.694	—	na	na	—
5	Modesto Irrigation District gage rooftop at Albers Road	37384112050480I	—	0.255	—	na	na	—
5	Modesto Irrigation District gage rooftop at Albers Road	37384112050480I	—	E0.018	—	na	na	—
5	Modesto Irrigation District gage rooftop at Albers Road	37384112050480I	—	E0.577	E0.03	na	na	—
5	Modesto Irrigation District gage rooftop at Albers Road	37384112050480I	—	E0.338	E0.02	na	na	—
5	Modesto Irrigation District gage rooftop at Albers Road	37384112050480I	—	E0.321	—	na	na	—
5	Modesto Irrigation District gage rooftop at Albers Road	37384112050480I	—	0.094	—	na	na	—
5	Modesto Irrigation District gage rooftop at Albers Road	37384112050480I	—	E0.260	—	na	na	—
5	Modesto Irrigation District gage rooftop at Albers Road	37384112050480I	—	0.561	—	na	na	na
16	Gridley High School precipitation gage at Gridley	39220512141020I	—	1.40	—	na	na	—
16	Gridley High School precipitation gage at Gridley	39220512141020I	—	0.880	—	na	na	—
16	Gridley High School precipitation gage at Gridley	39220512141020I	na	0.549	—	na	na	—
16	Gridley High School precipitation gage at Gridley	39220512141020I	na	—	—	na	na	—
16	Gridley High School precipitation gage at Gridley	39220512141020I	—	E0.055	—	na	na	—
16	Gridley High School precipitation gage at Gridley	39220512141020I	—	0.267	—	na	na	—
16	Gridley High School precipitation gage at Gridley	39220512141020I	—	E0.063	—	na	na	—
16	Gridley High School precipitation gage at Gridley	39220512141020I	—	E0.169	—	na	na	—
16	Gridley High School precipitation gage at Gridley	39220512141020I	—	—	—	na	na	—

Appendix 3. Calculated flux values presented as micrograms per square meter per day (μg/m²/day) for wet-deposition samples with at least one detection collected during the 2002–04 study period at eight sites in the Central Valley, California.—Continued

[All compounds were analyzed by gas chromatography/mass spectrometry (GCMS) at the U.S. Geological Survey (USGS) National Water Quality Laboratory (NWQL). The Chemical Abstract Service Number is given below each compound name in brackets. The five-digit number in parenthesis is a code used by the USGS to uniquely identify the given compound. Flux calculated as the analytical result in μg/L multiplied by the total sample volume collected in liters divided by the area of the respective sampler type (Funnel=0.0731 m², Autosampler= 0.0614 m²) and sample composite time in days. Abbreviations: E, estimated reported laboratory value; na, not applicable; (E), estimated compound due to known poor performance by applied method (Childress and others, 1999; Sandstrom and others, 2001); hh:mm, hour:minute; —, compound not detected; [-], no Chemical Abstract Service Number available for given compound]

Site number	Site name	Site identification number	Paraoxon-methyl (E) [950-35-6] (61664)	Pendimethalin [40487-42-1] (82683)	cis-Permethrin [54774-45-7] (82687)	cis-Propi-conazole [-] (79846)	trans-Propiconazole [-] (79847)	Phosmet [732-11-6] (61601)
16	Gridley High School precipitation gage at Gridley	39220512141020 1	E0.111	—	—	na	na	E0.078
16	Gridley High School precipitation gage at Gridley	39220512141020 1	—	—	—	na	na	—
16	Gridley High School precipitation gage at Gridley	39220512141020 1	—	—	—	na	na	na
16	Gridley High School precipitation gage at Gridley	39220512141020 1	—	E0.137	—	na	na	—
16	Gridley High School precipitation gage at Gridley	39220512141020 1	—	E0.062	—	na	na	—
16	Gridley High School precipitation gage at Gridley	39220512141020 1	—	0.141	—	na	na	—
16	Gridley High School precipitation gage at Gridley	39220512141020 1	—	0.212	—	na	na	—
16	Gridley High School precipitation gage at Gridley	39220512141020 1	—	E0.200	—	na	na	—
16	Gridley High School precipitation gage at Gridley	39220512141020 1	—	0.205	—	na	na	—
16	Gridley High School precipitation gage at Gridley	39220512141020 1	—	0.140	—	na	na	—
16	Gridley High School precipitation gage at Gridley	39220512141020 1	—	0.089	—	na	na	—
16	Gridley High School precipitation gage at Gridley	39220512141020 1	—	E0.103	—	na	na	—
16	Gridley High School precipitation gage at Gridley	39220512141020 1	—	E0.064	—	na	na	—
16	Gridley High School precipitation gage at Gridley	39220512141020 1	—	0.419	—	na	na	—
16	Gridley High School precipitation gage at Gridley	39220512141020 1	—	E0.488	—	na	na	—
16	Gridley High School precipitation gage at Gridley	39220512141020 1	—	E0.126	—	na	na	—
15	Oroville Dam precipitation gage at spillway	39323412129270 1	—	E0.260	—	na	na	—
15	Oroville Dam precipitation gage at spillway	39323412129270 1	—	—	—	na	na	—
15	Oroville Dam precipitation gage at spillway	39323412129270 1	—	—	—	na	na	—
15	Oroville Dam precipitation gage at spillway	39323412129270 1	na	0.179	—	na	na	—
15	Oroville Dam precipitation gage at spillway	39323412129270 1	—	0.254	—	na	na	—
15	Oroville Dam precipitation gage at spillway	39323412129270 1	—	—	—	na	na	—
15	Oroville Dam precipitation gage at spillway	39323412129270 1	—	E0.070	—	na	na	—
15	Oroville Dam precipitation gage at spillway	39323412129270 1	—	0.026	—	na	na	—
15	Oroville Dam precipitation gage at spillway	39323412129270 1	—	E0.104	—	na	na	—
15	Oroville Dam precipitation gage at spillway	39323412129270 1	—	—	0.04	na	na	—
15	Oroville Dam precipitation gage at spillway	39323412129270 1	—	E0.069	—	na	na	—
15	Oroville Dam precipitation gage at spillway	39323412129270 1	—	E0.278	—	na	na	—
15	Oroville Dam precipitation gage at spillway	39323412129270 1	—	E0.091	—	na	na	—

Appendix 3. Calculated flux values presented as micrograms per square meter per day (µg/m²/day) for wet-deposition samples with at least one detection collected during the 2002–04 study period at eight sites in the Central Valley, California.—Continued

[All compounds were analyzed by gas chromatography/mass spectrometry (GCMS) at the U.S. Geological Survey (USGS) National Water Quality Laboratory (NWQL). The Chemical Abstract Service Number is given below each compound name in brackets. The five-digit number in parenthesis is a code used by the USGS to uniquely identify the given compound. Flux calculated as the analytical result in µg/L multiplied by the total sample volume collected in liters divided by the area of the respective sampler type (Funnel=0.0731 m², Autosampler= 0.0614 m²) and sample composite time in days. Abbreviations: E, estimated reported laboratory value; na, not applicable; (E), estimated compound due to known poor performance by applied method (Childress and others, 1999; Sandstrom and others, 2001); hh:mm, hour:minute; –, compound not detected; [-], no Chemical Abstract Service Number available for given compound]

Site number	Site name	Site identification number	Phosmet oxon [3735-33-9] (61668)	Prometryn [7287-19-6] (04036)	Pronamide [23950-58-5] (82676)	Simazine [122-34-9] (04035)	Terbufos-oxon-sulfone [56070-15-6] (61674)	Trifluralin [1582-09-8] (82661)
11	Newman rain gage at wasteway levee near Draper Road	37173512103120l	–	0.053	–	0.482	–	–
11	Newman rain gage at wasteway levee near Draper Road	37173512103120l	–	–	–	0.104	–	0.049
11	Newman rain gage at wasteway levee near Draper Road	37173512103120l	–	–	–	0.217	E0.06	29.0
11	Newman rain gage at wasteway levee near Draper Road	37173512103120l	–	–	–	0.045	–	0.429
11	Newman rain gage at wasteway levee near Draper Road	37173512103120l	–	–	–	0.082	–	0.753
11	Newman rain gage at wasteway levee near Draper Road	37173512103120l	–	0.193	0.145	0.530	–	0.578
11	Newman rain gage at wasteway levee near Draper Road	37173512103120l	–	1.407	0.073	2.57	–	0.152
11	Newman rain gage at wasteway levee near Draper Road	37173512103120l	–	0.096	–	0.055	–	0.075
11	Newman rain gage at wasteway levee near Draper Road	37173512103120l	–	0.155	0.143	0.274	–	0.310
11	Newman rain gage at wasteway levee near Draper Road	37173512103120l	–	–	–	0.881	–	0.184
11	Newman rain gage at wasteway levee near Draper Road	37173512103120l	na	–	–	0.092	–	0.026
11	Newman rain gage at wasteway levee near Draper Road	37173512103120l	–	0.101	0.025	0.044	–	0.060
11	Newman rain gage at wasteway levee near Draper Road	37173512103120l	–	–	–	0.342	–	E0.096
11	Newman rain gage at wasteway levee near Draper Road	37173512103120l	E0.08	E0.042	0.042	1.84	–	0.195
11	Newman rain gage at wasteway levee near Draper Road	37173512103120l	–	–	0.019	0.055	–	0.011
11	Newman rain gage at wasteway levee near Draper Road	37173512103120l	–	E0.030	–	0.221	–	0.271
12	Turlock rain gage near Idaho Road	37271312053490l	–	0.043	–	0.324	–	E0.023
12	Turlock rain gage near Idaho Road	37271312053490l	–	0.020	–	0.193	–	–
12	Turlock rain gage near Idaho Road	37271312053490l	–	0.030	–	0.270	–	–
12	Turlock rain gage near Idaho Road	37271312053490l	–	–	–	0.351	–	0.043
12	Turlock rain gage near Idaho Road	37271312053490l	–	–	–	–	–	0.052
12	Turlock rain gage near Idaho Road	37271312053490l	–	–	–	–	–	E0.053
12	Turlock rain gage near Idaho Road	37271312053490l	–	–	–	0.242	E0.127	–
12	Turlock rain gage near Idaho Road	37271312053490l	–	0.285	–	0.209	–	E0.152
12	Turlock rain gage near Idaho Road	37271312053490l	–	0.581	0.059	0.256	–	E0.079
12	Turlock rain gage near Idaho Road	37271312053490l	–	0.086	–	0.057	–	E0.029
12	Turlock rain gage near Idaho Road	37271312053490l	–	0.090	E0.072	0.397	–	E0.09
12	Turlock rain gage near Idaho Road	37271312053490l	na	–	–	0.112	na	–
12	Turlock rain gage near Idaho Road	37271312053490l	–	–	–	0.226	–	0.080
12	Turlock rain gage near Idaho Road	37271312053490l	–	–	–	0.205	–	–
12	Turlock rain gage near Idaho Road	37271312053490l	–	–	–	0.078	–	–
12	Turlock rain gage near Idaho Road	37271312053490l	na	0.113	0.025	0.047	–	0.025

Appendix 3. Calculated flux values presented as micrograms per square meter per day ($\mu g/m^2/day$) for wet-deposition samples with at least one detection collected during the 2002–04 study period at eight sites in the Central Valley, California.—Continued

[All compounds were analyzed by gas chromatography/mass spectrometry (GCMS) at the U.S. Geological Survey (USGS) National Water Quality Laboratory (NWQL). The Chemical Abstract Service Number is given below each compound name in brackets. The five-digit number in parentheses is a code used by the USGS to uniquely identify the given compound. Flux calculated as the analytical result in $\mu g/L$ multiplied by the total sample volume collected in liters divided by the area of the respective sampler type (Funnel=0.0731 m^2, Autosampler= 0.0614 m^2) and sample composite time in days. Abbreviations: E, estimated reported laboratory value; na, not applicable; (E), estimated compound due to known poor performance by applied method (Childress and others, 1999; Sandstrom and others, 2001); hh:mm, hour:minute; —, compound not detected; [-], no Chemical Abstract Service Number available for given compound]

Site number	Site name	Site identification number	Phosmet oxon [3735-33-9] (61668)	Prometryn [7287-19-6] (04036)	Pronamide [23950-58-5] (82676)	Simazine [122-34-9] (04035)	Terbufos-oxon-sulfone [56070-15-6] (61674)	Trifluralin [1582-09-8] (82661)
12	Turlock rain gage near Idaho Road	37271312120534901	—	—	—	1.27	—	E0.165
12	Turlock rain gage near Idaho Road	37271312120534901	—	—	—	1.18	—	0.077
12	Turlock rain gage near Idaho Road	37271312120534901	—	—	—	0.064	—	E0.002
12	Turlock rain gage near Idaho Road	37271312120534901	—	—	—	0.292	—	0.053
12	Turlock rain gage near Idaho Road	37271312120534901	na	—	0.057	0.319	—	0.072
13	Turlock Airport rain gage	372857120414001	—	—	—	0.456	—	E0.045
13	Turlock Airport rain gage	372857120414001	—	—	—	0.442	—	—
13	Turlock Airport rain gage	372857120414001	—	0.038	—	0.256	—	0.043
13	Turlock Airport rain gage	372857120414001	—	—	—	0.845	—	—
13	Turlock Airport rain gage	372857120414001	—	—	—	0.199	—	—
13	Turlock Airport rain gage	372857120414001	—	—	—	0.074	—	—
13	Turlock Airport rain gage	372857120414001	—	—	—	0.167	—	—
13	Turlock Airport rain gage	372857120414001	—	0.197	0.141	0.902	—	E0.197
13	Turlock Airport rain gage	372857120414001	—	0.287	0.074	0.584	—	E0.064
13	Turlock Airport rain gage	372857120414001	—	0.048	—	0.075	—	—
13	Turlock Airport rain gage	372857120414001	—	—	—	0.427	—	E0.082
13	Turlock Airport rain gage	372857120414001	—	—	—	0.536	—	E0.016
13	Turlock Airport rain gage	372857120414001	—	—	—	0.123	—	0.029
13	Turlock Airport rain gage	372857120414001	—	—	—	48.9	—	0.073
13	Turlock Airport rain gage	372857120414001	—	—	—	1.59	—	E0.043
13	Turlock Airport rain gage	372857120414001	—	—	—	0.706	—	E0.039
13	Turlock Airport rain gage	372857120414001	na	0.071	0.027	0.188	—	E0.017
13	Turlock Airport rain gage	372857120414001	—	—	—	1.59	—	E0.076
13	Turlock Airport rain gage	372857120414001	—	—	—	0.180	—	E0.004
13	Turlock Airport rain gage	372857120414001	—	—	—	0.456	—	E0.03
13	Turlock Airport rain gage	372857120414001	—	—	0.081	1.06	—	0.151
10	Westley rain gage at pump building near lateral 6 North	373335121143001	—	0.054	—	0.076	—	0.020
10	Westley rain gage at pump building near lateral 6 North	373335121143001	—	0.026	0.021	0.239	—	0.104
10	Westley rain gage at pump building near lateral 6 North	373335121143001	—	—	—	0.157	—	—
10	Westley rain gage at pump building near lateral 6 North	373335121143001	—	—	—	0.123	—	—
10	Westley rain gage at pump building near lateral 6 North	373335121143001	—	—	—	0.031	—	E0.041

Appendix 3. Calculated flux values presented as micrograms per square meter per day (µg/m²/day) for wet-deposition samples with at least one detection collected during the 2002–04 study period at eight sites in the Central Valley, California.—Continued

[All compounds were analyzed by gas chromatography mass spectrometry (GCMS) at the U.S. Geological Survey (USGS) National Water Quality Laboratory (NWQL). The Chemical Abstract Service Number is given below each compound name in brackets. The five-digit number in parentheses is a code used by the USGS to uniquely identify the given compound. Flux calculated as the analytical result in µg/L multiplied by the total sample volume collected in liters divided by the area of the respective sampler type (Funnel=0.0731 m², Autosampler= 0.0614 m²) and sample composite time in days. Abbreviations: E, estimated reported laboratory value; na, not applicable; (E), estimated compound due to known poor performance by applied method (Childress and others, 1999; Sandstrom and others, 2001); hh:mm, hour:minute; –, compound not detected; [–], no Chemical Abstract Service Number available for given compound]

Site number	Site name	Site identification number	Phosmet oxon [3735-33-9] (61668)	Prometryn [7287-19-6] (04036)	Pronamide [23950-58-5] (82676)	Simazine [122-34-9] (04035)	Terbufos-oxon-sulfone [56070-15-6] (61674)	Trifluralin [1582-09-8] (82661)
10	Westley rain gage at pump building near lateral 6 North	373335121143001	–	0.059	–	0.143	–	0.059
10	Westley rain gage at pump building near lateral 6 North	373335121143001	–	0.211	0.169	0.443	–	0.358
10	Westley rain gage at pump building near lateral 6 North	373335121143001	–	0.756	0.089	0.330	–	0.113
10	Westley rain gage at pump building near lateral 6 North	373335121143001	–	0.050	–	0.079	–	E0.033
10	Westley rain gage at pump building near lateral 6 North	373335121143001	–	0.112	0.112	0.270	–	0.093
10	Westley rain gage at pump building near lateral 6 North	373335121143001	–	–	–	0.210	–	0.033
10	Westley rain gage at pump building near lateral 6 North	373335121143001	–	0.048	–	0.257	–	0.120
10	Westley rain gage at pump building near lateral 6 North	373335121143001	na	0.130	0.078	0.104	–	0.338
10	Westley rain gage at pump building near lateral 6 North	373335121143001	–	0.015	0.030	0.059	–	E0.002
10	Westley rain gage at pump building near lateral 6 North	373335121143001	–	–	–	0.057	–	0.218
4	Modesto Irrigation District gage rooftop at Modesto	373834121000601	–	–	–	0.201	–	0.019
4	Modesto Irrigation District gage rooftop at Modesto	373834121000601	–	–	–	0.314	–	–
4	Modesto Irrigation District gage rooftop at Modesto	373834121000601	–	–	–	–	–	–
4	Modesto Irrigation District gage rooftop at Modesto	373834121000601	na	na	–	0.051	na	E0.033
4	Modesto Irrigation District gage rooftop at Modesto	373834121000601	–	–	–	0.240	–	E0.020
4	Modesto Irrigation District gage rooftop at Modesto	373834121000601	–	–	–	0.044	–	0.080
4	Modesto Irrigation District gage rooftop at Modesto	373834121000601	–	0.253	–	0.281	–	–
4	Modesto Irrigation District gage rooftop at Modesto	373834121000601	–	0.166	–	0.469	–	0.253
4	Modesto Irrigation District gage rooftop at Modesto	373834121000601	–	0.438	0.066	0.219	–	E0.166
4	Modesto Irrigation District gage rooftop at Modesto	373834121000601	–	0.563	E0.047	0.141	–	0.099
4	Modesto Irrigation District gage rooftop at Modesto	373834121000601	–	0.052	0.030	0.039	–	0.07
4	Modesto Irrigation District gage rooftop at Modesto	373834121000601	–	0.058	–	0.107	–	E0.035
4	Modesto Irrigation District gage rooftop at Modesto	373834121000601	–	0.028	–	0.274	–	E0.011
4	Modesto Irrigation District gage rooftop at Modesto	373834121000601	–	–	–	0.269	–	E0.075
4	Modesto Irrigation District gage rooftop at Modesto	373834121000601	–	–	–	0.201	–	E0.064
4	Modesto Irrigation District gage rooftop at Modesto	373834121000601	–	–	–	0.110	–	0.020
4	Modesto Irrigation District gage rooftop at Modesto	373834121000601	–	–	–	0.073	–	0.031
4	Modesto Irrigation District gage rooftop at Modesto	373834121000601	–	–	–	0.378	–	0.029
4	Modesto Irrigation District gage rooftop at Modesto	373834121000601	–	–	–	0.225	–	0.112
4	Modesto Irrigation District gage rooftop at Modesto	373834121000601	–	–	–	0.060	–	E0.049

Appendix 3. Calculated flux values presented as micrograms per square meter per day (μg/m²/day) for wet-deposition samples with at least one detection collected during the 2002–04 study period at eight sites in the Central Valley, California.—Continued

[All compounds were analyzed by gas chromatography/mass spectrometry (GCMS) at the U.S. Geological Survey (USGS) National Water Quality Laboratory (NWQL). The Chemical Abstract Service Number is given below each compound name in brackets. The five-digit number in parenthesis is a code used by the USGS to uniquely identify the given compound. Flux calculated as the analytical result in μg/L multiplied by the total sample volume collected in liters divided by the area of the respective sampler type (Funnel=0.0731 m², Autosampler= 0.0614 m²) and sample composite time in days. Abbreviations: E, estimated reported laboratory value; na, not applicable; (E), estimated compound due to known poor performance by applied method (Childress and others, 1999; Sandstrom and others, 2001); hh:mm, hour:minute; –, compound not detected; [-], no Chemical Abstract Service Number available for given compound]

Site number	Site name	Site identification number	Phosmet oxon [3735-33-9] (61668)	Prometryn [7287-19-6] (04036)	Pronamide [23950-58-5] (82676)	Simazine [122-34-9] (04035)	Terbufos-oxon-sulfone [56070-15-6] (61674)	Trifluralin [1582-09-8] (82661)
4	Modesto Irrigation District gage rooftop at Modesto	373834121000601	–	–	–	0.057	–	–
4	Modesto Irrigation District gage rooftop at Modesto	373834121000601	–	–	–	0.066	–	0.028
4	Modesto Irrigation District gage rooftop at Modesto	373834121000601	na	0.125	–	0.172	–	–
4	Modesto Irrigation District gage rooftop at Modesto	373834121000601	–	0.052	–	0.045	–	–
4	Modesto Irrigation District gage rooftop at Modesto	373834121000601	–	0.098	–	0.033	–	–
4	Modesto Irrigation District gage rooftop at Modesto	373834121000601	–	0.500	0.167	1.06	–	E0.077
4	Modesto Irrigation District gage rooftop at Modesto	373834121000601	–	E0.015	–	0.108	–	E0.031
4	Modesto Irrigation District gage rooftop at Modesto	373834121000601	–	–	–	–	–	E0.015
4	Modesto Irrigation District gage rooftop at Modesto	373834121000601	–	E0.038	–	1.42	–	E0.077
4	Modesto Irrigation District gage rooftop at Modesto	373834121000601	–	–	–	1.047	–	E0.062
4	Modesto Irrigation District gage rooftop at Modesto	373834121000601	–	–	–	0.097	–	E0.005
4	Modesto Irrigation District gage rooftop at Modesto	373834121000601	–	–	–	0.382	–	E0.065
4	Modesto Irrigation District gage rooftop at Modesto	373834121000601	na	–	–	0.805	–	–
5	Modesto Irrigation District gage rooftop at Albers Road	373841120504801	–	–	–	0.225	–	E0.039
5	Modesto Irrigation District gage rooftop at Albers Road	373841120504801	–	–	–	0.196	–	–
5	Modesto Irrigation District gage rooftop at Albers Road	373841120504801	–	–	E0.025	0.623	–	–
5	Modesto Irrigation District gage rooftop at Albers Road	373841120504801	–	–	–	0.743	–	–
5	Modesto Irrigation District gage rooftop at Albers Road	373841120504801	–	–	–	0.398	–	0.031
5	Modesto Irrigation District gage rooftop at Albers Road	373841120504801	–	–	–	0.641	–	0.055
5	Modesto Irrigation District gage rooftop at Albers Road	373841120504801	–	0.143	–	0.486	–	E0.143
5	Modesto Irrigation District gage rooftop at Albers Road	373841120504801	–	E0.147	–	1.03	–	E0.147
5	Modesto Irrigation District gage rooftop at Albers Road	373841120504801	–	0.208	0.044	0.405	–	E0.055
5	Modesto Irrigation District gage rooftop at Albers Road	373841120504801	–	0.201	E0.042	0.328	–	E0.053
5	Modesto Irrigation District gage rooftop at Albers Road	373841120504801	–	0.056	–	0.052	–	–
5	Modesto Irrigation District gage rooftop at Albers Road	373841120504801	–	0.051	0.043	0.039	–	–
5	Modesto Irrigation District gage rooftop at Albers Road	373841120504801	–	0.023	0.013	0.067	–	E0.011
5	Modesto Irrigation District gage rooftop at Albers Road	373841120504801	–	–	E0.049	0.246	–	E0.066
5	Modesto Irrigation District gage rooftop at Albers Road	373841120504801	–	–	–	0.098	–	E0.07
5	Modesto Irrigation District gage rooftop at Albers Road	373841120504801	–	–	–	0.055	–	0.012
5	Modesto Irrigation District gage rooftop at Albers Road	373841120504801	na	–	–	0.192	na	–
5	Modesto Irrigation District gage rooftop at Albers Road	373841120504801	na	–	–	0.058	na	–
5	Modesto Irrigation District gage rooftop at Albers Road	373841120504801	–	–	–	1.40	–	E0.044

Appendix 3. Calculated flux values presented as micrograms per square meter per day (µg/m²/day) for wet-deposition samples with at least one detection collected during the 2002–04 study period at eight sites in the Central Valley, California.—Continued

[All compounds were analyzed by gas chromatography/mass spectrometry (GCMS) at the U.S. Geological Survey (USGS) National Water Quality Laboratory (NWQL). The Chemical Abstract Service Number is given below each compound name in brackets. The five-digit number in parenthesis is a code used by the USGS to uniquely identify the given compound. Flux calculated as the analytical result in µg/L multiplied by the total sample volume collected in liters divided by the area of the respective sampler type (Funnel=0.0731 m². Autosampler= 0.0614 m²) and sample composite time in days. Abbreviations: E, estimated compound due to known poor performance by applied method (Childress and others, 1999; Sandstrom and others, 2001); hh:mm, hour:minute; –, compound not detected; [-], no Chemical Abstract Service Number available for given compound]

Site number	Site name	Site identification number	Phosmet oxon [3735-33-9] (61668)	Prometryn [7287-19-6] (04036)	Pronamide [23950-58-5] (82676)	Simazine [122-34-9] (04035)	Terbufos-oxon-sulfone [56070-15-6] (61674)	Trifluralin [1582-09-8] (82661)
5	Modesto Irrigation District gage rooftop at Albers Road	373841120504801	–	–	–	0.455	–	E0.046
5	Modesto Irrigation District gage rooftop at Albers Road	373841120504801	–	–	–	0.141	–	–
5	Modesto Irrigation District gage rooftop at Albers Road	373841120504801	–	–	0.040	0.147	–	E0.033
5	Modesto Irrigation District gage rooftop at Albers Road	373841120504801	–	–	–	0.489	–	–
5	Modesto Irrigation District gage rooftop at Albers Road	373841120504801	–	–	–	0.118	–	–
5	Modesto Irrigation District gage rooftop at Albers Road	373841120504801	na	0.118	–	0.081	–	E0.011
5	Modesto Irrigation District gage rooftop at Albers Road	373841120504801	na	0.037	0.020	0.043	–	E0.020
5	Modesto Irrigation District gage rooftop at Albers Road	373841120504801	–	0.078	–	0.172	–	–
5	Modesto Irrigation District gage rooftop at Albers Road	373841120504801	–	0.060	–	0.104	–	–
5	Modesto Irrigation District gage rooftop at Albers Road	373841120504801	–	0.268	–	27.2	–	0.158
5	Modesto Irrigation District gage rooftop at Albers Road	373841120504801	–	–	–	0.309	–	–
5	Modesto Irrigation District gage rooftop at Albers Road	373841120504801	–	–	–	0.023	–	–
5	Modesto Irrigation District gage rooftop at Albers Road	373841120504801	–	–	–	2.31	–	E0.134
5	Modesto Irrigation District gage rooftop at Albers Road	373841120504801	–	–	–	0.518	–	–
5	Modesto Irrigation District gage rooftop at Albers Road	373841120504801	E0.047	–	–	1.01	–	0.051
5	Modesto Irrigation District gage rooftop at Albers Road	373841120504801	–	–	–	0.385	–	E0.002
5	Modesto Irrigation District gage rooftop at Albers Road	373841120504801	–	–	–	0.356	–	E0.041
5	Modesto Irrigation District gage rooftop at Albers Road	373841120504801	na	–	–	1.98	–	E0.045
16	Gridley High School precipitation gage at Gridley	392205121410201	–	–	–	0.794	–	0.958
16	Gridley High School precipitation gage at Gridley	392205121410201	–	–	–	1.43	–	E0.228
16	Gridley High School precipitation gage at Gridley	392205121410201	na	–	–	–	na	E0.093
16	Gridley High School precipitation gage at Gridley	392205121410201	na	–	–	0.069	na	E0.025
16	Gridley High School precipitation gage at Gridley	392205121410201	–	–	–	0.299	–	E0.019
16	Gridley High School precipitation gage at Gridley	392205121410201	–	–	–	0.089	–	0.105
16	Gridley High School precipitation gage at Gridley	392205121410201	–	–	–	–	–	0.058
16	Gridley High School precipitation gage at Gridley	392205121410201	–	–	–	–	–	E0.034
16	Gridley High School precipitation gage at Gridley	392205121410201	na	–	–	0.067	–	–
16	Gridley High School precipitation gage at Gridley	392205121410201	na	–	–	0.081	–	–
16	Gridley High School precipitation gage at Gridley	392205121410201	na	0.041	–	0.150	–	–
16	Gridley High School precipitation gage at Gridley	392205121410201	–	0.030	–	0.055	–	–
16	Gridley High School precipitation gage at Gridley	392205121410201	–	0.046	–	0.068	–	–

Appendix 3. Calculated flux values presented as micrograms per square meter per day (µg/m²/day) for wet-deposition samples with at least one detection collected during the 2002–04 study period at eight sites in the Central Valley, California.—Continued

[All compounds were analyzed by gas chromatography/mass spectrometry (GCMS) at the U.S. Geological Survey (USGS) National Water Quality Laboratory (NWQL). The Chemical Abstract Service Number is given below each compound name in brackets. The five-digit number in parenthesis is a code used by the USGS to uniquely identify the given compound. Flux calculated as the analytical result in µg/L multiplied by the total sample volume collected in liters divided by the area of the respective sampler type (Funnel=0.0731 m², Autosampler= 0.0614 m²) and sample composite time in days. Abbreviations: E, estimated reported laboratory value; na, not applicable; (E), estimated compound due to known poor performance by applied method (Childress and others, 1999; Sandstrom and others, 2001); hh:mm, hour:minute; -, compound not detected; [-], no Chemical Abstract Service Number available for given compound]

Site number	Site name	Site identification number	Phosmet oxon [3735-33-9] (61668)	Prometryn [7287-19-6] (04036)	Pronamide [23950-58-5] (82676)	Simazine [122-34-9] (04035)	Terbufos-oxon-sulfone [56070-15-6] (61674)	Trifluralin [1582-09-8] (82661)
16	Gridley High School precipitation gage at Gridley	392205121410201	-	0.034	-	0.037	-	-
16	Gridley High School precipitation gage at Gridley	392205121410201	-	0.054	-	0.049	-	-
16	Gridley High School precipitation gage at Gridley	392205121410201	-	-	-	0.106	-	-
16	Gridley High School precipitation gage at Gridley	392205121410201	-	-	-	E0.039	-	-
16	Gridley High School precipitation gage at Gridley	392205121410201	-	-	-	0.079	-	-
16	Gridley High School precipitation gage at Gridley	392205121410201	-	-	-	0.052	-	0.049
16	Gridley High School precipitation gage at Gridley	392205121410201	-	-	-	0.078	-	-
16	Gridley High School precipitation gage at Gridley	392205121410201	-	-	-	0.104	-	E0.016
16	Gridley High School precipitation gage at Gridley	392205121410201	-	-	0.078	0.667	-	E0.064
16	Gridley High School precipitation gage at Gridley	392205121410201	-	-	-	0.212	-	-
16	Gridley High School precipitation gage at Gridley	392205121410201	-	-	-	0.084	-	-
15	Oroville Dam precipitation gage at spillway	393234121292701	-	-	-	E0.055	-	E0.055
15	Oroville Dam precipitation gage at spillway	393234121292701	-	-	-	-	-	-
15	Oroville Dam precipitation gage at spillway	393234121292701	-	-	-	0.018	-	-
15	Oroville Dam precipitation gage at spillway	393234121292701	na	-	-	-	na	E0.024
15	Oroville Dam precipitation gage at spillway	393234121292701	-	-	-	0.164	-	0.082
15	Oroville Dam precipitation gage at spillway	393234121292701	-	-	-	-	-	-
15	Oroville Dam precipitation gage at spillway	393234121292701	-	0.026	-	E0.009	-	-
15	Oroville Dam precipitation gage at spillway	393234121292701	-	-	-	E0.016	-	-
15	Oroville Dam precipitation gage at spillway	393234121292701	-	-	-	E0.017	-	-
15	Oroville Dam precipitation gage at spillway	393234121292701	-	-	-	0.052	-	-
15	Oroville Dam precipitation gage at spillway	393234121292701	-	-	-	0.235	-	-
15	Oroville Dam precipitation gage at spillway	393234121292701	-	-	-	0.149	-	-

References Cited

Childress, C.J.O., Foreman, W.T., Connor, B.F., and Maloney, T.J., 1999, New reporting procedures based on long-term method detection levels and some considerations for interpretations of water-quality data provided by the U.S. Geological Survey National Water Quality Laboratory: U.S. Geological Survey Open-File Report 99–193, 19 p.

Sandstrom, M.W., Stroppel, M.E., Foreman, W.T., and Schroeder, M.P., 2001, Methods of analysis by the U.S. Geological Survey National Water Quality Laboratory—Determination of moderate-use pesticides and selected degradates in water by C-18 solid-phase extraction and capillary-column gas chromatography/mass spectrometry with selected-ion monitoring (methods 2002/2011): U.S. Geological Survey Water-Resources Investigations Report 01–4098, 70 p.

Appendix 4.   Calculated Flux Values Presented in Micrograms Per Square Meter Per Day (μg/m^2/day) for Bulk Wet-Deposition Samples With at Least One Detection Collected During the 2002–04 Study Period at Eight Sites in the Central Valley, California

Appendix 4. Calculated flux values presented in micrograms per square meter per day (µg/m²/day) for bulk wet-deposition samples with at least one detection collected during the 2002–04 study period at eight sites in the Central Valley, California.

[All compounds were analyzed by gas chromatography/mass spectrometry (GC/MS) at the U.S. Geological Survey (USGS) National Water Quality Laboratory (NWQL). The Chemical Abstract Service Number is given below each compound name in parentheses. The five digit number in parenthesis is a code used by the USGS to uniquely identify the given compound. Flux calculated as the analytical result in microgram per liter (µg/L), multiplied by the total sample volume collected in liters divided by the area of the respective sampler type (Funnel=0.0731 square meter [m²], Autosampler=0.061-4 m²) and sample composite time in days. Abbreviations: D-U, value deleted by NWQL because compound could not be determined due to matrix interference; E, estimated reported laboratory value; (E), estimated compound due to known poor performance by applied method (Childress and others, 1999; Sandstrom and others, 2001); hh:mm, hour:minute; [-], no Chemical Abstract Service Number available for given compound; <, less than laboratory reporting limit; –, not detected]

Site number	Site name	Site identification number	Start date	Start time (hh:mm)	End date	End time (hh:mm)	Composite sample time, days	Rain events during composite sample time, days	Sample type	Sampler type	Total sample volume collected, liters
11	Newman rain gage at wasteway levee near Draper Road	371735121031201	Dec 22, 2002	13:00	Jan 8, 2003	14:00	17	7	Bulk-wet	Funnel	0.78
11	Newman rain gage at wasteway levee near Draper Road	371735121031201	Jan 14, 2003	10:10	Feb 14, 2003	08:30	31	6	Bulk-wet	Funnel	0.81
11	Newman rain gage at wasteway levee near Draper Road	371735121031201	Feb 21, 2003	10:15	Mar 13, 2003	10:20	20	2	Bulk-wet	Funnel	1.3
11	Newman rain gage at wasteway levee near Draper Road	371735121031201	Mar 27, 2003	10:20	Apr 8, 2003	09:00	12	3	Bulk-wet	Funnel	0.7
11	Newman rain gage at wasteway levee near Draper Road	371735121031201	Apr 8, 2003	10:00	Apr 29, 2003	09:40	21	4	Bulk-wet	Funnel	0.76
11	Newman rain gage at wasteway levee near Draper Road	371735121031201	Apr 29, 2003	09:40	May 9, 2003	16:00	10	3	Bulk-wet	Funnel	0.71
12	Turlock rain gage near Idaho Road	372713120534901	Dec 22, 2002	13:30	Jan 8, 2003	14:40	17	7	Bulk-wet	Funnel	0.81
12	Turlock rain gage near Idaho Road	372713120534901	Feb 21, 2003	11:00	Mar 13, 2003	11:00	20	2	Bulk-wet	Funnel	0.45
12	Turlock rain gage near Idaho Road	372713120534901	Mar 27, 2003	11:20	Apr 8, 2003	09:40	12	3	Bulk-wet	Funnel	1.04
12	Turlock rain gage near Idaho Road	372713120534901	Apr 8, 2003	10:40	Apr 29, 2003	10:30	21	4	Bulk-wet	Funnel	1.46
13	Turlock Airport rain gage	372857120414001	Dec 22, 2002	14:00	Jan 9, 2003	13:50	18	7	Bulk-wet	Funnel	1.26
13	Turlock Airport rain gage	372857120414001	Feb 21, 2003	11:40	Mar 12, 2003	10:00	19	2	Bulk-wet	Funnel	0.52
13	Turlock Airport rain gage	372857120414001	Mar 18, 2003	11:50	Apr 8, 2003	10:20	21	5	Bulk-wet	Funnel	1.42
13	Turlock Airport rain gage	372857120414001	Apr 8, 2003	11:20	Apr 26, 2003	13:00	18	3	Bulk-wet	Funnel	0.76
13	Turlock Airport rain gage	372857120414001	Feb 4, 2004	10:30	Feb 19, 2004	10:10	15	5	Bulk-wet	Funnel	2.64
10	Westley rain gage at pump building near lateral 6 North	373335121143001	Dec 22, 2002	12:00	Jan 8, 2003	13:15	17	7	Bulk-wet	Funnel	1.05
10	Westley rain gage at pump building near lateral 6 North	373335121143001	Jan 14, 2003	09:20	Feb 15, 2003	11:45	32	6	Bulk-wet	Funnel	1.5
10	Westley rain gage at pump building near lateral 6 North	373335121143001	Mar 27, 2003	09:20	May 2, 2003	09:20	36	7	Bulk-wet	Funnel	1.26
10	Westley rain gage at pump building near lateral 6 North	373335121143001	Feb 4, 2004	13:30	Feb 19, 2004	14:20	15	5	Bulk-wet	Funnel	0.98

Appendix 4. Calculated flux values presented in micrograms per square meter per day (µg/m²/day) for bulk wet-deposition samples with at least one detection collected during the 2002–04 study period at eight sites in the Central Valley, California.—Continued

[All compounds were analyzed by gas chromatography/mass spectrometry (GC/MS) at the U.S. Geological Survey (USGS) National Water Quality Laboratory (NWQL). The Chemical Abstract Service Number is given below each compound name in brackets. The five digit number in parenthesis is a code used by the USGS to uniquely identify the given compound. Flux calculated as the analytical result in microgram per liter (µg/L) multiplied by the total sample volume collected in liters divided by the area of the respective sampler type (Funnel=0.0731 square meter [m²], Autosampler=0.0614 m²) and sample composite time in days. Abbreviations: D-U, value deleted by NWQL because compound could not be determined due to matrix interference; E, estimated reported laboratory value; (E), estimated compound due to known poor performance by applied method (Childress and others, 1999; Sandstrom and others, 2001); hh:mm, hour:minute; [-], no Chemical Abstract Service Number available for given compound; <, less than laboratory reporting limit; –, not detected]

Site number	Site name	Site identification number	Start date	Start time (hh:mm)	End date	End time (hh:mm)	Composite sample time, days	Rain events during composite sample time, days	Sample type	Sampler type	Total sample volume collected, liters
4	Modesto Irrigation District gage rooftop at Modesto	373834121000601	Dec 23, 2002	09:30	Jan 9, 2003	15:30	17	6	Bulk-wet	Funnel	1.3
4	Modesto Irrigation District gage rooftop at Modesto	373834121000601	Mar 27, 2003	14:50	Apr 8, 2003	13:10	12	3	Bulk-wet	Funnel	0.86
4	Modesto Irrigation District gage rooftop at Modesto	373834121000601	Apr 8, 2003	14:10	Apr 29, 2003	14:30	21	4	Bulk-wet	Funnel	1.45
4	Modesto Irrigation District gage rooftop at Modesto	373834121000601	Feb 4, 2004	08:00	Feb 19, 2004	08:10	15	5	Bulk-wet	Funnel	2.24
5	Modesto Irrigation District gage rooftop at Albers Road	373841120504801	Dec 23, 2002	10:30	Jan 9, 2003	14:30	17	6	Bulk-wet	Funnel	0.78
5	Modesto Irrigation District gage rooftop at Albers Road	373841120504801	Mar 27, 2003	13:50	Apr 8, 2003	11:50	12	3	Bulk-wet	Funnel	1.41
5	Modesto Irrigation District gage rooftop at Albers Road	373841120504801	Apr 8, 2003	12:50	Apr 29, 2003	13:30	21	4	Bulk-wet	Funnel	2.02
16	Gridley High School precipitation gage at Gridley	392205121410201	Mar 28, 2003	12:10	Apr 9, 2003	13:00	12	1	Bulk-wet	Funnel	0.85
16	Gridley High School precipitation gage at Gridley	392205121410201	Jul 21, 2003	09:00	Aug 6, 2003	11:30	16	1	Bulk-wet	Funnel	1.515
16	Gridley High School precipitation gage at Gridley	392205121410201	Aug 6, 2003	11:30	Aug 29, 2003	09:30	23	2	Bulk-wet	Funnel	0.26
15	Oroville Dam precipitation gage at spillway	393234121292701	Nov 14, 2002	12:00	Dec 19, 2002	11:00	35	8	Bulk-wet	Funnel	2.28
15	Oroville Dam precipitation gage at spillway	393234121292701	Dec 23, 2002	10:30	Jan 7, 2003	10:30	15	6	Bulk-wet	Funnel	4
15	Oroville Dam precipitation gage at spillway	393234121292701	Jan 13, 2003	09:50	Jan 23, 2003	11:30	10	5	Bulk-wet	Funnel	3.4
15	Oroville Dam precipitation gage at spillway	393234121292701	Jan 31, 2003	12:45	Feb 14, 2003	10:30	14	3	Bulk-wet	Funnel	1.68
15	Oroville Dam precipitation gage at spillway	393234121292701	Mar 28, 2003	13:20	Apr 9, 2003	11:30	12	1	Bulk-wet	Funnel	0.75
15	Oroville Dam precipitation gage at spillway	393234121292701	Jul 21, 2003	09:00	Aug 6, 2003	13:00	16	1	Bulk-wet	Funnel	0.69
15	Oroville Dam precipitation gage at spillway	393234121292701	Aug 6, 2003	13:00	Sep 2, 2003	11:15	27	2	Bulk-wet	Funnel	0.94
15	Oroville Dam precipitation gage at spillway	393234121292701	Jan 7, 2004	12:30	Jan 27, 2004	11:15	20	5	Bulk-wet	Funnel	1.01
15	Oroville Dam precipitation gage at spillway	393234121292701	Feb 6, 2004	11:50	Feb 17, 2004	15:00	11	4	Bulk-wet	Funnel	4.27

Appendix 4. Calculated flux values presented in micrograms per square meter per day (µg/m²/day) for bulk wet-deposition samples with at least one detection collected during the 2002–04 study period at eight sites in the Central Valley, California.—Continued

[All compounds were analyzed by gas chromatography/mass spectrometry (GC/MS) at the U.S. Geological Survey (USGS) National Water Quality Laboratory (NWQL). The Chemical Abstract Service Number is given below each compound name in parentheses. The five digit number in parenthesis is a code used by the USGS to uniquely identify the given compound. Flux calculated as the analytical result in microgram per liter (µg/L) multiplied by the total sample volume collected in liters divided by the area of the respective sampler type (Funnel=0.0731 square meter [m²], Autosampler=0.0614 m²) and sample composite time in days. Abbreviations: D-U, value deleted by NWQL because compound could not be determined due to matrix interference; E, estimated reported laboratory value; (E), estimated compound due to known poor performance by applied method (Childress and others, 1999; Sandstrom and others, 2001); hh:mm, hour:minute; [-], no Chemical Abstract Service Number available for given compound; <, less than laboratory reporting limit; –, not detected]

Site num-ber	Site name	Site identification number	1-Naphthol (E) [90-15-3] (49295)	2,6-Diethylaniline (E) [579-66-8] (82660)	3,4-Dichloroaniline (E) [95-76-1] (61625)	4-Chloro-2-methylphe-nol (E) [1570-64-5] (61633)	Alachlor [15972-60-8] (46342)	Atrazine [1912-24-9] (39632)	Azinphos-methyl (E) [86-50-0] (82686)
11	Newman rain gage at wasteway levee near Draper Road	371735121031201	–	–	E0.017	–	–	–	–
11	Newman rain gage at wasteway levee near Draper Road	371735121031201	–	–	E0.131	E0.017	–	–	–
11	Newman rain gage at wasteway levee near Draper Road	371735121031201	–	–	E0.116	–	–	–	–
11	Newman rain gage at wasteway levee near Draper Road	371735121031201	E0.03	–	–	–	–	–	–
11	Newman rain gage at wasteway levee near Draper Road	371735121031201	E0.05	–	–	–	–	–	–
11	Newman rain gage at wasteway levee near Draper Road	371735121031201	E0.002	–	–	–	0.023	–	E0.37
12	Turlock rain gage near Idaho Road	372713120534901	E0.02	–	–	–	–	–	–
12	Turlock rain gage near Idaho Road	372713120534901	–	–	–	–	–	–	–
12	Turlock rain gage near Idaho Road	372713120534901	E0.05	–	–	–	–	–	–
12	Turlock rain gage near Idaho Road	372713120534901	E0.05	–	–	–	–	–	–
13	Turlock Airport rain gage	372857120414001	E0.02	–	E0.017	–	–	–	–
13	Turlock Airport rain gage	372857120414001	–	–	–	–	–	–	–
13	Turlock Airport rain gage	372857120414001	E0.04	–	–	–	–	–	–
13	Turlock Airport rain gage	372857120414001	E0.07	–	–	–	–	–	E0.05
13	Turlock Airport rain gage	372857120414001	E0.04	–	E0.137	E0.014	–	–	–
10	Westley rain gage at pump building near lateral 6 North	373335121143001	E0.02	–	E0.043	–	–	–	–
10	Westley rain gage at pump building near lateral 6 North	373335121143001	E0.03	–	E0.072	E0.017	–	–	–
10	Westley rain gage at pump building near lateral 6 North	373335121143001	E0.12	–	–	–	–	–	E0.79
10	Westley rain gage at pump building near lateral 6 North	373335121143001	E0.01	–	E0.080	E0.003	–	–	–

Appendix 4. Calculated flux values presented in micrograms per square meter per day (µg/m²/day) for bulk wet-deposition samples with at least one detection collected during the 2002–04 study period at eight sites in the Central Valley, California.—Continued

[All compounds were analyzed by gas chromatography/mass spectrometry (GC/MS) at the U.S. Geological Survey (USGS) National Water Quality Laboratory (NWQL). The Chemical Abstract Service Number is given below each compound name in parentheses. The five digit number in brackets is a code used by the USGS to uniquely identify the given compound. Flux calculated as the analytical result in microgram per liter (µg/L) multiplied by the total sample volume collected in liters divided by the area of the respective sampler type (Funnel=0.0731 square meter [m²], Autosampler=0.0614 m²) and sample composite time in days. Abbreviations: D-U, value deleted by NWQL because compound could not be determined due to matrix interference; E, estimated reported laboratory value; (E), estimated compound due to known poor performance by applied method (Childress and others, 1999; Sandstrom and others, 2001); hh:mm, hour:minute; [–], no Chemical Abstract Service Number available for given compound; <, less than laboratory reporting limit; –, not detected]

Site number	Site name	Site identification number	1-Naphthol (E) [90-15-3] (49295)	2,6-Diethylaniline (E) [579-66-8] (82660)	3,4-Dichloroaniline (E) [95-76-1] (61625)	4-Chloro-2-methylphenol (E) [1570-64-5] (61633)	Alachlor [15972-60-8] (46342)	Atrazine [1912-24-9] (39632)	Azinphos-methyl (E) [86-50-0] (82686)
4	Modesto Irrigation District gage rooftop at Modesto	3738341210000601	E0.06	–	E0.033	E0.009	–	–	–
4	Modesto Irrigation District gage rooftop at Modesto	3738341210000601	E0.08	–	–	–	–	–	–
4	Modesto Irrigation District gage rooftop at Modesto	3738341210000601	E0.05	–	E0.055	E0.006	–	–	–
4	Modesto Irrigation District gage rooftop at Modesto	3738341210000601	E0.07	–	–	–	–	–	–
5	Modesto Irrigation District gage rooftop at Albers Road	3738411205048 01	E0.02	–	E0.630	–	–	–	–
5	Modesto Irrigation District gage rooftop at Albers Road	3738411205048 01	E0.06	–	–	–	–	–	–
5	Modesto Irrigation District gage rooftop at Albers Road	3738411205048 01	E0.14	–	E0.076	–	–	–	E0.08
16	Gridley High School precipitation gage at Gridley	3922051214102 01	E0.58	–	–	–	–	–	–
16	Gridley High School precipitation gage at Gridley	3922051214102 01	E0.62	–	–	–	–	–	E0.68
16	Gridley High School precipitation gage at Gridley	3922051214102 01	–	–	–	–	–	–	–
15	Oroville Dam precipitation gage at spillway	393234121292701	–	–	–	–	–	–	–
15	Oroville Dam precipitation gage at spillway	393234121292701	–	–	–	–	–	–	–
15	Oroville Dam precipitation gage at spillway	393234121292701	E0.08	E0.031	–	–	–	–	–
15	Oroville Dam precipitation gage at spillway	393234121292701	E0.31	–	–	–	–	–	–
15	Oroville Dam precipitation gage at spillway	393234121292701	–	–	–	–	–	–	–
15	Oroville Dam precipitation gage at spillway	393234121292701	–	–	–	–	–	–	–
15	Oroville Dam precipitation gage at spillway	393234121292701	–	–	–	–	–	–	E0.22
15	Oroville Dam precipitation gage at spillway	393234121292701	–	–	–	–	–	–	E0.21
15	Oroville Dam precipitation gage at spillway	393234121292701	–	–	–	–	–	E0.011	–
15	Oroville Dam precipitation gage at spillway	393234121292701	–	–	–	–	–	–	–

Appendix 4. Calculated flux values presented in micrograms per square meter per day (µg/m²/day) for bulk wet-deposition samples with at least one detection collected during the 2002–04 study period at eight sites in the Central Valley, California.—Continued

[All compounds were analyzed by gas chromatography/mass spectrometry (GC/MS) at the U.S. Geological Survey (USGS) National Water Quality Laboratory (NWQL). The Chemical Abstract Service Number is given below each compound name in parentheses. The five digit number in parenthesis is a code used by the USGS to uniquely identify the given compound. Flux calculated as the analytical result in microgram per liter (µg/L) multiplied by the total sample volume collected in liters divided by the area of the respective sampler type (Funnel=0.0731 square meter [m²]. Autosampler=0.0614 m²) and sample composite time in days. Abbreviations: D-U, value deleted by NWQL because compound could not be determined due to matrix interference; E, estimated reported laboratory value; (E), estimated compound due to known poor performance by applied method (Childress and others, 1999; Sandstrom and others, 2001); hh:mm, hour:minute; [-], no Chemical Abstract Service Number available for given compound; <, less than laboratory reporting limit, –, not detected]

Site number	Site name	Site identification number	Azinphos-methyl-oxon (E) [-] 61635	Carbaryl (E) [63-25-2] (82680)	Chlorpyrifos [2921-88-2] (38933)	Chlorpyrifos oxon (E) [5598-15-2] (61636)	Cypermethrin (E) [52315-07-8] (61586)	Dacthal (DCPA) [1861-32-1] (82682)	Diazinon [333-41-5] (39572)
11	Newman rain gage at wasteway levee near Draper Road	37173512103120l	–	–	–	–	–	0.017	0.050
11	Newman rain gage at wasteway levee near Draper Road	37173512103120l	–	E0.015	0.024	–	–	0.013	0.375
11	Newman rain gage at wasteway levee near Draper Road	37173512103120l	–	E0.036	0.080	–	–	0.187	0.712
11	Newman rain gage at wasteway levee near Draper Road	37173512103120l	–	E0.172	–	–	–	0.029	–
11	Newman rain gage at wasteway levee near Draper Road	37173512103120l	–	E0.681	–	–	–	0.065	0.052
11	Newman rain gage at wasteway levee near Draper Road	37173512103120l	–	–	E0.010	–	–	0.016	0.032
12	Turlock rain gage near Idaho Road	37271312053490l	–	E0.010	0.030	–	–	0.017	0.041
12	Turlock rain gage near Idaho Road	37271312053490l	–	E0.028	0.065	–	–	0.034	0.206
12	Turlock rain gage near Idaho Road	37271312053490l	–	E0.289	0.171	–	–	0.081	0.138
12	Turlock rain gage near Idaho Road	37271312053490l	–	E1.19	E0.025	–	–	0.045	0.075
13	Turlock Airport rain gage	37285712041400l	–	E0.020	0.052	E0.02	–	0.025	0.271
13	Turlock Airport rain gage	37285712041400l	–	–	0.025	–	–	0.032	0.085
13	Turlock Airport rain gage	37285712041400l	–	E0.074	0.035	–	–	0.019	0.043
13	Turlock Airport rain gage	37285712041400l	–	E0.770	–	–	–	0.031	0.059
13	Turlock Airport rain gage	37285712041400l	–	–	0.159	–	–	E0.022	0.361
10	Westley rain gage at pump building near lateral 6 North	37333512114300l	–	E0.018	0.012	–	–	0.045	0.144
10	Westley rain gage at pump building near lateral 6 North	37333512114300l	–	E0.031	0.024	E0.03	–	0.027	2.48
10	Westley rain gage at pump building near lateral 6 North	37333512114300l	E0.02	E1.86	–	–	–	0.020	0.296
10	Westley rain gage at pump building near lateral 6 North	37333512114300l	–	–	0.054	E0.05	–	0.027	0.231

Appendix 4. Calculated flux values presented in micrograms per square meter per day (µg/m²/day) for bulk wet-deposition samples with at least one detection collected during the 2002–04 study period at eight sites in the Central Valley, California.—Continued

[All compounds were analyzed by gas chromatography/mass spectrometry (GC/MS) at the U.S. Geological Survey (USGS) National Water Quality Laboratory (NWQL). The Chemical Abstract Service Number is given below each compound name in brackets. The five digit number in parenthesis is a code used by the USGS to uniquely identify the given compound. Flux calculated as the analytical result in microgram per liter (µg/L) multiplied by the total sample volume collected in liters divided by the area of the respective sampler type (Funnel=0.0731 square meter [m²], Autosampler=0.0614 m²) and sample composite time in days. Abbreviations: D-U, value deleted by NWQL because compound could not be determined due to matrix interference; E, estimated reported laboratory value; (E), estimated compound due to known poor performance by applied method (Childress and others, 1999; Sandstrom and others, 2001); hh:mm, hour:minute; [–], no Chemical Abstract Service Number available for given compound; <, less than laboratory reporting limit; –, not detected]

Site number	Site name	Site identification number	Azinphos-methyl-oxon (E) [–] 61635	Carbaryl (E) [63-25-2] (82680)	Chlorpyrifos [2921-88-2] (38933)	Chlorpyrifos oxon (E) [5598-15-2] (61636)	Cypermethrin (E) [52315-07-8] (61586)	Dacthal (DCPA) [1861-32-1] (82682)	Diazinon [333-41-5] (39572)
4	Modesto Irrigation District gage rooftop at Modesto	373834121000601	–	E0.288	0.166	E0.09	E0.030	0.042	2.33
4	Modesto Irrigation District gage rooftop at Modesto	373834121000601	–	E0.757	0.090	E0.04	–	0.039	0.122
4	Modesto Irrigation District gage rooftop at Modesto	373834121000601	–	E1.20	E0.020	–	–	0.084	0.119
4	Modesto Irrigation District gage rooftop at Modesto	373834121000601	–	E0.061	0.208	E0.12	–	0.043	0.423
5	Modesto Irrigation District gage rooftop at Albers Road	373841120504801	–	E0.009	0.948	E0.11	–	0.012	0.694
5	Modesto Irrigation District gage rooftop at Albers Road	373841120504801	–	E0.367	0.129	–	–	0.045	0.109
5	Modesto Irrigation District gage rooftop at Albers Road	373841120504801	–	E1.26	0.035	–	–	0.097	0.117
16	Gridley High School precipitation gage at Gridley	392205121410201	–		E0.058			0.047	0.163
16	Gridley High School precipitation gage at Gridley	392205121410201	–	E5.76	3.73	E1.66	–	0.207	8.63
16	Gridley High School precipitation gage at Gridley	392205121410201	–	E0.562	0.043	E0.12	–	0.020	0.08
15	Oroville Dam precipitation gage at spillway	393234121292701	–		E0.016		–	0.020	E0.012
15	Oroville Dam precipitation gage at spillway	393234121292701	–				–	0.036	0.046
15	Oroville Dam precipitation gage at spillway	393234121292701	–		E0.028	E0.09	–	0.037	0.158
15	Oroville Dam precipitation gage at spillway	393234121292701	–		E0.023		–	0.023	0.184
15	Oroville Dam precipitation gage at spillway	393234121292701	–				–	0.051	
15	Oroville Dam precipitation gage at spillway	393234121292701	–	E0.699	0.123		–	0.094	–
15	Oroville Dam precipitation gage at spillway	393234121292701	–				–	0.058	–
15	Oroville Dam precipitation gage at spillway	393234121292701	–		E0.008	–	–	0.008	0.047
15	Oroville Dam precipitation gage at spillway	393234121292701	–	E0.073	E0.058	E0.15	–	0.058	0.321

Appendix 4. Calculated flux values presented in micrograms per square meter per day (µg/m²/day) for bulk wet-deposition samples with at least one detection collected during the 2002–04 study period at eight sites in the Central Valley, California.—Continued

[All compounds were analyzed by gas chromatography mass spectrometry (GC/MS) at the U.S. Geological Survey (USGS) National Water Quality Laboratory (NWQL). The Chemical Abstract Service Number is given below each compound name in parentheses. The five digit number in parenthesis is a code used by the USGS to uniquely identify the given compound. Flux calculated as the analytical result in microgram per liter (µg/L) multiplied by the total sample volume collected in liters divided by the area of the respective sampler type (Funnel=0.0731 square meter [m²], Autosampler=0.0614 m²) and sample composite time in days. Abbreviations: D-U, value deleted by NWQL because compound could not be determined due to matrix interference; E, estimated compound due to known poor performance by applied method (Childress and others, 1999; Sandstrom and others, 2001); hh:mm, hour:minute; [-], no Chemical Abstract Service Number available for given compound; <, less than laboratory reporting limit; —, not detected]

Site number	Site name	Site identification number	Diazoxon (E) [962-58-3] (61638)	Dichlorvos (E) [62-73-7] (38775)	Iprodione (E) [36734-19-7] (61593)	Malathion [121-75-5] (39532)	Malaoxon (E) [1634-78-2] (61652)	Metalaxyl [57837-19-1] (61596)	Methidathion [950-37-8] (61598)	Metolachlor [51218-45-2] (39415)
11	Newman rain gage at wasteway levee near Draper Road	371735121031201	—	—	—	—	—	—	—	0.038
11	Newman rain gage at wasteway levee near Draper Road	371735121031201	—	—	E1.12	—	—	—	—	E0.017
11	Newman rain gage at wasteway levee near Draper Road	371735121031201	—	—	E8.90	—	—	—	0.053	0.036
11	Newman rain gage at wasteway levee near Draper Road	371735121031201	—	—	E0.319	E0.029	E0.096	—	—	E0.016
11	Newman rain gage at wasteway levee near Draper Road	371735121031201	—	—	E0.117	—	—	—	—	0.104
11	Newman rain gage at wasteway levee near Draper Road	371735121031201	—	—	—	—	—	—	—	0.149
12	Turlock rain gage near Idaho Road	372713120534901	—	—	—	—	—	—	0.025	E0.008
12	Turlock rain gage near Idaho Road	372713120534901	—	—	E3.26	E0.040	—	—	0.028	E0.031
12	Turlock rain gage near Idaho Road	372713120534901	E0.05	—	E0.864	0.180	E0.256	—	—	E0.038
12	Turlock rain gage near Idaho Road	372713120534901	—	—	E0.165	—	E0.095	—	—	0.075
13	Turlock Airport rain gage	372857120414001	—	—	—	E0.015	—	—	0.478	E0.007
13	Turlock Airport rain gage	372857120414001	—	—	E13.1	—	—	—	0.039	—
13	Turlock Airport rain gage	372857120414001	—	—	E0.882	E0.074	E0.109	—	—	E0.019
13	Turlock Airport rain gage	372857120414001	E0.03	—	E0.243	E0.028	E0.087	—	—	0.055
13	Turlock Airport rain gage	372857120414001	E0.07	—	E0.513	—	—	—	0.072	—
10	Westley rain gage at pump building near lateral 6 North	373335121143001	E0.44	—	—	—	—	—	E0.010	0.035
10	Westley rain gage at pump building near lateral 6 North	373335121143001	—	—	E2.23	—	E0.062	—	0.027	E0.038
10	Westley rain gage at pump building near lateral 6 North	373335121143001	—	—	E0.126	—	E0.039	—	—	0.468
10	Westley rain gage at pump building near lateral 6 North	373335121143001	E0.08	—	E0.094	—	—	—	—	E0.024

Appendix 4. Calculated flux values presented in micrograms per square meter per day ($\mu g/m^2/day$) for bulk wet-deposition samples with at least one detection collected during the 2002–04 study period at eight sites in the Central Valley, California.—Continued

[All compounds were analyzed by gas chromatography/mass spectrometry (GC/MS) at the U.S. Geological Survey (USGS) National Water Quality Laboratory (NWQL). The Chemical Abstract Service Number is given below each compound name in brackets. The five digit number in parenthesis is a code used by the USGS to uniquely identify the given compound. Flux calculated as the analytical result in microgram per liter ($\mu g/L$) multiplied by the total sample volume collected in liters divided by the area of the respective sampler type (Funnel=0.0731 square meter [m^2]. Autosampler=0.0614 m^2) and sample composite time in days. Abbreviations: D-U, value deleted by NWQL because compound could not be determined due to matrix interference; E, estimated reported laboratory value; (E), estimated compound due to known poor performance by applied method (Childress and others, 1999; Sandstrom and others, 2001); hh:mm, hour:minute; [–], no Chemical Abstract Service Number available for given compound; <, less than laboratory reporting limit; –, not detected]

Site number	Site name	Site identification number	Diazoxon (E) [962-58-3] (61638)	Dichlorvos (E) [62-73-7] (38775)	Iprodione (E) [36734-19-7] (61593)	Malathion [121-75-5] (39532)	Malaoxon (E) [1634-78-2] (61652)	Metalaxyl [57837-19-1] (61596)	Methidathion [950-37-8] (61598)	Metolachlor [51218-45-2] (39415)
4	Modesto Irrigation District gage rooftop at Modesto	37383412100060 1	–	–	–	E0.039	–	–	0.305	E0.021
4	Modesto Irrigation District gage rooftop at Modesto	37383412100060 1	E0.04	–	E0.851	1.50	E0.145	–	–	E0.027
4	Modesto Irrigation District gage rooftop at Modesto	37383412100060 1	–	–	E0.104	E0.074	E0.134	0.089	–	0.139
4	Modesto Irrigation District gage rooftop at Modesto	37383412100060 1	E0.12	–	E0.472	E0.049	–	–	0.049	E0.025
5	Modesto Irrigation District gage rooftop at Albers Road	37384112050480 1	–	–	–	–	–	–	0.101	–
5	Modesto Irrigation District gage rooftop at Albers Road	37384112050480 1	–	–	E0.482	0.965	E0.161	–	–	E0.032
5	Modesto Irrigation District gage rooftop at Albers Road	37384112050480 1	–	–	E0.145	E0.090	E0.145	–	0.055	E0.076
16	Gridley High School precipitation gage at Gridley	39220512141020 1	–	–	E1.49	E0.174	E1.908	–	–	E0.140
16	Gridley High School precipitation gage at Gridley	39220512141020 1	E0.62	E0.41	–	4.15	–	–	–	–
16	Gridley High School precipitation gage at Gridley	39220512141020 1	–	–	–	0.685	0.471	–	–	–
15	Oroville Dam precipitation gage at spillway	39323412129270 1	–	–	–	–	–	–	–	–
15	Oroville Dam precipitation gage at spillway	39323412129270 1	–	–	–	–	–	–	–	–
15	Oroville Dam precipitation gage at spillway	39323412129270 1	–	–	–	–	–	–	–	–
15	Oroville Dam precipitation gage at spillway	39323412129270 1	–	–	E0.628	–	–	–	–	–
15	Oroville Dam precipitation gage at spillway	39323412129270 1	–	–	–	–	–	–	–	E0.072
15	Oroville Dam precipitation gage at spillway	39323412129270 1	–	–	–	0.331	–	–	–	–
15	Oroville Dam precipitation gage at spillway	39323412129270 1	–	–	–	–	–	–	–	–
15	Oroville Dam precipitation gage at spillway	39323412129270 1	–	–	–	–	–	–	–	–
15	Oroville Dam precipitation gage at spillway	39323412129270 1	E0.008	–	–	–	–	–	E0.011	–
15	Oroville Dam precipitation gage at spillway	39323412129270 1	E0.15	–	–	–	–	–	–	–

Appendix 4. Calculated flux values presented in micrograms per square meter per day (µg/m²/day) for bulk wet-deposition samples with at least one detection collected during the 2002–04 study period at eight sites in the Central Valley, California.—Continued

[All compounds were analyzed by gas chromatography/mass spectrometry (GC/MS) at the U.S. Geological Survey (USGS) National Water Quality Laboratory (NWQL). The Chemical Abstract Service Number is given below each compound name in parentheses. The five digit number in parenthesis is a code used by the USGS to uniquely identify the given compound. Flux calculated as the analytical result in microgram per liter (µg/L) multiplied by the total sample volume collected in liters divided by the area of the respective sampler type (Funnel=0.0731 square meter [m²], Autosampler=0.0614 m²) and sample composite time in days. Abbreviations: D-U, value deleted by NWQL because compound could not be determined due to matrix interference; E, estimated reported laboratory value; (E), estimated compound due to known poor performance by applied method (Childress and others, 1999; Sandstrom and others, 2001); hh:mm, hour:minute; [–], no Chemical Abstract Service Number available for given compound; <, less than laboratory reporting limit; —, not detected]

Site number	Site name	Site identification number	Metribuzin [21087-64-9] (82630)	Myclobutanil (E) [88671-89-0] (61599)	Parathion-methyl [298-00-0] (61664)	Paraoxon-methyl (E) [950-35-6] (61664)	Pendimethalin [40487-42-1] (82683)	Phosmet [732-11-6] (61601)	Prometryn [7287-19-6] (04036)
11	Newman rain gage at wasteway levee near Draper Road	37173512103120 1	—	—	—	—	0.043	—	—
11	Newman rain gage at wasteway levee near Draper Road	37173512103120 1	—	E0.070	—	—	0.052	—	0.011
11	Newman rain gage at wasteway levee near Draper Road	37173512103120 1	—	E0.703	—	—	0.338	E0.032	—
11	Newman rain gage at wasteway levee near Draper Road	37173512103120 1	—	E0.051	—	—	—	—	—
11	Newman rain gage at wasteway levee near Draper Road	37173512103120 1	—	E0.075	—	—	—	—	—
11	Newman rain gage at wasteway levee near Draper Road	37173512103120 1	—	E0.110	—	—	E0.045	—	—
12	Turlock rain gage near Idaho Road	37271312053490 1	—	—	—	—	0.055	—	0.033
12	Turlock rain gage near Idaho Road	37271312053490 1	—	E0.487	—	—	0.120	—	—
12	Turlock rain gage near Idaho Road	37271312053490 1	—	E0.085	—	—	0.133	—	—
12	Turlock rain gage near Idaho Road	37271312053490 1	—	E0.255	—	—	E0.085	E0.045	—
13	Turlock Airport rain gage	37285712041400 1	—	—	—	—	0.096	—	0.042
13	Turlock Airport rain gage	37285712041400 1	—	E0.747	—	—	0.093	E0.031	—
13	Turlock Airport rain gage	37285712041400 1	—	E0.210	—	—	0.109	E0.083	—
13	Turlock Airport rain gage	37285712041400 1	—	E0.270	—	—	—	—	—
13	Turlock Airport rain gage	37285712041400 1	—	E0.058	—	—	E0.766	—	—
10	Westley rain gage at pump building near lateral 6 North	37333512114300 1	0.041	E0.010	—	—	0.177	—	0.039
10	Westley rain gage at pump building near lateral 6 North	37333512114300 1	0.192	E0.229	—	—	0.920	—	0.089
10	Westley rain gage at pump building near lateral 6 North	37333512114300 1	0.111	E0.406	—	—	—	—	0.044
10	Westley rain gage at pump building near lateral 6 North	37333512114300 1	—	E0.027	—	—	E1.221	—	0.032

Appendix 4. Calculated flux values presented in micrograms per square meter per day (μg/m²/day) for bulk wet-deposition samples with at least one detection collected during the 2002–04 study period at eight sites in the Central Valley, California.—Continued

[All compounds were analyzed by gas chromatography/mass spectrometry (GC/MS) at the U.S. Geological Survey (USGS) National Water Quality Laboratory (NWQL). The Chemical Abstract Service Number is given below each compound name in brackets. The five digit number in parenthesis is a code used by the USGS to uniquely identify the given compound. Flux calculated as the analytical result in micrograms per liter (μg/L) multiplied by the total sample volume collected in liters divided by the area of the respective sampler type (Funnel=0.0731 square meter [m²], Autosampler=0.0614 m²) and sample composite time in days. Abbreviations: D-U, value deleted by NWQL because compound could not be determined due to matrix interference; E, estimated reported laboratory value; (E), estimated compound due to known poor performance by applied method (Childress and others, 1999; Sandstrom and others, 2001); hh:mm, hour:minute; [–], no Chemical Abstract Service Number available for given compound; <, less than laboratory reporting limit; –, not detected]

Site number	Site name	Site identification number	Metribuzin [21087-64-9] (82630)	Myclobutanil (E) [88671-89-0] (61599)	Parathion-methyl [298-00-0] (61664)	Paraoxon-methyl (E) [950-35-6] (61664)	Pendimethalin [40487-42-1] (82683)	Phosmet [732-11-6] (61601)	Prometryn [7287-19-6] (04036)
4	Modesto Irrigation District gage rooftop at Modesto	37383412100060 1	–	–	–	–	0.211	–	0.033
4	Modesto Irrigation District gage rooftop at Modesto	37383412100060 1	–	E0.114	–	–	0.243	–	–
4	Modesto Irrigation District gage rooftop at Modesto	37383412100060 1	–	E0.198	–	–	0.546	–	–
4	Modesto Irrigation District gage rooftop at Modesto	37383412100060 1	–	E0.472	–	–	E0.300	–	E0.025
5	Modesto Irrigation District gage rooftop at Albers Road	37384112050480 1	–	E0.009	–	–	0.093	–	0.016
5	Modesto Irrigation District gage rooftop at Albers Road	37384112050480 1	–	E0.148	–	–	0.206	–	–
5	Modesto Irrigation District gage rooftop at Albers Road	37384112050480 1	–	E0.387	–	–	E0.104	E0.069	–
16	Gridley High School precipitation gage at Gridley	39220512141020 1	–	–	–	–	–	–	–
16	Gridley High School precipitation gage at Gridley	39220512141020 1	–	E0.228	10.1	E2.49	–	–	–
16	Gridley High School precipitation gage at Gridley	39220512141020 1	–	–	E1.39	–	–	–	–
15	Oroville Dam precipitation gage at spillway	39323412129270 1	–	–	–	–	E0.078	–	–
15	Oroville Dam precipitation gage at spillway	39323412129270 1	–	–	–	–	E0.100	–	–
15	Oroville Dam precipitation gage at spillway	39323412129270 1	–	–	–	–	–	–	–
15	Oroville Dam precipitation gage at spillway	39323412129270 1	–	E0.169	–	–	E0.138	–	–
15	Oroville Dam precipitation gage at spillway	39323412129270 1	–	–	–	–	–	–	–
15	Oroville Dam precipitation gage at spillway	39323412129270 1	–	–	0.737	–	–	–	–
15	Oroville Dam precipitation gage at spillway	39323412129270 1	–	–	–	–	–	–	–
15	Oroville Dam precipitation gage at spillway	39323412129270 1	–	–	–	–	–	–	–
15	Oroville Dam precipitation gage at spillway	39323412129270 1	–	–	–	–	–	–	–
15	Oroville Dam precipitation gage at spillway	39323412129270 1	–	–	–	–	–	–	–

Appendix 4. Calculated flux values presented in micrograms per square meter per day (µg/m²/day) for bulk wet-deposition samples with at least one detection collected during the 2002–04 study period at eight sites in the Central Valley, California.—Continued

[All compounds were analyzed by gas chromatography/mass spectrometry (GC/MS) at the U.S. Geological Survey (USGS) National Water Quality Laboratory (NWQL). The Chemical Abstract Service Number is given below each compound name in brackets. The five digit number in parenthesis is a code used by the USGS to uniquely identify the given compound. Flux calculated as the analytical result in microgram per liter (µg/L) multiplied by the total sample volume collected in liters divided by the area of the respective sampler type (Funnel=0.0731 square meter [m²]. Autosampler=0.0614 m²) and sample composite time in days. Abbreviations: D-U, value deleted by NWQL because compound could not be determined due to matrix interference; E, estimated reported laboratory value; (E), estimated compound due to known poor performance by applied method (Childress and others, 1999; Sandstrom and others, 2001); hh:mm, hour:minute; [-], no Chemical Abstract Service Number available for given compound; <, less than laboratory reporting limit; —, not detected]

Site number	Site name	Site identification number	Pronamide [23950-58-5] (82676)	Simazine [122-34-9] (04035)	Trifluralin [1582-09-8] (82661)
11	Newman rain gage at wasteway levee near Draper Road	371735121031201	—	E0.006	0.015
11	Newman rain gage at wasteway levee near Draper Road	371735121031201	—	0.349	0.033
11	Newman rain gage at wasteway levee near Draper Road	371735121031201	—	0.801	0.107
11	Newman rain gage at wasteway levee near Draper Road	371735121031201	—	0.141	0.032
11	Newman rain gage at wasteway levee near Draper Road	371735121031201	—	0.169	0.029
11	Newman rain gage at wasteway levee near Draper Road	371735121031201	—	0.071	E0.016
12	Turlock rain gage near Idaho Road	372713120534901	0.014	0.082	E0.010
12	Turlock rain gage near Idaho Road	372713120534901	—	0.394	0.034
12	Turlock rain gage near Idaho Road	372713120534901	—	0.859	E0.043
12	Turlock rain gage near Idaho Road	372713120534901	—	0.320	E0.030
13	Turlock Airport rain gage	372857120414001	0.027	0.190	E0.012
13	Turlock Airport rain gage	372857120414001	—	1.38	—
13	Turlock Airport rain gage	372857120414001	—	7.46	—
13	Turlock Airport rain gage	372857120414001	—	4.68	E0.021
13	Turlock Airport rain gage	372857120414001	0.029	0.932	0.072
10	Westley rain gage at pump building near lateral 6 North	373335121143001	0.053	0.084	0.021
10	Westley rain gage at pump building near lateral 6 North	373335121143001	—	0.684	0.038
10	Westley rain gage at pump building near lateral 6 North	373335121143001	—	0.148	0.032
10	Westley rain gage at pump building near lateral 6 North	373335121143001	0.105	0.424	0.064

Appendix 4. Calculated flux values presented in micrograms per square meter per day (μg/m²/day) for bulk wet-deposition samples with at least one detection collected during the 2002–04 study period at eight sites in the Central Valley, California.—Continued

[All compounds were analyzed by gas chromatography/mass spectrometry (GC/MS) at the U.S. Geological Survey (USGS) National Water Quality Laboratory (NWQL). The Chemical Abstract Service Number is given below each compound name in brackets. The five digit number in parentheses is a code used by the USGS to uniquely identify the given compound. Flux calculated as the analytical result in microgram per liter (μg/L) multiplied by the total sample volume collected in liters divided by the area of the respective sampler type (Funnel=0.0731 square meter [m²], Autosampler=0.0614 m²) and sample composite time in days. Abbreviations: D-U, value deleted by NWQL because compound could not be determined due to matrix interference; E, estimated reported laboratory value; (E), estimated compound due to known poor performance by applied method (Childress and others, 1999; Sandstrom and others, 2001); hh:mm, hour:minute; [-], no Chemical Abstract Service Number available for given compound; <, less than laboratory reporting limit; –, not detected]

Site number	Site name	Site identification number	Pronamide [23950-58-5] (82676)	Simazine [122-34-9] (04035)	Trifluralin [1582-09-8] (82661)
4	Modesto Irrigation District gage rooftop at Modesto	373834121000601	–	0.119	E0.021
4	Modesto Irrigation District gage rooftop at Modesto	373834121000601	–	E0.180	E0.031
4	Modesto Irrigation District gage rooftop at Modesto	373834121000601	–	0.228	E0.040
4	Modesto Irrigation District gage rooftop at Modesto	373834121000601	0.031	0.527	0.067
5	Modesto Irrigation District gage rooftop at Albers Road	373841120504801	0.016	0.144	E0.009
5	Modesto Irrigation District gage rooftop at Albers Road	373841120504801	–	0.270	0.071
5	Modesto Irrigation District gage rooftop at Albers Road	373841120504801	0.035	0.698	E0.028
16	Gridley High School precipitation gage at Gridley	392205121410201	–	E0.151	E0.070
16	Gridley High School precipitation gage at Gridley	392205121410201	–	–	–
16	Gridley High School precipitation gage at Gridley	392205121410201	–	–	–
15	Oroville Dam precipitation gage at spillway	393234121292701	–	–	–
15	Oroville Dam precipitation gage at spillway	393234121292701	–	–	–
15	Oroville Dam precipitation gage at spillway	393234121292701	–	–	–
15	Oroville Dam precipitation gage at spillway	393234121292701	–	0.176	–
15	Oroville Dam precipitation gage at spillway	393234121292701	–	–	–
15	Oroville Dam precipitation gage at spillway	393234121292701	–	–	–
15	Oroville Dam precipitation gage at spillway	393234121292701	–	–	–
15	Oroville Dam precipitation gage at spillway	393234121292701	–	0.025	–
15	Oroville Dam precipitation gage at spillway	393234121292701	–	0.278	–

References Cited

Childress, C.J.O., Foreman, W.T., Connor, B.F., and Maloney, T.J., 1999, New reporting procedures based on long-term method detection levels and some considerations for interpretations of water-quality data provided by the U.S. Geological Survey National Water Quality Laboratory: U.S. Geological Survey Open-File Report 99–193, 19 p.

Sandstrom, M.W., Stroppel, M.E., Foreman, W.T., and Schroeder, M.P., 2001, Methods of analysis by the U.S. Geological Survey National Water Quality Laboratory—Determination of moderate-use pesticides and selected degradates in water by C-18 solid-phase extraction and capillary-column gas chromatography/mass spectrometry with selected-ion monitoring (methods 2002/2011): U.S. Geological Survey Water-Resources Investigations Report 01–4098, 70 p.

Appendix 5.   Calculated Flux Values Presented in Micrograms Per Square Meter Per Day (µg/m^2/day) for Dry-Deposition Samples With at Least One Detection Collected During the 2002–04 Study Period at Seven Sites in the Central Valley, California

Appendix 5. Calculated flux values presented in micrograms per square meter per day (µg/m²/day) for dry-deposition samples with at least one detection collected during the 2002–04 study period at seven sites in the Central Valley, California.

[All compounds were analyzed by gas chromatography/mass spectrometry (GC/MS) at the U.S. Geological Survey (USGS) National Water Quality Laboratory (NWQL). The Chemical Abstract Service Number is given below each compound name in brackets. Flux calculated as the analytical result in microgram per sample (µg/sample) divided by the area of the respective sampler type (Funnel=0.0731 square meter [m²], Autosampler=0.0614 m²) and sample composite time in days. Abbreviations: E, estimated reported laboratory value; hh:mm, hour:minute; [-], no Chemical Abstract Service Number available for given compound; –, not detected]

Site num-ber	Site name	Site identification number	Start date	Start time (hh:mm)	End date	End time (hh:mm)	Sample composite time, days	Sampler type	1-Naphthol[1] [90-15-3]	3,4-Dichloroaniline[1] [95-76-1]	Alachlor [15972-60-8]
11	Newman rain gage at wasteway levee near Draper Road	371735121031201	Apr 9, 2002	08:50	May 21, 2002	10:30	42	Funnel	–	–	–
11	Newman rain gage at wasteway levee near Draper Road	371735121031201	May 21, 2002	10:30	Oct 3, 2002	11:30	135	Funnel	–	–	–
11	Newman rain gage at wasteway levee near Draper Road	371735121031201	Oct 3, 2002	11:30	Nov 6, 2002	10:45	34	Funnel	–	–	–
11	Newman rain gage at wasteway levee near Draper Road	371735121031201	May 9, 2003	16:00	Jul 18, 2003	09:30	70	Funnel	E0.004	–	–
11	Newman rain gage at wasteway levee near Draper Road	371735121031201	Mar 8, 2004	15:40	Apr 16, 2004	13:30	39	Funnel	E0.002	E0.001	–
12	Turlock rain gage near Idaho Road	372713120534901	Apr 25, 2002	12:00	May 21, 2002	11:05	26	Funnel	–	–	E0.004
12	Turlock rain gage near Idaho Road	372713120534901	May 21, 2002	11:05	Oct 3, 2002	12:15	135	Funnel	–	–	–
12	Turlock rain gage near Idaho Road	372713120534901	Oct 3, 2002	12:15	Nov 6, 2002	11:30	34	Funnel	E0.005	–	–
12	Turlock rain gage near Idaho Road	372713120534901	May 9, 2003	17:00	Jul 18, 2003	10:30	70	Funnel	–	–	–
12	Turlock rain gage near Idaho Road	372713120534901	Mar 8, 2004	12:10	Apr 16, 2004	12:30	39	Funnel	E0.004	–	–
13	Turlock Airport rain gage	372857120414001	May 21, 2002	12:00	Oct 3, 2002	13:00	135	Funnel	–	–	–
13	Turlock Airport rain gage	372857120414001	Oct 3, 2002	13:00	Nov 6, 2002	12:15	34	Funnel	–	–	–
13	Turlock Airport rain gage	372857120414001	May 5, 2003	10:40	Jul 22, 2003	11:30	78	Funnel	–	–	–
13	Turlock Airport rain gage	372857120414001	Mar 8, 2004	11:20	Apr 16, 2004	11:10	39	Funnel	E0.002	E0.003	–
10	Westley rain gage at pump building near lateral 6 North	373335121143001	Apr 9, 2002	08:10	May 21, 2002	09:30	42	Funnel	–	–	–
10	Westley rain gage at pump building near lateral 6 North	373335121143001	May 21, 2002	09:30	Oct 3, 2002	10:30	135	Funnel	–	–	–
10	Westley rain gage at pump building near lateral 6 North	373335121143001	Oct 3, 2002	10:30	Nov 6, 2002	09:45	34	Funnel	–	–	–
10	Westley rain gage at pump building near lateral 6 North	373335121143001	May 22, 2003	09:30	Jul 18, 2003	08:45	57	Funnel	E0.003	–	–
10	Westley rain gage at pump building near lateral 6 North	373335121143001	Mar 8, 2004	16:30	Mar 23, 2004	15:15	15	Funnel	E0.005	–	–
10	Westley rain gage at pump building near lateral 6 North	373335121143001	Mar 25, 2004	15:30	Apr 16, 2004	14:50	22	Funnel	E0.036	E0.003	E0.003

Appendix 5. Calculated flux values presented in micrograms per square meter per day ($\mu g/m^2/day$) for dry-deposition samples with at least one detection collected during the 2002–04 study period at seven sites in the Central Valley, California.—Continued

[All compounds were analyzed by gas chromatography/mass spectrometry (GC/MS) at the U.S. Geological Survey (USGS) National Water Quality Laboratory (NWQL). The Chemical Abstract Service Number is given below each compound name in brackets. Flux calculated as the analytical result in microgram per sample (μg/sample) divided by the area of the respective sampler type (Funnel=0.0731 square meter [m^2], Autosampler=0.0614 m^2) and sample composite time in days. Abbreviations: E, estimated reported laboratory value; hh:mm, hour:minute; [-], no Chemical Abstract Service Number available for given compound; –, not detected]

Site number	Site name	Site identification number	Start date	Start time (hh:mm)	End date	End time (hh:mm)	Sample composite time, days	Sampler type	1-Naphthol[1] [90-15-3]	3,4-Dichloroaniline[1] [95-76-1]	Alachlor [15972-60-8]
4	Modesto Irrigation District gage rooftop at Modesto	37383412100601	Apr 9, 2002	10:40	May 21, 2002	13:30	42	Autosampler	–	–	–
4	Modesto Irrigation District gage rooftop at Modesto	37383412100601	May 21, 2002	12:30	Oct 3, 2002	11:00	135	Autosampler	–	–	–
4	Modesto Irrigation District gage rooftop at Modesto	37383412100601	Oct 4, 2002	12:00	Nov 6, 2002	11:50	33	Autosampler	–	–	–
4	Modesto Irrigation District gage rooftop at Modesto	37383412100601	Oct 4, 2002	12:10	Nov 6, 2002	12:00	33	Funnel	–	–	–
4	Modesto Irrigation District gage rooftop at Modesto	37383412100601	May 5, 2003	09:15	Jul 22, 2003	09:30	78	Funnel	E0.003	–	–
4	Modesto Irrigation District gage rooftop at Modesto	37383412100601	May 5, 2003	09:20	Jul 22, 2003	09:40	78	Autosampler	E0.003	–	–
4	Modesto Irrigation District gage rooftop at Modesto	37383412100601	Mar 8, 2004	10:40	Apr 16, 2004	09:00	39	Funnel	E0.003	E0.002	–
5	Modesto Irrigation District gage rooftop at Albers Road	373841120504801	Apr 9, 2002	11:20	May 21, 2002	12:30	42	Autosampler	–	–	–
5	Modesto Irrigation District gage rooftop at Albers Road	373841120504801	May 21, 2002	12:00	Oct 3, 2002	12:00	136	Autosampler	–	–	–
5	Modesto Irrigation District gage rooftop at Albers Road	373841120504801	Oct 4, 2002	11:00	Nov 6, 2002	13:10	33	Autosampler	–	–	–
5	Modesto Irrigation District gage rooftop at Albers Road	373841120504801	Oct 4, 2002	11:10	Nov 6, 2002	13:00	33	Funnel	–	–	–
5	Modesto Irrigation District gage rooftop at Albers Road	373841120504801	May 9, 2003	17:45	Jul 22, 2003	10:00	74	Autosampler	–	–	–
5	Modesto Irrigation District gage rooftop at Albers Road	373841120504801	May 9, 2003	17:50	Jul 22, 2003	10:10	74	Funnel	–	–	–
5	Modesto Irrigation District gage rooftop at Albers Road	373841120504801	Mar 8, 2004	11:00	Apr 16, 2004	09:50	39	Funnel	E0.003	E0.005	–
16	Gridley High School precipitation gage at Gridley	392205121410201	Feb 20, 2003	14:00	Mar 11, 2003	11:40	19	Autosampler	–	–	–
16	Gridley High School precipitation gage at Gridley	392205121410201	May 9, 2003	10:30	Jul 21, 2003	09:00	73	Funnel	E0.006	–	–
16	Gridley High School precipitation gage at Gridley	392205121410201	May 9, 2003	10:40	Jul 21, 2003	09:15	73	Autosampler	E0.002	–	–

Appendix 5. Calculated flux values presented in micrograms per square meter per day (µg/m²/day) for dry-deposition samples with at least one detection collected during the 2002–04 study period at seven sites in the Central Valley, California.—Continued

[All compounds were analyzed by gas chromatography/mass spectrometry (GC/MS) at the U.S. Geological Survey (USGS) National Water Quality Laboratory (NWQL). The Chemical Abstract Service Number is given below each compound name in brackets. Flux calculated as the analytical result in microgram per sample (µg/sample) divided by the area of the respective sampler type (Funnel=0.0731 square meter [m²], Autosampler=0.0614 m²) and sample composite time in days. Abbreviations: E, estimated reported laboratory value; hh:mm, hour:minute; [-], no Chemical Abstract Service Number available for given compound; -, not detected]

Site number	Site name	Site identification number	Azinphos-methyl [86-50-0]	Carbaryl [63-25-2]	Chlorpyrifos [2921-88-2]	Chlorpyrifos oxon [5598-15-2]	Cypermethrin [52315-07-8]	Dacthal (DCPA) [1861-32-1]	Diazinon [333-41-5]	Diazoxon [962-58-3]
11	Newman rain gage at wasteway levee near Draper Road	3717355121031201	-	E0.003	-	-	-	E0.002	-	-
11	Newman rain gage at wasteway levee near Draper Road	3717355121031201	-	-	E0.002	-	-	E0.0005	-	-
11	Newman rain gage at wasteway levee near Draper Road	3717355121031201	-	-	-	-	-	-	-	-
11	Newman rain gage at wasteway levee near Draper Road	3717355121031201	E0.116	E0.004	E0.024	-	-	E0.0004	E0.027	E0.006
11	Newman rain gage at wasteway levee near Draper Road	3717355121031201	-	E0.015	-	-	-	E0.001	-	-
12	Turlock rain gage near Idaho Road	3727131205334901	E0.108	E0.009	-	-	-	E0.002	-	-
12	Turlock rain gage near Idaho Road	3727131205334901	E0.009	-	E0.004	-	-	E0.002	E0.001	-
12	Turlock rain gage near Idaho Road	3727131205334901	-	-	E0.006	-	-	E0.003	E0.005	-
12	Turlock rain gage near Idaho Road	3727131205334901	-	-	E0.233	-	-	-	-	-
12	Turlock rain gage near Idaho Road	3727131205334901	-	E0.029	E0.002	-	-	E0.001	-	-
13	Turlock Airport rain gage	3728571204144001	E0.004	-	E0.001	-	-	E0.0003	-	-
13	Turlock Airport rain gage	3728571204144001	-	-	E0.004	-	-	E0.004	E0.008	-
13	Turlock Airport rain gage	3728571204144001	E0.006	-	E0.132	E0.140	-	E0.0004	-	-
13	Turlock Airport rain gage	3728571204144001	-	E0.019	-	-	-	E0.0005	-	-
10	Westley rain gage at pump building near lateral 6 North	3733351211143001	E0.070	E0.046	-	-	-	E0.001	-	-
10	Westley rain gage at pump building near lateral 6 North	3733351211143001	E0.003	E0.002	E0.001	-	-	E0.001	-	-
10	Westley rain gage at pump building near lateral 6 North	3733351211143001	E0.017	E0.004	E0.002	-	-	E0.004	E0.001	-
10	Westley rain gage at pump building near lateral 6 North	3733351211143001	E0.010	E0.013	-	-	-	E0.001	E0.007	E0.002
10	Westley rain gage at pump building near lateral 6 North	3733351211143001	-	E0.036	-	-	-	-	-	-
10	Westley rain gage at pump building near lateral 6 North	3733351211143001	-	E0.186	-	-	-	E0.001	E0.010	E0.007

Appendix 5. Calculated flux values presented in micrograms per square meter per day ($\mu g/m^2/day$) for dry-deposition samples with at least one detection collected during the 2002–04 study period at seven sites in the Central Valley, California.—Continued

[All compounds were analyzed by gas chromatography/mass spectrometry (GC/MS) at the U.S. Geological Survey (USGS) National Water Quality Laboratory (NWQL). The Chemical Abstract Service Number is given below each compound name in brackets. Flux calculated as the analytical result in microgram per sample (μg/sample) divided by the area of the respective sampler type (Funnel=0.0731 square meter [m^2], Autosampler=0.0614 m^2) and sample composite time in days. Abbreviations: E, estimated reported laboratory value; hh:mm, hour:minute; [–], no Chemical Abstract Service Number available for given compound; –, not detected]

Site number	Site name	Site identification number	Azinphos-methyl [86-50-0]	Carbaryl [63-25-2]	Chlorpyrifos [2921-88-2]	Chlorpyrifos oxon [5598-15-2]	Cypermethrin [52315-07-8]	Dacthal (DCPA) [1861-32-1]	Diazinon [333-41-5]	Diazoxon [962-58-3]
4	Modesto Irrigation District gage rooftop at Modesto	3738341210000601	E0.043	E0.014	E0.011	–	–	E0.002	–	–
4	Modesto Irrigation District gage rooftop at Modesto	3738341210000601	E0.003	E0.003	E0.001	–	–	E0.001	E0.001	–
4	Modesto Irrigation District gage rooftop at Modesto	3738341210000601	–	E0.010	E0.020	–	–	E0.006	E0.006	E0.001
4	Modesto Irrigation District gage rooftop at Modesto	3738341210000601	–	–	E0.013	–	E0.050	E0.008	E0.016	–
4	Modesto Irrigation District gage rooftop at Modesto	3738341210000601	–	–	E0.009	–	–	E0.001	E0.001	–
4	Modesto Irrigation District gage rooftop at Modesto	3738341210000601	–	–	E0.016	E0.021	–	E0.002	E0.002	–
4	Modesto Irrigation District gage rooftop at Modesto	3738341210000601	–	E0.030	–	–	–	E0.001	–	–
5	Modesto Irrigation District gage rooftop at Albers Road	3738411205048 01	–	E0.048	E0.018	–	–	E0.002	–	–
5	Modesto Irrigation District gage rooftop at Albers Road	3738411205048 01	E0.008	E0.003	E0.004	–	–	E0.0004	–	–
5	Modesto Irrigation District gage rooftop at Albers Road	3738411205048 01	E0.014	E0.013	E0.009	–	–	E0.009	E0.007	–
5	Modesto Irrigation District gage rooftop at Albers Road	3738411205048 01	E0.010	E0.009	E0.006	–	–	E0.005	E0.004	–
5	Modesto Irrigation District gage rooftop at Albers Road	3738411205048 01	E0.015	E0.004	E0.054	E0.104	–	E0.001	E0.001	–
5	Modesto Irrigation District gage rooftop at Albers Road	3738411205048 01	–	–	E0.039	E0.031	–	E0.001	E0.002	–
5	Modesto Irrigation District gage rooftop at Albers Road	3738411205048 01	–	E0.027	E0.002	–	–	E0.001	–	–
16	Gridley High School precipitation gage at Gridley	3922051214102 01	–	–	–	–	–	–	E0.008	–
16	Gridley High School precipitation gage at Gridley	3922051214102 01	–	–	E0.004	–	–	E0.001	E0.001	–
16	Gridley High School precipitation gage at Gridley	3922051214102 01	E0.002	–	E0.004	E0.009	–	–	E0.002	–

Appendix 5. Calculated flux values presented in micrograms per square meter per day (µg/m²/day) for dry-deposition samples with at least one detection collected during the 2002–04 study period at seven sites in the Central Valley, California.—Continued

[All compounds were analyzed by gas chromatography/mass spectrometry (GC/MS) at the U.S. Geological Survey (USGS) National Water Quality Laboratory (NWQL). The Chemical Abstract Service Number is given below each compound name in brackets. Flux calculated as the analytical result in microgram per sample (µg/sample) divided by the area of the respective sampler type (Funnel=0.0731 square meter [m²], Autosampler=0.0614 m²) and sample composite time in days. Abbreviations: E, estimated reported laboratory value; lh·mm, hour/minute; [-], no Chemical Abstract Service Number available for given compound; –, not detected]

Site number	Site name	Site identification number	Dichlorvos¹ [62-73-7]	Dimethoate¹ [60-51-5]	Hexazinone¹ [51235-04-2]	Iprodione¹ [36734-19-7]	Malathion [121-75-5]	Malaoxon [1634-78-2]	Methidathion [950-37-8]
11	Newman rain gage at wasteway levee near Draper Road	371735121031201	–	–	–	–	–	–	–
11	Newman rain gage at wasteway levee near Draper Road	371735121031201	–	–	–	–	–	–	–
11	Newman rain gage at wasteway levee near Draper Road	371735121031201	E0.017	–	–	–	–	–	–
11	Newman rain gage at wasteway levee near Draper Road	371735121031201	–	–	–	E0.010	–	E0.008	–
11	Newman rain gage at wasteway levee near Draper Road	371735121031201	–	–	–	–	–	–	–
12	Turlock rain gage near Idaho Road	37273131205334901	–	–	–	–	–	–	–
12	Turlock rain gage near Idaho Road	37273131205334901	E0.003	–	–	–	–	–	–
12	Turlock rain gage near Idaho Road	37273131205334901	–	–	–	–	–	–	–
12	Turlock rain gage near Idaho Road	37273131205334901	–	E0.092	–	–	–	–	–
12	Turlock rain gage near Idaho Road	37273131205334901	–	E0.018	–	E0.074	–	E0.004	–
13	Turlock Airport rain gage	372857120414001	–	–	–	–	–	–	–
13	Turlock Airport rain gage	372857120414001	–	–	–	–	–	E0.022	–
13	Turlock Airport rain gage	372857120414001	–	E0.003	–	E0.011	–	E0.003	–
13	Turlock Airport rain gage	372857120414001	–	–	–	E0.032	–	–	–
10	Westley rain gage at pump building near lateral 6 North	373335121143001	–	–	–	–	–	–	–
10	Westley rain gage at pump building near lateral 6 North	373335121143001	–	E0.001	–	–	E0.002	E0.013	–
10	Westley rain gage at pump building near lateral 6 North	373335121143001	–	–	–	–	E0.003	E0.015	–
10	Westley rain gage at pump building near lateral 6 North	373335121143001	–	E0.013	–	–	–	E0.045	–
10	Westley rain gage at pump building near lateral 6 North	373335121143001	–	–	E0.007	E0.017	–	–	–
10	Westley rain gage at pump building near lateral 6 North	373335121143001	–	E0.014	–	E0.032	–	E0.012	–

Appendix 5. Calculated flux values presented in micrograms per square meter per day (μg/m²/day) for dry-deposition samples with at least one detection collected during the 2002–04 study period at seven sites in the Central Valley, California.—Continued

[All compounds were analyzed by gas chromatography/mass spectrometry (GC/MS) at the U.S. Geological Survey (USGS) National Water Quality Laboratory (NWQL). The Chemical Abstract Service Number is given below each compound name in brackets. Flux calculated as the analytical result in microgram per sample (μg/sample) divided by the area of the respective sampler type (Funnel=0.0731 square meter [m²], Autosampler=0.0614 m²) and sample composite time in days. Abbreviations: E, estimated reported laboratory value; hh:mm, hour:minute; [-], no Chemical Abstract Service Number available for given compound; -, not detected]

Site number	Site name	Site identification number	Dichlorvos¹ [62-73-7]	Dimethoate¹ [60-51-5]	Hexazinone¹ [51235-04-2]	Iprodione¹ [36734-19-7]	Malathion [121-75-5]	Malaoxon [1634-78-2]	Methidathion [950-37-8]
4	Modesto Irrigation District gage rooftop at Modesto	37383412100000601	-	-	-	-	-	-	-
4	Modesto Irrigation District gage rooftop at Modesto	37383412100000601	-	-	-	-	E0.002	E0.006	-
4	Modesto Irrigation District gage rooftop at Modesto	37383412100000601	-	-	-	-	E0.009	-	-
4	Modesto Irrigation District gage rooftop at Modesto	37383412100000601	-	-	-	E0.114	0.058	-	-
4	Modesto Irrigation District gage rooftop at Modesto	37383412100000601	E0.002	-	-	-	-	E0.016	-
4	Modesto Irrigation District gage rooftop at Modesto	37383412100000601	E0.004	-	-	E0.009	-	E0.009	-
4	Modesto Irrigation District gage rooftop at Modesto	37383412100000601	-	-	-	E0.045	-	E0.009	-
5	Modesto Irrigation District gage rooftop at Albers Road	37384112050504801	-	-	-	-	-	-	-
5	Modesto Irrigation District gage rooftop at Albers Road	37384112050504801	-	-	-	-	0.004	E0.006	-
5	Modesto Irrigation District gage rooftop at Albers Road	37384112050504801	-	-	-	-	E0.004	-	-
5	Modesto Irrigation District gage rooftop at Albers Road	37384112050504801	-	-	-	-	-	-	-
5	Modesto Irrigation District gage rooftop at Albers Road	37384112050504801	E0.006	E0.003	-	E0.009	-	-	-
5	Modesto Irrigation District gage rooftop at Albers Road	37384112050504801	E0.003	E0.065	-	-	-	-	-
5	Modesto Irrigation District gage rooftop at Albers Road	37384112050504801	-	-	-	E0.032	-	-	-
16	Gridley High School precipitation gage at Gridley	39220512141410201	-	-	-	E0.078	-	E0.014	E0.018
16	Gridley High School precipitation gage at Gridley	39220512141410201	-	-	-	-	-	-	-
16	Gridley High School precipitation gage at Gridley	39220512141410201	-	-	-	-	-	E0.003	-

Appendix 5. Calculated flux values presented in micrograms per square meter per day (µg/m²/day) for dry-deposition samples with at least one detection collected during the 2002–04 study period at seven sites in the Central Valley, California.—Continued

[All compounds were analyzed by gas chromatography/mass spectrometry (GC/MS) at the U.S. Geological Survey (USGS) National Water Quality Laboratory (NWQL). The Chemical Abstract Service Number is given below each compound name in brackets. Flux calculated as the analytical result in microgram per sample (µg/sample) divided by the area of the respective sampler type (Funnel=0.0731 square meter [m²], Autosampler=0.0614 m²) and sample composite time in days. Abbreviations: E, estimated reported laboratory value; hh:mm, hour:minute; [-], no Chemical Abstract Service Number available for given compound. -, not detected]

Site number	Site name	Site identification number	Metolachlor [51218-45-2]	Metribuzin[1] [21087-64-9]	Myclobutanil [88671-89-0]	Parathion-methyl[1] [298-00-0]	Paraoxon-methyl[1] [950-35-6]	Pendimethalin[1] [40487-42-1]	cis-Permethrin [54774-45-7]
11	Newman rain gage at wasteway levee near Draper Road	371735121031201	E0.016	-	-	-	-	-	-
11	Newman rain gage at wasteway levee near Draper Road	371735121031201	E0.004	E0.004	-	E0.004	-	-	-
11	Newman rain gage at wasteway levee near Draper Road	371735121031201	E0.004	-	-	-	-	-	-
11	Newman rain gage at wasteway levee near Draper Road	371735121031201	E0.065	-	-	E0.010	E0.006	-	E0.014
11	Newman rain gage at wasteway levee near Draper Road	371735121031201	E0.003	-	E0.003	-	-	-	-
12	Turlock rain gage near Idaho Road	372713120534901	E0.038	-	E0.020	-	-	-	-
12	Turlock rain gage near Idaho Road	372713120534901	E0.002	-	-	-	-	-	-
12	Turlock rain gage near Idaho Road	372713120534901	-	-	-	E0.005	-	-	E0.047
12	Turlock rain gage near Idaho Road	372713120534901	-	-	E0.004	-	-	-	-
13	Turlock Airport rain gage	372857120414001	-	-	-	-	-	-	-
13	Turlock Airport rain gage	372857120414001	-	-	-	-	-	-	-
13	Turlock Airport rain gage	372857120414001	E0.004	-	E0.005	E0.008	E0.006	-	E0.124
13	Turlock Airport rain gage	372857120414001	E0.001	-	-	-	-	-	-
10	Westley rain gage at pump building near lateral 6 North	373335121143001	E0.153	-	E0.033	-	-	-	-
10	Westley rain gage at pump building near lateral 6 North	373335121143001	E0.007	-	E0.002	-	-	E0.001	E0.001
10	Westley rain gage at pump building near lateral 6 North	373335121143001	E0.018	E0.004	E0.003	-	-	E0.020	-
10	Westley rain gage at pump building near lateral 6 North	373335121143001	E0.035	-	E0.004	E0.074	E0.022	-	E0.003
10	Westley rain gage at pump building near lateral 6 North	373335121143001	-	-	E0.007	-	-	-	-
10	Westley rain gage at pump building near lateral 6 North	373335121143001	E0.043	-	E0.028	-	-	E0.012	E0.027

Appendix 5.   Calculated flux values presented in micrograms per square meter per day (μg/m²/day) for dry-deposition samples with at least one detection collected during the 2002–04 study period at seven sites in the Central Valley, California.—Continued

[All compounds were analyzed by gas chromatography/mass spectrometry (GC/MS) at the U.S. Geological Survey (USGS) National Water Quality Laboratory (NWQL). The Chemical Abstract Service Number is given below each compound name in brackets. Flux calculated as the analytical result in microgram per sample (μg/sample) divided by the area of the respective sampler type (Funnel=0.0731 square meter [m²], Autosampler=0.0614 m²) and sample composite time in days. Abbreviations: E, estimated reported laboratory value; hh:mm, hour:minute; [–], no Chemical Abstract Service Number available for given compound; –, not detected]

Site number	Site name	Site identification number	Metolachlor [51218-45-2]	Metribuzin¹ [21087-64-9]	Myclobutanil [88671-89-0]	Parathion-methyl¹ [298-00-0]	Paraoxon-methyl¹ [950-35-6]	Pendimethalin¹ [40487-42-1]	cis-Permethrin [54774-45-7]
4	Modesto Irrigation District gage rooftop at Modesto	37383412100060 1	E0.009	–	E0.028	–	–	–	–
4	Modesto Irrigation District gage rooftop at Modesto	37383412100060 1	E0.001	–	E0.002	E0.005	–	E0.001	E0.001
4	Modesto Irrigation District gage rooftop at Modesto	37383412100060 1	–	–	–	–	–	E0.021	–
4	Modesto Irrigation District gage rooftop at Modesto	37383412100060 1	–	–	–	–	–	–	–
4	Modesto Irrigation District gage rooftop at Modesto	37383412100060 1	E0.002	–	–	E0.009	E0.006	–	E0.004
4	Modesto Irrigation District gage rooftop at Modesto	37383412100060 1	E0.002	–	E0.005	E0.007	E0.006	–	E0.003
4	Modesto Irrigation District gage rooftop at Modesto	37383412100060 1	E0.003	–	E0.003	–	–	–	–
5	Modesto Irrigation District gage rooftop at Albers Road	37384112050480 1	E0.009	–	E0.031	E0.012	–	–	–
5	Modesto Irrigation District gage rooftop at Albers Road	37384112050480 1	E0.001	–	E0.002	E0.006	–	–	–
5	Modesto Irrigation District gage rooftop at Albers Road	37384112050480 1	E0.002	–	E0.003	E0.005	–	E0.005	E0.004
5	Modesto Irrigation District gage rooftop at Albers Road	37384112050480 1	E0.002	–	E0.003	E0.007	–	E0.013	–
5	Modesto Irrigation District gage rooftop at Albers Road	37384112050480 1	E0.006	–	E0.009	E0.032	E0.017	–	E0.004
5	Modesto Irrigation District gage rooftop at Albers Road	37384112050480 1	E0.011	–	–	E0.017	E0.005	–	E0.006
5	Modesto Irrigation District gage rooftop at Albers Road	37384112050480 1	E0.002	–	E0.004	–	–	–	–
16	Gridley High School precipitation gage at Gridley	39220512141020 1	–	–	E0.013	–	–	–	–
16	Gridley High School precipitation gage at Gridley	39220512141020 1	–	–	–	E0.015	E0.011	–	–
16	Gridley High School precipitation gage at Gridley	39220512141020 1	–	–	–	E0.022	E0.016	–	E0.002

Appendix 5. Calculated flux values presented in micrograms per square meter per day (µg/m²/day) for dry-deposition samples with at least one detection collected during the 2002–04 study period at seven sites in the Central Valley, California.—Continued

[All compounds were analyzed by gas chromatography/mass spectrometry (GC/MS) at the U.S. Geological Survey (USGS) National Water Quality Laboratory (NWQL). The Chemical Abstract Service Number is given below each compound name in brackets. Flux calculated as the analytical result in microgram per sample (µg/sample) divided by the area of the respective sampler type (Funnel=0.0731 square meter [m²], Autosampler=0.0614 m²) and sample composite time in days. Abbreviations: E, estimated reported laboratory value; lih:min, hour:minute; [–], no Chemical Abstract Service Number available for given compound; –, not detected]

Site number	Site name	Site identification number	trans-Permethrin [51877-74-8]	Phosmet [732-11-6]	Phosmet oxon [3735-33-9]	Prometryn [7287-19-6]	Pronamide [23950-58-5]	Simazine [122-34-9]	Trifluralin [1582-09-8]
11	Newman rain gage at wasteway levee near Draper Road	37173512103120l	–	E0.019	–	–	–	–	E0.180
11	Newman rain gage at wasteway levee near Draper Road	37173512103120l	E0.004	–	–	–	–	–	E0.003
11	Newman rain gage at wasteway levee near Draper Road	37173512103120l	–	–	–	–	–	–	E0.005
11	Newman rain gage at wasteway levee near Draper Road	37173512103120l	E0.016	E0.007	–	–	–	–	E0.004
11	Newman rain gage at wasteway levee near Draper Road	37173512103120l	–	–	–	–	–	E0.013	–
12	Turlock rain gage near Idaho Road	37271312053490l	–	E0.007	–	–	–	E0.023	E0.001
12	Turlock rain gage near Idaho Road	37271312053490l	E0.008	–	–	–	–	–	E0.0004
12	Turlock rain gage near Idaho Road	37271312053490l	–	–	–	E0.004	–	–	E0.001
12	Turlock rain gage near Idaho Road	37271312053490l	E0.051	–	–	–	–	–	E0.001
12	Turlock rain gage near Idaho Road	37271312053490l	–	E0.015	–	–	–	E0.038	E0.001
13	Turlock Airport rain gage	37285712041400l	–	E0.018	–	–	–	–	E0.0005
13	Turlock Airport rain gage	37285712041400l	–	E0.056	–	–	–	–	E0.002
13	Turlock Airport rain gage	37285712041400l	E0.175	E0.010	E0.018	–	–	E0.049	E0.001
13	Turlock Airport rain gage	37285712041400l	–	–	–	–	–	E0.011	E0.0004
10	Westley rain gage at pump building near lateral 6 North	37333512114300l	–	E0.060	–	–	–	–	–
10	Westley rain gage at pump building near lateral 6 North	37333512114300l	–	E0.003	–	–	–	E0.0004	E0.001
10	Westley rain gage at pump building near lateral 6 North	37333512114300l	–	E0.009	–	–	–	E0.044	E0.003
10	Westley rain gage at pump building near lateral 6 North	37333512114300l	E0.007	E0.024	E0.020	–	–	–	E0.002
10	Westley rain gage at pump building near lateral 6 North	37333512114300l	–	–	–	–	–	E0.078	–
10	Westley rain gage at pump building near lateral 6 North	37333512114300l	E0.031	E0.019	E0.013	–	E0.004	E0.023	E0.008

Appendix 5. Calculated flux values presented in micrograms per square meter per day ($\mu g/m^2/day$) for dry-deposition samples with at least one detection collected during the 2002–04 study period at seven sites in the Central Valley, California.—Continued

[All compounds were analyzed by gas chromatography/mass spectrometry (GC/MS) at the U.S. Geological Survey (USGS) National Water Quality Laboratory (NWQL). The Chemical Abstract Service Number is given below each compound name in brackets. Flux calculated as the analytical result in microgram per sample (μg/sample) divided by the area of the respective sampler type (Funnel=0.0731 square meter [m^2], Autosampler=0.0614 m^2) and sample composite time in days. Abbreviations: E, estimated reported laboratory value; hh:mm, hour:minute; [–], no Chemical Abstract Service Number available for given compound; –, not detected]

Site number	Site name	Site identification number	trans-Permethrin [51877-74-8]	Phosmet[1] [732-11-6]	Phosmet oxon[1] [3735-33-9]	Prometryn [7287-19-6]	Pronamide [23950-58-5]	Simazine [122-34-9]	Trifluralin[1] [1582-09-8]
4	Modesto Irrigation District gage rooftop at Modesto	373834121000601	–	E0.068	–	–	–	E0.033	–
4	Modesto Irrigation District gage rooftop at Modesto	373834121000601	E0.003	E0.009	–	–	–	E0.012	E0.0004
4	Modesto Irrigation District gage rooftop at Modesto	373834121000601	–	–	–	–	–	E0.079	E0.002
4	Modesto Irrigation District gage rooftop at Modesto	373834121000601	–	–	–	E0.004	–	E0.077	E0.002
4	Modesto Irrigation District gage rooftop at Modesto	373834121000601	E0.006	E0.007	E0.015	–	–	E0.007	–
4	Modesto Irrigation District gage rooftop at Modesto	373834121000601	E0.005	E0.011	E0.017	–	–	E0.005	–
4	Modesto Irrigation District gage rooftop at Modesto	373834121000601	–	–	–	–	–	E0.013	–
5	Modesto Irrigation District gage rooftop at Albers Road	373841120504801	–	E0.188	–	–	–	E0.067	E0.001
5	Modesto Irrigation District gage rooftop at Albers Road	373841120504801	–	E0.010	–	–	–	E0.059	–
5	Modesto Irrigation District gage rooftop at Albers Road	373841120504801	E0.013	E0.021	–	–	–	E0.245	–
5	Modesto Irrigation District gage rooftop at Albers Road	373841120504801	E0.011	E0.022	–	–	–	E0.218	E0.001
5	Modesto Irrigation District gage rooftop at Albers Road	373841120504801	E0.006	E0.014	E0.008	–	–	E0.064	–
5	Modesto Irrigation District gage rooftop at Albers Road	373841120504801	E0.005	–	–	–	–	–	–
5	Modesto Irrigation District gage rooftop at Albers Road	373841120504801	–	–	–	–	–	E0.032	–
16	Gridley High School precipitation gage at Gridley	392205121410201	–	–	–	–	–	E0.025	–
16	Gridley High School precipitation gage at Gridley	392205121410201	–	–	–	–	–	–	–
16	Gridley High School precipitation gage at Gridley	392205121410201	E0.002	E0.009	E0.004	–	–	E0.002	–

[1]Compound is biased low (B-L) based on low and variable lab reagent spike recoveries (less than 50 percent).

Appendix 6.   Calculated Flux Values Presented in
Micrograms Per Square Meter Per Day ($\mu g/m^2/day$) for
Bulk Dry-Deposition Samples With at Least One Detection
Collected During the 2002–04 Study Period at Eight Sites in
the Central Valley, California

Appendix 6.   Calculated flux values presented in micrograms per square meter per day (μg/m²/day) for bulk dry-deposition samples with at least one detection collected during the 2002–04 study period at eight sites in the Central Valley, California

[All compounds were analyzed by gas chromatography/mass spectrometry (GC/MS) at the U.S. Geological Survey (USGS) National Water Quality Laboratory (NWQL). The Chemical Abstract Service Number is given below each compound name in brackets. Flux calculated as the analytical result in μg/sample divided by the area of the respective sampler type (Funnel=0.0731 square meter [m²]. Autosampler=0.0614 m²) and sample composite time in days. Abbreviations: E, estimated reported laboratory value; hh:mm, hour:minute; [-], no Chemical Abstract Service Number available for given compound; -, not detected]

Site number	Site name	Site identification number	Start date	Start time (hh:mm)	End date	End time (hh:mm)	Sample composite time, days	Sampler type	1-Naphthol[1] [90-15-3]	3,4-dichloroaniline[1] [95-76-1]
11	Newman rain gage at wasteway levee near Draper Road	371735121031201	Jul 18, 2003	09:30	Aug 4, 2003	12:15	17	Funnel	-	-
11	Newman rain gage at wasteway levee near Draper Road	371735121031201	Nov 10, 2003	10:30	Dec 11, 2003	11:50	31	Funnel	E0.007	-
11	Newman rain gage at wasteway levee near Draper Road	371735121031201	Dec 16, 2003	11:00	Jan 29, 2004	14:00	44	Funnel	E0.002	E0.009
12	Turlock rain gage near Idaho Road	372713120534901	Jul 18, 2003	10:30	Aug 4, 2003	13:00	17	Funnel	-	-
12	Turlock rain gage near Idaho Road	372713120534901	Nov 10, 2003	11:10	Dec 11, 2003	12:40	31	Funnel	E0.003	E0.003
12	Turlock rain gage near Idaho Road	372713120534901	Jan 6, 2004	12:15	Jan 29, 2004	13:00	23	Funnel	E0.004	E0.015
13	Turlock Airport rain gage	372857120414001	Nov 10, 2003	11:40	Dec 11, 2003	13:20	31	Funnel	E0.004	E0.008
13	Turlock Airport rain gage	372857120414001	Jan 6, 2004	11:00	Jan 29, 2004	12:10	23	Funnel	E0.004	E0.011
10	Westley rain gage at pump building near lateral 6 North	373335121143001	Mar 1, 2003	11:00	Mar 27, 2003	09:20	26	Funnel	-	E0.004
10	Westley rain gage at pump building near lateral 6 North	373335121143001	Nov 11, 2003	10:30	Dec 11, 2003	10:50	30	Funnel	E0.003	E0.005
4	Modesto Irrigation District gage rooftop at Modesto	373834121000601	Feb 21, 2003	14:40	Mar 27, 2003	14:50	34	Funnel	E0.003	E0.008
4	Modesto Irrigation District gage rooftop at Modesto	373834121000601	Jul 22, 2003	09:30	Aug 4, 2003	15:15	13	Funnel	E0.008	-
4	Modesto Irrigation District gage rooftop at Modesto	373834121000601	Nov 10, 2003	13:15	Dec 11, 2003	15:00	31	²Autosampler	E0.003	E0.003
4	Modesto Irrigation District gage rooftop at Modesto	373834121000601	Jan 6, 2004	08:40	Jan 29, 2004	09:50	23	Funnel	E0.008	E0.010
5	Modesto Irrigation District gage rooftop at Albers Road	373841120504801	Feb 21, 2003	13:20	Mar 27, 2003	13:50	34	Funnel	-	E0.005
5	Modesto Irrigation District gage rooftop at Albers Road	373841120504801	Jul 22, 2003	10:00	Aug 4, 2003	14:30	13	Funnel	E0.011	-
5	Modesto Irrigation District gage rooftop at Albers Road	373841120504801	Nov 10, 2003	12:10	Dec 11, 2003	14:05	31	²Autosampler	E0.002	-
5	Modesto Irrigation District gage rooftop at Albers Road	373841120504801	Nov 10, 2003	12:30	Dec 11, 2003	14:10	31	Funnel	E0.004	E0.007
5	Modesto Irrigation District gage rooftop at Albers Road	373841120504801	Jan 6, 2004	09:35	Jan 29, 2004	10:50	23	Funnel	E0.003	E0.036
5	Modesto Irrigation District gage rooftop at Albers Road	373841120504801	Jan 6, 2004	09:35	Jan 29, 2004	10:50	23	²Autosampler	E0.003	E0.042
16	Gridley High School precipitation gage at Gridley	392205121410201	Feb 20, 2003	14:00	Mar 28, 2003	12:10	36	Funnel	E0.007	E0.006
16	Gridley High School precipitation gage at Gridley	392205121410201	Jan 7, 2004	11:00	Jan 27, 2004	10:10	20	Funnel	E0.004	E0.003
16	Gridley High School precipitation gage at Gridley	392205121410201	Mar 2, 2004	10:40	Mar 27, 2004	14:10	25	Funnel	E0.003	E0.003
16	Gridley High School precipitation gage at Gridley	392205121410201	Mar 2, 2004	10:50	Mar 27, 2004	14:10	25	²Autosampler	E0.005	E0.005
15	Oroville Dam precipitation gage at spillway	393234121292701	Feb 20, 2003	15:20	Mar 28, 2003	13:20	36	Funnel	E0.015	-
15	Oroville Dam precipitation gage at spillway	393234121292701	Mar 2, 2004	11:40	Mar 27, 2004	15:30	25	Funnel	E0.005	-

Appendix 6. Calculated flux values presented in micrograms per square meter per day (µg/m²/day) for bulk dry-deposition samples with at least one detection collected during the 2002–04 study period at eight sites in the Central Valley, California—Continued

[All compounds were analyzed by gas chromatography/mass spectrometry (GC/MS) at the U.S. Geological Survey (USGS) National Water Quality Laboratory (NWQL). The Chemical Abstract Service Number is given below each compound name in brackets. Flux calculated as the analytical result in µg/sample divided by the area of the respective sampler type (Funnel=0.0731 square meter [m²], Autosampler=0.0614 m²) and sample composite time in days. Abbreviations: E, estimated reported laboratory value; hh:mm, hour:minute; [-], no Chemical Abstract Service Number available for given compound; -, not detected]

Site number	Site name	Site identification number	Azinphos-methyl[1] [86-50-0]	Carbaryl [63-25-2]	Chlorpyrifos [2921-88-2]	Chlorpyrifos oxon [5598-15-2]	Deethylatrazine, CIAT[1] [6190-65-4]	Cyfluthrin [68359-37-5]	Dacthal (DCPA) [1861-32-1]	Diazinon [333-41-5]
11	Newman rain gage at wasteway levee near Draper Road	371735121031201	E0.047	-	E0.011	-	-	-	-	-
11	Newman rain gage at wasteway levee near Draper Road	371735121031201	-	E0.005	E0.011	-	-	-	E0.007	-
11	Newman rain gage at wasteway levee near Draper Road	371735121031201	-	-	-	-	-	-	E0.011	E0.411
12	Turlock rain gage near Idaho Road	372713120534901	E0.044	E0.021	E0.036	-	-	-	E0.003	E0.011
12	Turlock rain gage near Idaho Road	372713120534901	-	E0.006	E0.005	-	E0.004	-	E0.012	-
12	Turlock rain gage near Idaho Road	372713120534901	-	E0.011	E0.105	E0.048	-	-	E0.004	E1.21
13	Turlock Airport rain gage	372857120414001	-	E0.016	E0.003	-	-	-	E0.009	E0.008
13	Turlock Airport rain gage	372857120414001	-	E0.014	E0.208	E0.067	-	-	E0.006	E0.690
10	Westley rain gage at pump building near lateral 6 North	373335121143001	-	-	E0.007	-	-	-	E0.006	E0.003
10	Westley rain gage at pump building near lateral 6 North	373335121143001	-	-	E0.002	-	-	E0.043	E0.006	E0.004
4	Modesto Irrigation District gage rooftop at Modesto	373834121000601	-	-	E0.014	-	-	-	E0.005	E0.026
4	Modesto Irrigation District gage rooftop at Modesto	373834121000601	-	-	E0.010	-	-	-	E0.003	-
4	Modesto Irrigation District gage rooftop at Modesto	373834121000601	-	E0.014	E0.004	-	-	-	E0.005	E0.009
4	Modesto Irrigation District gage rooftop at Modesto	373834121000601	-	E0.033	E0.143	E0.061	-	-	E0.008	E1.73
5	Modesto Irrigation District gage rooftop at Albers Road	373841120504801	-	-	E0.004	-	-	-	E0.001	E0.005
5	Modesto Irrigation District gage rooftop at Albers Road	373841120504801	E0.052	-	E0.219	-	-	-	E0.006	-
5	Modesto Irrigation District gage rooftop at Albers Road	373841120504801	-	-	E0.005	-	-	-	E0.007	E0.005
5	Modesto Irrigation District gage rooftop at Albers Road	373841120504801	-	E0.011	E0.009	-	-	-	E0.009	E0.012
5	Modesto Irrigation District gage rooftop at Albers Road	373841120504801	-	E0.011	E0.225	E0.057	-	-	E0.006	E1.52
5	Modesto Irrigation District gage rooftop at Albers Road	373841120504801	-	E0.013	E0.268	E0.068	-	-	E0.007	E1.81
16	Gridley High School precipitation gage at Gridley	392205121410201	-	-	E0.008	-	-	-	E0.001	E0.092
16	Gridley High School precipitation gage at Gridley	392205121410201	-	E0.009	E0.235	E0.075	-	-	E0.005	E5.26
16	Gridley High School precipitation gage at Gridley	392205121410201	-	E0.012	E0.011	-	-	-	E0.001	E0.003
16	Gridley High School precipitation gage at Gridley	392205121410201	-	E0.025	E0.012	E0.014	-	-	E0.001	E0.009
15	Oroville Dam precipitation gage at spillway	393234121292701	-	-	E0.005	-	-	-	E0.002	E0.002
15	Oroville Dam precipitation gage at spillway	393234121292701	-	E0.017	-	-	-	-	E0.001	-

Appendix 6. Calculated flux values presented in micrograms per square meter per day (µg/m²/day) for bulk dry-deposition samples with at least one detection collected during the 2002–04 study period at eight sites in the Central Valley, California—Continued

[All compounds were analyzed by gas chromatography/mass spectrometry (GC/MS) at the U.S. Geological Survey (USGS) National Water Quality Laboratory (NWQL). The Chemical Abstract Service Number is given below each compound name in brackets. Flux calculated as the analytical result in µg/sample divided by the area of the respective sampler type (Funnel=0.0731 square meter [m²], Autosampler=0.0614 m²) and sample composite time in days. Abbreviations: E, estimated reported laboratory value; hh:mm, hour:minute; [-], no Chemical Abstract Service Number available for given compound; -, not detected]

Site number	Site name	Site identification number	Diazoxon[1] [962-58-3]	Dichlorvos[1] [62-73-7]	Dimethoate[1] [60-51-5]	Fipronil sulfone[1] [120068-36-2]	Hexazinone[1] [51235-04-2]	Iprodione[1] [36734-19-7]	Malathion[1] [121-75-5]	Malaoxon[1] [1634-78-2]
11	Newman rain gage at wasteway levee near Draper Road	371735121031201	-	-	E0.079	-	-	-	-	E0.103
11	Newman rain gage at wasteway levee near Draper Road	371735121031201	-	-	-	-	E0.057	E0.019	-	-
11	Newman rain gage at wasteway levee near Draper Road	371735121031201	E0.050	-	-	E0.002	E0.071	E0.007	-	-
12	Turlock rain gage near Idaho Road	372713120534901	-	E0.012	E0.044	-	-	-	-	E0.031
12	Turlock rain gage near Idaho Road	372713120534901	-	-	-	-	E0.026	-	-	E0.012
12	Turlock rain gage near Idaho Road	372713120534901	E0.126	E0.005	-	-	E0.086	-	-	-
13	Turlock Airport rain gage	372857120414001	-	-	E0.013	-	E0.026	E0.015	E0.009	E0.009
13	Turlock Airport rain gage	372857120414001	E0.092	-	-	-	E0.073	E0.015	E0.006	E0.007
10	Westley rain gage at pump building near lateral 6 North	373335121143001	-	-	-	-	-	E0.352	-	E0.010
10	Westley rain gage at pump building near lateral 6 North	373335121143001	-	-	-	-	E0.028	E0.006	-	E0.022
4	Modesto Irrigation District gage rooftop at Modesto	373841210000601	-	-	-	-	-	E0.237	E0.004	E0.007
4	Modesto Irrigation District gage rooftop at Modesto	373841210000601	-	-	-	-	-	-	-	E0.052
4	Modesto Irrigation District gage rooftop at Modesto	373841210000601	-	-	-	-	E0.012	-	E0.007	-
4	Modesto Irrigation District gage rooftop at Modesto	373841210000601	E0.202	E0.004	-	-	E0.111	-	0.061	E0.031
5	Modesto Irrigation District gage rooftop at Albers Road	373841120504801	-	-	E0.023	-	-	E0.700	-	E0.047
5	Modesto Irrigation District gage rooftop at Albers Road	373841120504801	-	E0.029	-	-	-	-	-	-
5	Modesto Irrigation District gage rooftop at Albers Road	373841120504801	-	-	-	-	-	-	-	E0.013
5	Modesto Irrigation District gage rooftop at Albers Road	373841120504801	E0.091	-	-	-	E0.031	E0.006	E0.004	-
5	Modesto Irrigation District gage rooftop at Albers Road	373841120504801	E0.108	-	-	-	-	E0.006	E0.005	-
5	Modesto Irrigation District gage rooftop at Albers Road	373841120504801	-	-	-	-	-	E0.007	-	-
16	Gridley High School precipitation gage at Gridley	392205121410201	E0.005	-	-	-	-	E0.043	E0.008	-
16	Gridley High School precipitation gage at Gridley	392205121410201	E0.203	E0.004	-	-	E0.049	-	E0.009	-
16	Gridley High School precipitation gage at Gridley	392205121410201	E0.006	-	-	-	-	E0.016	-	-
16	Gridley High School precipitation gage at Gridley	392205121410201	-	-	-	-	-	E0.031	-	E0.008
15	Oroville Dam precipitation gage at spillway	393234121292701	-	-	-	-	-	E0.046	-	-
15	Oroville Dam precipitation gage at spillway	393234121292701	-	-	-	-	-	E0.010	-	-

Appendix 6. Calculated flux values presented in micrograms per square meter per day (μg/m²/day) for bulk dry-deposition samples with at least one detection collected during the 2002–04 study period at eight sites in the Central Valley, California—Continued

[All compounds were analyzed by gas chromatography/mass spectrometry (GCMS) at the U.S. Geological Survey (USGS) National Water Quality Laboratory (NWQL). The Chemical Abstract Service Number is given below each compound name in brackets. Flux calculated as the analytical result in μg/sample divided by the area of the respective sampler type (Funnel=0.0731 square meter [m²], Autosampler=0.0614 m²) and sample composite time in days. Abbreviations: E, estimated reported laboratory value; hh:mm, hour:minute; [-], no Chemical Abstract Service Number available for given compound; -, not detected]

Site number	Site name	Site identification number	Methidathion [950-37-8]	Metolachlor [51218-45-2]	Metribuzin[1] [21087-64-9]	Myclobutanil [88671-89-0]	Parathion-methyl[1] [298-00-0]	Paraoxon-methyl[1] [950-35-6]	Pendimethilan[1] [40487-42-1]
11	Newman rain gage at wasteway levee near Draper Road	371735121031201	-	E0.065	-	-	E0.015	-	-
11	Newman rain gage at wasteway levee near Draper Road	371735121031201	-	-	-	-	-	-	-
11	Newman rain gage at wasteway levee near Draper Road	371735121031201	0.179	E0.012	-	E0.005	-	-	E0.079
12	Turlock rain gage near Idaho Road	372713120534901	-	-	-	-	E0.014	-	-
12	Turlock rain gage near Idaho Road	372713120534901	-	E0.005	-	E0.006	E0.007	-	E0.044
12	Turlock rain gage near Idaho Road	372713120534901	E0.103	E0.004	-	-	-	-	E0.063
13	Turlock Airport rain gage	372857120414001	E0.003	E0.003	-	E0.007	E0.008	-	E0.044
13	Turlock Airport rain gage	372857120414001	E0.092	E0.003	E0.010	-	-	-	E0.044
10	Westley rain gage at pump building near lateral 6 North	373335121143001	-	E0.004	-	E0.037	E0.010	-	E0.145
10	Westley rain gage at pump building near lateral 6 North	373335121143001	-	E0.014	E0.044	-	-	-	-
4	Modesto Irrigation District gage rooftop at Modesto	373834121000601	E0.004	E0.004	-	0.048	-	-	E0.010
4	Modesto Irrigation District gage rooftop at Modesto	373834121000601	-	E0.005	-	E0.022	E0.015	E0.017	-
4	Modesto Irrigation District gage rooftop at Modesto	373834121000601	E0.008	E0.003	-	E0.003	E0.007	-	E0.033
4	Modesto Irrigation District gage rooftop at Modesto	373834121000601	E0.426	E0.004	-	-	-	-	E0.084
5	Modesto Irrigation District gage rooftop at Albers Road	373841120504801	-	E0.001	-	E0.034	-	-	-
5	Modesto Irrigation District gage rooftop at Albers Road	373841120504801	-	-	-	-	E0.062	E0.026	E0.033
5	Modesto Irrigation District gage rooftop at Albers Road	373841120504801	E0.005	E0.002	-	-	E0.010	-	E0.057
5	Modesto Irrigation District gage rooftop at Albers Road	373841120504801	E0.006	E0.003	-	E0.005	E0.024	-	E0.061
5	Modesto Irrigation District gage rooftop at Albers Road	373841120504801	E0.157	E0.002	-	E0.002	-	-	E0.073
5	Modesto Irrigation District gage rooftop at Albers Road	373841120504801	E0.187	E0.002	-	E0.002	-	-	-
16	Gridley High School precipitation gage at Gridley	392205121410201	E0.013	-	-	E0.012	-	-	-
16	Grndley High School precipitation gage at Gridley	392205121410201	E0.049	-	-	-	E0.009	-	E0.068
16	Gridley High School precipitation gage at Gridley	392205121410201	E0.007	E0.002	-	-	-	-	-
16	Gridley High School precipitation gage at Gridley	392205121410201	E0.019	E0.003	-	E0.006	-	-	-
15	Oroville Dam precipitation gage at spillway	393234121292701	-	-	-	-	-	-	-
15	Oroville Dam precipitation gage at spillway	393234121292701	-	-	-	-	-	-	-

Appendix 6. Calculated flux values presented in micrograms per square meter per day (μg/m²/day) for bulk dry-deposition samples with at least one detection collected during the 2002–04 study period at eight sites in the Central Valley, California—Continued

[All compounds were analyzed by gas chromatography/mass spectrometry (GC/MS) at the U.S. Geological Survey (USGS) National Water Quality Laboratory (NWQL). The Chemical Abstract Service Number is given below each compound name in brackets. Flux calculated as the analytical result in μg/sample divided by the area of the respective sampler type (Funnel=0.0731 square meter [m²], Autosampler=0.0614 m²) and sample composite time in days. Abbreviations: E, estimated reported laboratory value; hh:mm, hour:minute; [–], no Chemical Abstract Service Number available for given compound; –, not detected]

Site number	Site name	Site identification number	cis-Permethrin [54774-45-7]	trans-Permethrin [51877-74-8]	Phosmet[1] [732-11-6]	Phosmet oxon[1] [3735-33-9]	Prometryn [7287-19-6]	Pronamide [23950-58-5]	Simazine [122-34-9]	Trifluralin[1] [1582-09-8]
11	Newman rain gage at wasteway levee near Draper Road	371735121031201	E0.061	E0.059	E0.050	–	–	–	–	E0.010
11	Newman rain gage at wasteway levee near Draper Road	371735121031201	–	–	–	–	–	E0.007	–	–
11	Newman rain gage at wasteway levee near Draper Road	371735121031201	E0.005	E0.005	–	–	–	E0.009	E0.047	E0.032
12	Turlock rain gage near Idaho Road	372713120534901	E0.025	E0.012	E0.065	–	–	–	E0.014	–
12	Turlock rain gage near Idaho Road	372713120534901	E0.005	E0.005	–	–	E0.079	E0.005	E0.051	E0.009
12	Turlock rain gage near Idaho Road	372713120534901	E0.030	E0.036	–	–	–	–	E0.067	E0.007
13	Turlock Airport rain gage	372857120414001	E0.044	E0.054	–	–	E0.025	–	E0.078	E0.006
13	Turlock Airport rain gage	372857120414001	–	–	–	–	–	–	E0.097	E0.008
10	Westley rain gage at pump building near lateral 6 North	373335121143001	E0.011	E0.010	–	–	–	–	E0.018	–
10	Westley rain gage at pump building near lateral 6 North	373335121143001	–	–	–	–	E0.043	–	E0.042	E0.007
4	Modesto Irrigation District gage rooftop at Modesto	373834121000601	E0.014	E0.023	E0.024	E0.045	–	–	E0.034	–
4	Modesto Irrigation District gage rooftop at Modesto	373834121000601	E0.006	E0.009	–	–	–	–	E0.011	–
4	Modesto Irrigation District gage rooftop at Modesto	373834121000601	–	–	–	–	E0.018	–	E0.030	–
4	Modesto Irrigation District gage rooftop at Modesto	373834121000601	E0.032	E0.043	–	–	–	–	E0.099	E0.009
5	Modesto Irrigation District gage rooftop at Albers Road	373841120504801	E0.019	E0.017	E0.068	–	–	–	E0.046	–
5	Modesto Irrigation District gage rooftop at Albers Road	373841120504801	–	–	–	–	–	–	E0.069	–
5	Modesto Irrigation District gage rooftop at Albers Road	373841120504801	–	–	–	–	E0.006	–	E0.009	–
5	Modesto Irrigation District gage rooftop at Albers Road	373841120504801	–	–	–	–	E0.029	–	E0.066	E0.007
5	Modesto Irrigation District gage rooftop at Albers Road	373841120504801	E0.009	E0.011	–	–	–	–	E0.071	E0.004
5	Modesto Irrigation District gage rooftop at Albers Road	373841120504801	E0.011	E0.013	–	–	–	–	E0.085	E0.004
16	Gridley High School precipitation gage at Gridley	392205121410201	–	–	–	–	–	–	E0.038	E0.003
16	Gridley High School precipitation gage at Gridley	392205121410201	–	–	–	–	–	–	E0.030	E0.007
16	Gridley High School precipitation gage at Gridley	392205121410201	–	–	E0.010	E0.006	–	–	E0.024	–
16	Gridley High School precipitation gage at Gridley	392205121410201	–	–	–	–	–	–	E0.104	–
15	Oroville Dam precipitation gage at spillway	393234121292701	–	–	–	–	–	–	E0.008	–
15	Oroville Dam precipitation gage at spillway	393234121292701	–	–	–	–	–	–	E0.008	–

[1] Compound is biased low (B-L) based on low and variable lab reagent spike recoveries (less than 50 percent).

[2] The autosampler for this sampling period failed to cover the dry deposition collection side at the onset of a rain event and therefore is a composite sample of dry and wet deposition, and was analyzed using NWQL research method LS8054.

Appendix 7.   Analytical Results Presented in Micrograms Per Liter (µg/L) for Compounds With at Least One Detection Collected as Soil-Box Runoff (Aqueous Phase) Samples From Two Sites in the San Joaquin Valley and One Site in the Sacramento Valley, California, November 2002–March 2004

Appendix 7. Analytical results presented in micrograms per liter (µg/L) for compounds with at least one detection collected as soil-box runoff (aqueous phase) samples from two sites in the San Joaquin Valley and one site in the Sacramento Valley, California, November 2002–March 2004.

[All compounds were analyzed by gas chromatography/mass spectrometry (GC/MS) at the U.S. Geological Survey (USGS) National Water Quality Laboratory (NWQL). The Chemical Abstract Service Number is given below each compound name in brackets. The five digit alphanumeric number in parenthesis is a code used by the USGS to uniquely identify the given compound. Abbreviations: D-U, value deleted by NWQL because compound could not be determined due to matrix interference; E, estimated reported laboratory valued; hh:mm, hour:minute; (E), estimated compound due to known poor performance by applied method (Childress and others, 1999; Sandstrom and others, 2001); –, compound not detected; <, less than laboratory reporting limit]

Site number	Site name	Site identification number	Sample begin date	Start time (hh:mm)	Sample end date	End time (hh:mm)	Volume collected	1-Naphthol (E) [90-15-3] (49295)	3,4-Dichloroaniline (E) [95-76-1] (61625)	Azinphos-methyl (E) [86-50-0] (82686)
4	Modesto Irrigation District gage rooftop at Modesto	3738341210000601	Nov 7, 2002	02:00	Nov 10, 2002	03:10	0.422	<0.09	E0.04	<0.05
4	Modesto Irrigation District gage rooftop at Modesto	3738341210000601	Dec 23, 2002	09:30	Jan 9, 2003	15:35	0.7	<0.09	E0.014	<0.05
4	Modesto Irrigation District gage rooftop at Modesto	3738341210000601	Jan 9, 2003	15:30	Jan 14, 2003	10:40	0.66	<0.09	E0.013	<0.05
4	Modesto Irrigation District gage rooftop at Modesto	3738341210000601	Jan 14, 2003	10:30	Mar 18, 2003	16:35	0.21	<0.09	E0.034	<0.05
4	Modesto Irrigation District gage rooftop at Modesto	3738341210000601	Mar 18, 2003	16:35	Dec 11, 2003	13:20	0.23	<0.09	E0.023	<0.05
4	Modesto Irrigation District gage rooftop at Modesto	3738341210000601	Dec 11, 2003	13:20	Dec 16, 2003	15:20	0.12	<0.09	<0.004	<0.05
4	Modesto Irrigation District gage rooftop at Modesto	3738341210000601	Dec 23, 2003	12:20	Dec 31, 2003	09:40	0.325	<0.09	<0.004	<0.05
4	Modesto Irrigation District gage rooftop at Modesto	3738341210000601	Dec 31, 2003	09:40	Feb 4, 2004	08:20	0.235	<0.09	<0.025	<0.05
5	Modesto Irrigation District gage rooftop at Albers Road	3738411205004801	Nov 7, 2002	02:00	Nov 10, 2002	03:10	0.389	<0.09	E0.043	E0.02
5	Modesto Irrigation District gage rooftop at Albers Road	3738411205004801	Nov 14, 2002	09:10	Dec 16, 2002	12:15	0.625	<0.09	E0.045	E0.021
5	Modesto Irrigation District gage rooftop at Albers Road	3738411205004801	Dec 16, 2002	12:00	Dec 23, 2002	10:40	0.45	<0.09	E0.041	<0.05
5	Modesto Irrigation District gage rooftop at Albers Road	3738411205004801	Jan 9, 2003	14:30	Jan 14, 2003	11:50	1.07	<0.09	E0.033	<0.05
5	Modesto Irrigation District gage rooftop at Albers Road	3738411205004801	Jan 14, 2003	11:50	Feb 21, 2003	13:25	0.075	<0.09	E0.138	<0.05
5	Modesto Irrigation District gage rooftop at Albers Road	3738411205004801	Feb 21, 2003	13:25	Apr 8, 2003	12:10	0.485	<0.09	E0.075	<0.05
5	Modesto Irrigation District gage rooftop at Albers Road	3738411205004801	May 13, 2003	10:00	Nov 10, 2003	12:40	0.09	<0.09		<0.05
5	Modesto Irrigation District gage rooftop at Albers Road	3738411205004801	Nov 10, 2003	12:40	Dec 11, 2003	14:00	0.63	<0.09	E0.015	<0.05
5	Modesto Irrigation District gage rooftop at Albers Road	3738411205004801	Dec 11, 2003	14:00	Dec 16, 2003	14:30	0.64	<0.09	E0.027	<0.05
5	Modesto Irrigation District gage rooftop at Albers Road	3738411205004801	Feb 4, 2004	09:30	Mar 26, 2004	11:00	0.048	<0.09	E0.149	<0.05
16	Gridley High School precipitation gage at Gridley	3922051214110201	Nov 4, 2003	10:15	Nov 12, 2003	10:20	1.03	E0.01	E0.01	E0.009
16	Gridley High School precipitation gage at Gridley	3922051214110201	Nov 12, 2003	10:20	Dec 3, 2003	12:30	3.01	<0.09	<0.006	<0.05
16	Gridley High School precipitation gage at Gridley	3922051214110201	Dec 3, 2003	12:30	Dec 12, 2003	11:00	4.2	<0.09	E0.008	<0.05
16	Gridley High School precipitation gage at Gridley	3922051214110201	Dec 12, 2003	11:00	Dec 18, 2003	10:10	2	D-U	E0.009	<0.05
16	Gridley High School precipitation gage at Gridley	3922051214110201	Dec 18, 2003	10:00	Dec 23, 2003	08:00	1.22	D-U	<0.004	<0.05
16	Gridley High School precipitation gage at Gridley	3922051214110201	Dec 23, 2003	08:00	Dec 30, 2003	09:45	3	<0.09	<0.005	<0.05
16	Gridley High School precipitation gage at Gridley	3922051214110201	Dec 30, 2003	09:45	Jan 27, 2004	10:20	0.17	<0.09	<0.004	<0.05
16	Gridley High School precipitation gage at Gridley	3922051214110201	Jan 27, 2004	10:20	Feb 6, 2004	10:30	0.18	E0.02	<0.004	<0.05
16	Gridley High School precipitation gage at Gridley	3922051214110201	Feb 6, 2004	10:30	Feb 17, 2004	14:20	0.11	<0.09	<0.004	<0.05
16	Gridley High School precipitation gage at Gridley	3922051214110201	Feb 17, 2004	14:20	Feb 20, 2004	10:30	1.28	<0.09	<0.004	<0.05
16	Gridley High School precipitation gage at Gridley	3922051214110201	Feb 20, 2004	10:30	Mar 2, 2004	11:00	4.18	<0.09	<0.004	<0.05
16	Gridley High School precipitation gage at Gridley	3922051214110201	Mar 2, 2004	11:00	Mar 27, 2004	14:10	0.24	E0.01	E0.012	<0.05

Appendix 7. Analytical results presented in micrograms per liter (µg/L) for compounds with at least one detection collected as soil-box runoff (aqueous phase) samples from two sites in the San Joaquin Valley and one site in the Sacramento Valley, California, November 2002–March 2004.—Continued

[All compounds were analyzed by gas chromatography/mass spectrometry (GC/MS) at the U.S. Geological Survey (USGS) National Water Quality Laboratory (NWQL). The Chemical Abstract Service Number is given below each compound name in brackets. The five digit alphanumeric number in parenthesis is a code used by the USGS to uniquely identify the given compound. Abbreviations: D-U, value deleted by NWQL because compound could not be determined due to matrix interference; E, estimated reported laboratory valued; hh:mm, hour:minute; (E), estimated compound due to known poor performance by applied method (Childress and others, 1999; Sandstrom and others, 2001); –, compound not detected; <, less than laboratory reporting limit]

Site number	Site name	Site identification number	Carbaryl (E) [63-25-2] (82680)	Chlorpyrifos [2921-88-2] (38933)	Dacthal, DCPA [1861-32-1] (826820)	Diazinon [333-41-5] (39572)	Diazoxon (E) [962-58-3] (61638)	Dimethoate [60-51-5] (82662)	Hexazinone [51235-04-2] (04025)	Iprodione (E) [36734-19-7] (61593)
4	Modesto Irrigation District gage rooftop at Modesto	373834121000601	E0.012	0.009	0.018	0.026	<0.04	<0.006	D-U	<1.42
4	Modesto Irrigation District gage rooftop at Modesto	373834121000601	E0.007	0.016	0.009	0.154	E0.04	<0.006	D-U	<1.42
4	Modesto Irrigation District gage rooftop at Modesto	373834121000601	<0.041	0.039	0.009	0.304	E0.04	<0.006	D-U	<1.42
4	Modesto Irrigation District gage rooftop at Modesto	373834121000601	<0.041	0.061	0.013	0.1	<0.04	<0.006	D-U	E0.085
4	Modesto Irrigation District gage rooftop at Modesto	373834121000601	<0.041	<0.005	0.016	0.02	<0.01	<0.006	D-U	<1.42
4	Modesto Irrigation District gage rooftop at Modesto	373834121000601	<0.041	0.053	0.031	0.101	<0.01	<0.006	<0.013	<1.42
4	Modesto Irrigation District gage rooftop at Modesto	373834121000601	<0.041	0.015	0.009	0.124	E0.03	<0.006	<0.025	<1.42
4	Modesto Irrigation District gage rooftop at Modesto	373834121000601	E0.018	0.083	0.015	0.874	E0.09	<0.006	0.023	<1.42
5	Modesto Irrigation District gage rooftop at Albers Road	373841120504801	<0.041	0.011	0.009	0.012	<0.04	<0.006	D-U	<1.42
5	Modesto Irrigation District gage rooftop at Albers Road	373841120504801	<0.041	0.024	0.015	0.026	<0.04	E0.027	D-U	<1.42
5	Modesto Irrigation District gage rooftop at Albers Road	373841120504801	<0.041	0.028	0.01	0.043	<0.04	<0.006	D-U	<1.42
5	Modesto Irrigation District gage rooftop at Albers Road	373841120504801	E0.004	0.544	0.005	1.62	E0.09	<0.006	D-U	<1.42
5	Modesto Irrigation District gage rooftop at Albers Road	373841120504801	<0.041	0.06	0.021	0.37	E0.08	<0.006	D-U	E3.52
5	Modesto Irrigation District gage rooftop at Albers Road	373841120504801	<0.041	0.017	0.013	0.042	0.12	D-U	D-U	D-U
5	Modesto Irrigation District gage rooftop at Albers Road	373841120504801	<0.041	<0.005	<0.003	<0.005	<0.01	<0.006	<0.013	<1.42
5	Modesto Irrigation District gage rooftop at Albers Road	373841120504801	<0.041	0.009	0.01	0.011	<0.01	<0.006	<0.013	<1.42
5	Modesto Irrigation District gage rooftop at Albers Road	373841120504801	<0.041	0.015	0.011	0.096	E0.02	<0.006	<0.013	<1.42
5	Modesto Irrigation District gage rooftop at Albers Road	373841120504801	<0.041	<0.005	0.047	<0.005	<0.01	<0.006	<0.013	E0.356
16	Gridley High School precipitation gage at Gridley	392205121410201	E0.023	0.005	0.008	0.012	<0.01	<0.006	E0.007	<1.42
16	Gridley High School precipitation gage at Gridley	392205121410201	E0.011	E0.003	E0.003	0.005	<0.01	<0.006	<0.013	<1.42
16	Gridley High School precipitation gage at Gridley	392205121410201	E0.024	0.005	0.006	0.01	<0.01	<0.006	<0.013	<1.42
16	Gridley High School precipitation gage at Gridley	392205121410201	<0.055	E0.004	0.004	0.013	<0.01	<0.006	<0.013	<1.42
16	Gridley High School precipitation gage at Gridley	392205121410201	E0.034	0.009	0.005	0.031	<0.01	<0.006	<0.013	<1.42
16	Gridley High School precipitation gage at Gridley	392205121410201	E0.012	0.018	E0.003	0.022	E0.01	<0.006	<0.013	<1.42
16	Gridley High School precipitation gage at Gridley	392205121410201	<0.041	0.025	0.007	0.078	<0.01	<0.006	<0.013	<1.42
16	Gridley High School precipitation gage at Gridley	392205121410201	<0.041	0.057	0.01	1.75	E0.09	<0.006	<0.013	<1.42
16	Gridley High School precipitation gage at Gridley	392205121410201	E0.043	0.042	0.014	0.796	E0.09	<0.006	<0.013	<1.42
16	Gridley High School precipitation gage at Gridley	392205121410201	<0.041	0.014	0.004	0.131	E0.01	<0.006	<0.013	<1.42
16	Gridley High School precipitation gage at Gridley	392205121410201	<0.041	0.007	0.004	0.07	<0.01	<0.006	<0.013	E0.016
16	Gridley High School precipitation gage at Gridley	392205121410201	E0.045	0.028	0.01	0.046	<0.01	<0.006	<0.013	E0.046

Appendix 7. Analytical results presented in micrograms per liter (μg/L) for compounds with at least one detection collected as soil-box runoff (aqueous phase) samples from two sites in the San Joaquin Valley and one site in the Sacramento Valley, California, November 2002–March 2004.—Continued

[All compounds were analyzed by gas chromatography/mass spectrometry (GC/MS) at the U.S. Geological Survey (USGS) National Water Quality Laboratory (NWQL). The Chemical Abstract Service Number is given below each compound name in brackets. The five digit alphanumeric number in parenthesis is a code used by the USGS to uniquely identify the given compound. Abbreviations: D-U, value deleted by NWQL because compound could not be determined due to matrix interference; E, estimated reported laboratory valued; hh:mm, hour:minute; (E), estimated compound due to known poor performance by applied method (Childress and others, 1999; Sandstrom and others, 2001); –, compound not detected; <, less than laboratory reporting limit]

Site number	Site name	Site identification number	Malathion [121-75-5] (39532)	Malaoxon (E) [1634-78-2] (61652)	Methidathion [950-37-8] (61598)	Metolachlor [51218-45-2] (39415)	Metribuzin [21087-64-9] (82630)	Myclobutanil (E) [88671-89-0] (61599)	Parathion-methyl [298-00-0] (82667)
4	Modesto Irrigation District gage rooftop at Modesto	373834121000601	E0.02	0.12	<0.006	0.309	0.028	E0.064	0.02
4	Modesto Irrigation District gage rooftop at Modesto	373834121000601	<0.027	<0.008	0.009	0.193	0.09	E0.047	<0.006
4	Modesto Irrigation District gage rooftop at Modesto	373834121000601	<0.027	<0.008	0.022	0.143	0.112	E0.049	<0.006
4	Modesto Irrigation District gage rooftop at Modesto	373834121000601	<0.027	<0.008	<0.006	0.046	0.025	E0.104	<0.006
4	Modesto Irrigation District gage rooftop at Modesto	373834121000601	<0.027	<0.008	<0.006	0.02	<0.006	E0.052	<0.006
4	Modesto Irrigation District gage rooftop at Modesto	373834121000601	<0.027	<0.008	<0.006	0.05	<0.006	<0.008	<0.015
4	Modesto Irrigation District gage rooftop at Modesto	373834121000601	<0.027	<0.008	0.013	E0.01	<0.015	E0.026	<0.015
4	Modesto Irrigation District gage rooftop at Modesto	373834121000601	<0.027	<0.008	0.045	E0.009	<0.015	E0.021	<0.015
5	Modesto Irrigation District gage rooftop at Albers Road	373841120504801	<0.027	<0.008	<0.006	0.132	0.034	E0.067	0.042
5	Modesto Irrigation District gage rooftop at Albers Road	373841120504801	<0.027	<0.008	<0.006	0.309	0.116	E0.075	<0.006
5	Modesto Irrigation District gage rooftop at Albers Road	373841120504801	<0.027	<0.008	0.016	0.154	0.043	E0.053	<0.006
5	Modesto Irrigation District gage rooftop at Albers Road	373841120504801	<0.027	<0.008	0.054	0.066	0.069	E0.039	<0.006
5	Modesto Irrigation District gage rooftop at Albers Road	373841120504801	<0.027	<0.008	0.064	0.089	0.069	E1.02	<0.006
5	Modesto Irrigation District gage rooftop at Albers Road	373841120504801	E0.015	<0.125	<0.006	0.029	<0.125	E D-U	<0.006
5	Modesto Irrigation District gage rooftop at Albers Road	373841120504801	<0.027	<0.008	<0.006	0.046	<0.006	E0.117	0.085
5	Modesto Irrigation District gage rooftop at Albers Road	373841120504801	<0.027	<0.008	0.01	E0.012	<0.006	E0.028	E0.014
5	Modesto Irrigation District gage rooftop at Albers Road	373841120504801	<0.027	<0.008	0.044	0.023	0.021	E0.034	<0.015
5	Modesto Irrigation District gage rooftop at Albers Road	373841120504801	0.089	<0.008	0.115	<0.013	<0.006	E0.151	<0.015
16	Gridley High School precipitation gage at Gridley	392205121410201	<0.027	E0.012	<0.006	<0.013	<0.006	<0.008	0.041
16	Gridley High School precipitation gage at Gridley	392205121410201	<0.027	<0.008	<0.006	<0.013	<0.006	<0.008	0.017
16	Gridley High School precipitation gage at Gridley	392205121410201	<0.027	<0.008	<0.006	<0.013	<0.006	<0.008	E0.013
16	Gridley High School precipitation gage at Gridley	392205121410201	<0.027	<0.008	<0.006	<0.013	<0.006	<0.008	E0.011
16	Gridley High School precipitation gage at Gridley	392205121410201	<0.027	<0.008	<0.006	<0.013	<0.006	<0.008	E0.011
16	Gridley High School precipitation gage at Gridley	392205121410201	<0.027	<0.008	<0.006	<0.013	<0.006	<0.008	E0.008
16	Gridley High School precipitation gage at Gridley	392205121410201	<0.027	<0.008	0.043	<0.013	<0.006	<0.008	<0.015
16	Gridley High School precipitation gage at Gridley	392205121410201	<0.027	<0.008	0.038	<0.013	<0.006	<0.008	<0.015
16	Gridley High School precipitation gage at Gridley	392205121410201	<0.027	<0.008	<0.006	<0.013	<0.006	<0.008	<0.015
16	Gridley High School precipitation gage at Gridley	392205121410201	<0.027	<0.008	<0.006	E0.002	<0.006	E0.003	<0.015
16	Gridley High School precipitation gage at Gridley	392205121410201	<0.027	<0.008	<0.006	<0.013	<0.006	E0.008	E0.003
16	Gridley High School precipitation gage at Gridley	392205121410201	<0.027	<0.008	<0.006	<0.013	<0.006	E0.008	E0.008
16	Gridley High School precipitation gage at Gridley	392205121410201	E0.022	<0.008	0.019	0.015	<0.006	E0.022	E0.015

Appendix 7. Analytical results presented in micrograms per liter (µg/L) for compounds with at least one detection collected as soil-box runoff (aqueous phase) samples from two sites in the San Joaquin Valley and one site in the Sacramento Valley, California, November 2002–March 2004.—Continued

[All compounds were analyzed by gas chromatography/mass spectrometry (GC/MS) at the U.S. Geological Survey (USGS) National Water Quality Laboratory (NWQL). The Chemical Abstract Service Number is given below each compound name in brackets. The five digit alphanumeric number in parentheses is a code used by the USGS to uniquely identify the given compound. Abbreviations: D-U, value deleted by NWQL because compound could not be determined due to matrix interference; E, estimated reported laboratory valued; hh:mm, hour/minute; (E), estimated compound due to known poor performance by applied method (Childress and others 1999; Sandstrom and others, 2001); –, compound not detected; <, less than laboratory reporting limit]

Site number	Site name	Site identification number	Paraoxon-methyl (E) [950-35-6] (61664)	Pendimethalin [40487-42-1] (82683)	Prometon (E) [1610-18-0] (04037)	Prometryn [7287-19-6] (04036)	Pronamide [23950-58-5] (82676)	Simazine [122-34-9] (04035)	Trifluralin [1582-09-8] (82661)
4	Modesto Irrigation District gage rooftop at Modesto	37383412100060	<0.03	0.094	<0.01	<0.005	<0.004	0.765	0.021
4	Modesto Irrigation District gage rooftop at Modesto	37383412100060	<0.03	0.029	<0.01	0.009	0.006	0.297	E0.009
4	Modesto Irrigation District gage rooftop at Modesto	37383412100060	<0.03	0.03	<0.01	0.009	<0.004	0.246	E0.008
4	Modesto Irrigation District gage rooftop at Modesto	37383412100060	<0.03	0.07	<0.01	<0.005	<0.004	0.25	<0.009
4	Modesto Irrigation District gage rooftop at Modesto	37383412100060	<0.03	<0.035	<0.01	0.03	<0.025	0.083	<0.009
4	Modesto Irrigation District gage rooftop at Modesto	37383412100060	<0.03	<0.095	<0.01	<0.065	<0.004	<0.005	<0.009
4	Modesto Irrigation District gage rooftop at Modesto	37383412100060	<0.03	0.039	<0.01	<0.005	<0.004	0.027	<0.009
4	Modesto Irrigation District gage rooftop at Modesto	37383412100060	<0.03	E0.095	<0.01	0.007	<0.004	0.077	<0.015
5	Modesto Irrigation District gage rooftop at Albers Road	37384112050480	<0.03	0.071	<0.01	<0.005	<0.004	0.388	0.015
5	Modesto Irrigation District gage rooftop at Albers Road	37384112050480	<0.03	0.065	E0.01	0.088	0.017	0.934	<0.009
5	Modesto Irrigation District gage rooftop at Albers Road	37384112050480	<0.03	0.06	<0.01	0.023	<0.004	0.324	<0.009
5	Modesto Irrigation District gage rooftop at Albers Road	37384112050480	<0.03	0.033	<0.01	<0.006	0.004	0.146	E0.005
5	Modesto Irrigation District gage rooftop at Albers Road	37384112050480	<0.03	<0.022	<0.01	<0.005	<0.004	1.01	<0.009
5	Modesto Irrigation District gage rooftop at Albers Road	37384112050480	D-U	0.039	D-U	D-U	<0.004	E0.082	<0.009
5	Modesto Irrigation District gage rooftop at Albers Road	37384112050480	<0.03	<0.022	<0.01	<0.005	<0.004	0.556	<0.009
5	Modesto Irrigation District gage rooftop at Albers Road	37384112050480	<0.03	0.024	<0.01	0.017	<0.004	0.056	<0.009
5	Modesto Irrigation District gage rooftop at Albers Road	37384112050480	<0.03	0.032	<0.01	0.017	<0.009	0.076	<0.009
5	Modesto Irrigation District gage rooftop at Albers Road	37384112050480	<0.03	<0.022	<0.01	<0.005	<0.004	0.534	<0.009
16	Gridley High School precipitation gage at Gridley	39220512141020	E0.02	E0.014	E0.01	<0.005	<0.004	0.025	<0.009
16	Gridley High School precipitation gage at Gridley	39220512141020	<0.03	E0.009	<0.01	E0.003	<0.004	0.017	<0.009
16	Gridley High School precipitation gage at Gridley	39220512141020	<0.03	E0.013	D-U	0.007	<0.004	0.019	<0.009
16	Gridley High School precipitation gage at Gridley	39220512141020	<0.03	0.023	E0.01	0.007	<0.004	0.026	<0.009
16	Gridley High School precipitation gage at Gridley	39220512141020	<0.03	0.022	E0.01	<0.005	<0.004	0.024	<0.009
16	Gridley High School precipitation gage at Gridley	39220512141020	<0.03	E0.015	<0.01	<0.005	<0.004	0.028	<0.009
16	Gridley High School precipitation gage at Gridley	39220512141020	<0.03	<0.022	<0.01	<0.005	<0.004	<0.005	<0.009
16	Gridley High School precipitation gage at Gridley	39220512141020	<0.03	<0.022	<0.01	<0.005	<0.004	0.039	<0.009
16	Gridley High School precipitation gage at Gridley	39220512141020	<0.03	<0.022	<0.01	<0.005	<0.004	0.06	<0.009
16	Gridley High School precipitation gage at Gridley	39220512141020	E0.03	E0.022	<0.01	<0.005	<0.004	0.014	<0.009
16	Gridley High School precipitation gage at Gridley	39220512141020	<0.03	<0.022	<0.03	<0.005	<0.004	0.021	<0.009
16	Gridley High School precipitation gage at Gridley	39220512141020	<0.03	0.029	E0.01	<0.005	<0.004	0.141	<0.009

References Cited

Childress, C.J.O., Foreman, W.T., Connor, B.F., and Maloney, T.J., 1999, New reporting procedures based on long-term method detection levels and some considerations for interpretations of water-quality data provided by the U.S. Geological Survey National Water Quality Laboratory: U.S. Geological Survey Open-File Report 99–193, 19 p.

Sandstrom, M.W., Stroppel, M.E., Foreman, W.T., and Schroeder, M.P., 2001, Methods of analysis by the U.S. Geological Survey National Water Quality Laboratory–Determination of moderate-use pesticides and selected degradates in water by C-18 solid-phase extraction and capillary-column gas chromatography/mass spectrometry with selected-ion monitoring (methods 2002/2011): U.S. Geological Survey Water-Resources Investigations Report 01–4098, 70 p.

Appendix 8.   Analytical Results Presented in Micrograms Per Liter (µg/L) for Compounds With at Least One Detection From the Soil Box as Soil-Box Runoff (Suspended Sediment) Samples Collected at Two Sites in the San Joaquin Valley and One Site in the Sacramento Valley, California, November 2002–March 2004

Appendix 8. Analytical results presented in micrograms per liter (µg/L) for compounds with at least one detection from the soil box as soil-box runoff (suspended sediment) samples collected at two sites in the San Joaquin Valley and one site in the Sacramento Valley, California, November 2002–March 2004.

[All compounds were analyzed by gas chromatography/mass spectrometry (GC/MS) at the U.S. Geological Survey (USGS) National Water Quality Laboratory (NWQL). The Chemical Abstract Service Number is given below each compound name in brackets. The five digit alphanumeric number in parenthesis is a code used by the USGS to uniquely identify the given compound. Abbreviations: D-U, value deleted by NWQL because compound could not be determined due to matrix interference; E, estimated reported laboratory valued; GFF, glass fiber filters; hh:mm, hour:minute; g, gram; L, liter; NA, not applicable; [-], no Chemical Abstract Service Number available for given compound; <, less than laboratory reporting limit; (E), estimated compound due to known poor performance by applied method (Childress and others, 1999; Sandstrom and others, 2001); –, compound not detected]

Site number	Site name	Site identification number	Sample begin date	Start time (hh:mm)	Sample end date	End time (hh:mm)	Number of GFF per sample	Volume filtered (L)	Estimated suspended sediment mass (g)	1-Naphthol[1] [E] [90-15-3] (S0282)
4	Modesto Irrigation District gage rooftop at Modesto	37383412100060601	Dec 23, 2002	09:30	Jan 9, 2003	15:35	1	0.7	0.23	<0.79
4	Modesto Irrigation District gage rooftop at Modesto	37383412100060601	Jan 9, 2003	15:30	Jan 14, 2003	1040	1	0.66	0.22	E0.73
4	Modesto Irrigation District gage rooftop at Modesto	37383412100060601	Jan 14, 2003	10:30	Mar 18, 2003	16:35	1	0.21	0.31	<0.79
4	Modesto Irrigation District gage rooftop at Modesto	37383412100060601	Mar 18, 2003	16:35	Dec 11, 2003	13:20	1	0.23	0.25	<0.79
4	Modesto Irrigation District gage rooftop at Modesto	37383412100060601	Dec 11, 2003	13:20	Dec 16, 2003	15:20	1	0.12	0.68	<0.79
4	Modesto Irrigation District gage rooftop at Modesto	37383412100060601	Dec 23, 2003	12:20	Dec 31, 2003	09:40	1	0.31	0.26	<0.79
4	Modesto Irrigation District gage rooftop at Modesto	37383412100060601	Dec 31, 2003	09:40	Feb 4, 2004	08:20	1	0.235	0.19	<0.79
5	Modesto Irrigation District gage rooftop at Albers Road	37384112050480801	Nov 7, 2002	02:00	Nov 10, 2002	03:10	1	0.389	3.88	<0.79
5	Modesto Irrigation District gage rooftop at Albers Road	37384112050480801	Jan 9, 2003	14:30	Jan 14, 2003	11:50	1	1.7	0.48	<0.79
5	Modesto Irrigation District gage rooftop at Albers Road	37384112050480801	Feb 21, 2003	13:25	Apr 8, 2003	12:10	1	0.475	0.63	<0.79
5	Modesto Irrigation District gage rooftop at Albers Road	37384112050480801	May 13, 2003	10:00	Nov 10, 2003	12:40	1	0.9	0.17	<0.79
5	Modesto Irrigation District gage rooftop at Albers Road	37384112050480801	Nov 10, 2003	12:40	Dec 11, 2003	14:00	1	0.61	0.35	<0.79
5	Modesto Irrigation District gage rooftop at Albers Road	37384112050480801	Dec 11, 2003	14:00	Dec 16, 2003	14:30	2	0.64	0.32	<0.79
5	Modesto Irrigation District gage rooftop at Albers Road	37384112050480801	Feb 4, 2004	09:30	Mar 26, 2004	11:00	1	0.48	0.20	<0.79
16	Gridley High School precipitation gage at Gridley	39220512141020201	Nov 4, 2003	10:15	Nov 12, 2003	10:20	5	0.989	4.44	<0.79
16	Gridley High School precipitation gage at Gridley	39220512141020201	Nov 12, 2003	10:20	Dec 3, 2003	12:30	6	2.8	2.36	<0.79
16	Gridley High School precipitation gage at Gridley	39220512141020201	Dec 3, 2003	12:30	Dec 12, 2003	11:00	9	3.745	4.73	<0.79
16	Gridley High School precipitation gage at Gridley	39220512141020201	Dec 12, 2003	11:00	Dec 18, 2003	10:10	5	2	3.56	<0.79
16	Gridley High School precipitation gage at Gridley	39220512141020201	Dec 18, 2003	10:00	Dec 23, 2003	08:00	2	1.22	0.72	<0.79
16	Gridley High School precipitation gage at Gridley	39220512141020201	Dec 23, 2003	08:00	Dec 30, 2003	09:45	5	3	1.65	<0.79
16	Gridley High School precipitation gage at Gridley	39220512141020201	Dec 30, 2003	09:45	Jan 27, 2004	10:20	1	0.17	0.37	<0.79
16	Gridley High School precipitation gage at Gridley	39220512141020201	Jan 27, 2004	10:20	Feb 6, 2004	10:30	1	0.18	-0.66	<0.79
16	Gridley High School precipitation gage at Gridley	39220512141020201	Feb 6, 2004	10:30	Feb 17, 2004	14:20	1	0.11	0.200	<0.79
16	Gridley High School precipitation gage at Gridley	39220512141020201	Feb 17, 2004	14:20	Feb 20, 2004	10:30	4	1.16	1.63	<0.79
16	Gridley High School precipitation gage at Gridley	39220512141020201	Feb 20, 2004	10:30	Mar 2, 2004	11:00	38	4.18	21.82	<0.79
16	Gridley High School precipitation gage at Gridley	39220512141020201	Mar 2, 2004	11:00	Mar 27, 2004	14:10	1	0.24	0.76	<0.79
Samples ruined during laboratory prep										
4	Modesto Irrigation District gage rooftop at Modesto	37383412100060601	Nov 7, 2002	02:00	Nov 10, 2002	03:10	NA	NA	NA	NA
5	Modesto Irrigation District gage rooftop at Albers Road	37384112050480801	Nov 14, 2002	09:10	Dec 16, 2002	12:15	NA	NA	NA	NA
5	Modesto Irrigation District gage rooftop at Albers Road	37384112050480801	Dec 16, 2002	12:00	Dec 23, 2002	10:40	NA	NA	NA	NA
5	Modesto Irrigation District gage rooftop at Albers Road	37384112050480801	Jan 14, 2003	11:50	Feb 21, 2003	13:25	NA	NA	NA	NA

Appendix 8. Analytical results presented in micrograms per liter (µg/L) for compounds with at least one detection from the soil box as soil-box runoff (suspended sediment) samples collected at two sites in the San Joaquin Valley and one site in the Sacramento Valley, California, November 2002–March 2004.—Continued

[All compounds were analyzed by gas chromatography/mass spectrometry (GC/MS) at the U.S. Geological Survey (USGS) National Water Quality Laboratory (NWQL). The Chemical Abstract Service Number is given below each compound name in brackets. The five digit alphanumeric number in parenthesis is a code used by the USGS to uniquely identify the given compound. Abbreviations: D-U, value deleted by NWQL because compound could not be determined due to matrix interference; E, estimated reported laboratory valued; GFF, glass fiber filters; hh:mm, hour:minute; g, gram; L, liter; [-], no Chemical Abstract Service Number available for given compound; <, less than laboratory reporting limit; (E), estimated compound due to known poor performance by applied method (Childress and others, 1999; Sandstrom and others, 2001); -, compound not detected]

Site num-ber	Site name	Site identification number	3,4-dichloro-aniline[1] (E) [95-76-1] (S0287)	Carbaryl[1] (E) [63-25-2] (S0295)	Chlorpyrifos[1] [2921-88-2] (S0296)	Dacthal (DCPA) [1861-32-1] (S0300)	Diazinon [333-41-5] (S0302)	Fipronil [120068-37-3] (S0314)	Iprodione[1] (E) [36734-19-7] (S0322)
4	Modesto Irrigation District gage rooftop at Modesto	3738341210000601	E0.69	<0.16	<0.08	<0.08	0.53	<0.08	<0.79
4	Modesto Irrigation District gage rooftop at Modesto	3738341210000601	E0.40	<0.16	0.75	E0.37	0.81	<0.08	<0.79
4	Modesto Irrigation District gage rooftop at Modesto	3738341210000601	E0.36	<0.16	<0.08	<0.08	0.18	<0.08	<0.79
4	Modesto Irrigation District gage rooftop at Modesto	3738341210000601	<5.9	<0.16	<0.08	<0.08	<0.08	E0.13	<0.79
4	Modesto Irrigation District gage rooftop at Modesto	3738341210000601	<5.9	<0.16	<0.08	<0.08	<0.08	<0.08	<0.79
4	Modesto Irrigation District gage rooftop at Modesto	3738341210000601	<5.9	<0.16	<0.08	<0.08	<0.08	<0.08	<0.79
4	Modesto Irrigation District gage rooftop at Modesto	3738341210000601	<5.9	<0.16	<0.08	<0.08	0.16	<0.08	<0.79
5	Modesto Irrigation District gage rooftop at Albers Road	3738411205048S01	E0.31	<0.16	<0.08	0.69	<0.08	<0.08	<0.79
5	Modesto Irrigation District gage rooftop at Albers Road	3738411205048S01	E0.78	<0.16	0.86	E0.21	0.37	<0.08	<0.79
5	Modesto Irrigation District gage rooftop at Albers Road	3738411205048S01	E0.87	<0.16	<0.08	<0.08	0.58	<0.08	E0.11
5	Modesto Irrigation District gage rooftop at Albers Road	3738411205048S01	<5.9	<0.16	<0.08	<0.08	<0.08	<0.08	<0.79
5	Modesto Irrigation District gage rooftop at Albers Road	3738411205048S01	E0.14	<0.16	<0.08	<0.08	<0.08	<0.08	<0.79
5	Modesto Irrigation District gage rooftop at Albers Road	3738411205048S01	<5.9	<0.16	<0.08	<0.08	0.64	<0.08	<0.79
5	Modesto Irrigation District gage rooftop at Albers Road	3738411205048S01	<5.9	<0.16	<0.08	<0.08	<0.08	<0.08	<0.79
16	Gridley High School precipitation gage at Gridley	3922051214102O1	<5.9	E0.14	0.57	E0.49	<0.08	<0.08	<0.79
16	Gridley High School precipitation gage at Gridley	3922051214102O1	<5.9	<0.16	E0.15	E0.10	<0.08	E0.30	<0.79
16	Gridley High School precipitation gage at Gridley	3922051214102O1	<5.9	E0.35	E0.21	E0.15	<0.08	<0.08	<0.79
16	Gridley High School precipitation gage at Gridley	3922051214102O1	E0.14	E0.65	E0.31	E0.18	<0.08	<0.08	<0.79
16	Gridley High School precipitation gage at Gridley	3922051214102O1	<5.9	E0.40	E0.49	E0.26	<0.08	<0.08	<0.79
16	Gridley High School precipitation gage at Gridley	3922051214102O1	<5.9	<0.16	E0.50	E0.10	<0.08	<0.08	<0.79
16	Gridley High School precipitation gage at Gridley	3922051214102O1	<5.9	<0.16	<0.08	<0.08	<0.08	<0.08	<0.79
16	Gridley High School precipitation gage at Gridley	3922051214102O1	<5.9	<0.16	<0.08	<0.08	0.52	<0.08	<0.79
16	Gridley High School precipitation gage at Gridley	3922051214102O1	<5.9	<0.16	<0.08	<0.08	0.31	<0.08	<0.79
16	Gridley High School precipitation gage at Gridley	3922051214102O1	E0.22	E0.37	0.72	E0.26	0.13	<0.08	<0.79
16	Gridley High School precipitation gage at Gridley	3922051214102O1	<5.9	E0.16	0.14	E0.25	0.29	<0.08	E0.16
16	Gridley High School precipitation gage at Gridley	3922051214102O1	E0.17	<0.16	0.14	<0.08	<0.08	<0.08	<0.79

Appendix 8. Analytical results presented in micrograms per liter (µg/L) for compounds with at least one detection from the soil box as soil-box runoff (suspended sediment) samples collected at two sites in the San Joaquin Valley and one site in the Sacramento Valley, California, November 2002–March 2004.—Continued

[All compounds were analyzed by gas chromatography/mass spectrometry (GC/MS) at the U.S. Geological Survey (USGS) National Water Quality Laboratory (NWQL). The Chemical Abstract Service Number is given below each compound name in brackets. The five digit alphanumeric number in parenthesis is a code used by the USGS to uniquely identify the given compound. Abbreviations: D-U, value deleted by NWQL because compound could not be determined due to matrix interference; E, estimated reported laboratory valued; GFF, glass fiber filters; hh:mm, hour:minute; g, gram; L, liter; [-], no Chemical Abstract Service Number available for given compound; <, less than laboratory reporting limit; (E), estimated compound due to known poor performance by applied method (Childress and others, 1999; Sandstrom and others, 2001); -, compound not detected]

Site number	Site name	Site identification number	Malaoxon[1] (E) [1634-78-2] (S0324)	Myclobutanil (E) [88671-89-0] (S0330)	Parathion-methyl [298-00-0] (S0332)	Pendimethalin [40487-42-1] (S0333)	Simazine [122-34-9] (S0343)	Trifluralin [1582-09-8] (S0348)
4	Modesto Irrigation District gage rooftop at Modesto	37383412000601	D-U	E0.69	<0.16	<0.11	E0.93	<0.08
4	Modesto Irrigation District gage rooftop at Modesto	37383412000601	D-U	<0.08	<0.16	<0.11	E0.84	<0.08
4	Modesto Irrigation District gage rooftop at Modesto	37383412000601	D-U	<0.08	<0.16	<0.11	0.26	<0.08
4	Modesto Irrigation District gage rooftop at Modesto	37383412000601	D-U	<0.08	<0.16	<0.11	<0.16	<0.08
4	Modesto Irrigation District gage rooftop at Modesto	37383412000601	D-U	<0.08	<0.16	<0.11	<0.16	<0.08
4	Modesto Irrigation District gage rooftop at Modesto	37383412000601	D-U	<0.08	<0.16	<0.11	<0.16	<0.08
4	Modesto Irrigation District gage rooftop at Modesto	37383412000601	D-U	<0.08	<0.16	<0.11	<0.16	<0.08
5	Modesto Irrigation District gage rooftop at Albers Road	37384112050480	D-U	E0.89	<0.16	0.96	0.13	0.81
5	Modesto Irrigation District gage rooftop at Albers Road	37384112050480	D-U	<0.08	<0.16	0.17	E0.68	E0.1
5	Modesto Irrigation District gage rooftop at Albers Road	37384112050480	D-U	E0.12	<0.16	0.15	0.11	<0.08
5	Modesto Irrigation District gage rooftop at Albers Road	37384112050480	D-U	<0.08	<0.16	<0.11	0.68	<0.08
5	Modesto Irrigation District gage rooftop at Albers Road	37384112050480	D-U	<0.08	<0.16	<0.11	E0.75	<0.08
5	Modesto Irrigation District gage rooftop at Albers Road	37384112050480	<0.40	E0.5	<0.16	<0.9	0.11	<0.08
5	Modesto Irrigation District gage rooftop at Albers Road	37384112050480	D-U	<0.08	<0.16	<0.11	<0.16	<0.08
16	Gridley High School precipitation gage at Gridley	39220512141020	<0.40	<0.08	E0.27	0.13	E0.96	<0.08
16	Gridley High School precipitation gage at Gridley	39220512141020	<0.40	<0.08	E0.26	<0.11	E0.21	<0.08
16	Gridley High School precipitation gage at Gridley	39220512141020	<0.40	<0.08	E0.80	0.7	E0.47	<0.08
16	Gridley High School precipitation gage at Gridley	39220512141020	<0.40	<0.08	0.92	E0.46	E0.73	<0.08
16	Gridley High School precipitation gage at Gridley	39220512141020	E0.23	<0.08	<0.16	E0.37	E0.59	<0.08
16	Gridley High School precipitation gage at Gridley	39220512141020	<0.40	<0.08	E0.20	E0.24	E0.26	<0.08
16	Gridley High School precipitation gage at Gridley	39220512141020	D-U	<0.08	<0.16	<0.11	<0.16	<0.08
16	Gridley High School precipitation gage at Gridley	39220512141020	D-U	<0.08	<0.16	<0.11	<0.16	<0.08
16	Gridley High School precipitation gage at Gridley	39220512141020	D-U	<0.08	<0.16	<0.11	<0.16	<0.08
16	Gridley High School precipitation gage at Gridley	39220512141020	D-U	<0.08	<0.16	<0.11	<0.16	<0.08
16	Gridley High School precipitation gage at Gridley	39220512141020	<0.40	<0.08	<0.16	<0.11	E0.72	<0.08
16	Gridley High School precipitation gage at Gridley	39220512141020	<0.40	E0.86	E0.86	0.16	0.22	E0.18
16	Gridley High School precipitation gage at Gridley	39220512141020	D-U	<0.08	<0.16	<0.11	0.24	<0.08

[1]Compound is biased low (B-L) because of low laboratory matrix and reagent spike recoveries, see table 10. Compounds with non-detections are not presented in the appendix but concentrations may be present in low concentrations in the environment.

References Cited

Childress, C.J.O., Foreman, W.T., Connor, B.F., and Maloney, T.J., 1999, New reporting procedures based on long-term method detection levels and some considerations for interpretations of water-quality data provided by the U.S. Geological Survey National Water Quality Laboratory: U.S. Geological Survey Open-File Report 99–193, 19 p.

Sandstrom, M.W., Stroppel, M.E., Foreman, W.T., and Schroeder, M.P., 2001, Methods of analysis by the U.S. Geological Survey National Water Quality Laboratory—Determination of moderate-use pesticides and selected degradates in water by C-18 solid-phase extraction and capillary-column gas chromatography/mass spectrometry with selected-ion monitoring (methods 2002/2011): U.S. Geological Survey Water-Resources Investigations Report 01–4098, 70 p.

Appendix 9.   Analytical Results Presented in Micrograms Per Kilogram (µg/Kg) for Compounds With at Least One Detection From the Soil Box as Surficial-Soil Samples Collected at Two Sites in the San Joaquin Valley and One Site in the Sacramento Valley, California, November 2002–March 2004

Appendix 9. Analytical results presented in micrograms per kilogram (µg/Kg) for compounds with at least one detection from the soil box as surficial-soil samples collected at two sites in the San Joaquin Valley and one site in the Sacramento Valley, California, November 2002–March 2004.

[The background concentrations and control samples are considered quality assurance samples but are presented with the environmental samples for comparative purposes. All compounds were analyzed by gas chromatography/mass spectrometry (GC/MS) at the U.S. Geological Survey (USGS) National Water Quality Laboratory (NWQL). The Chemical Abstract Service Number is given below each compound name in brackets. The five digit alphanumeric number in parenthesis is a code used by the USGS to uniquely identify the given compound. Mix 1 collected near site 10 (Westly Raingage). Mix 2 collected near site 11 (Newman Raingage). Mix 3 collected near site 5 (Modesto Irrigation District at Albers Road). Mix 4 collected near site 16 (Gridley High School). Sample weight was wet weight in grams. Total area of soil box is 0.374 square meters. Abbreviations: CSUS, California State University, Sacramento; E, estimated reported laboratory value; NA, not applicable; [-], no Chemical Abstract Service Number available for given compound; <, less than laboratory reporting value]

Site number	Site name	Site identification number	Start date	Start time (hh:mm)	End date	End time (hh:mm)	1-Napthol[1] (E) [90-15-3] (S0282)	2,6-Diethylaniline[1] (E) [579-66-8] (S0283)	2-Chloro-2,6-diethylacetanilide[1] [6967-29-9] (S0285)
			Environmental surficial-soil samples						
4	Modesto Irrigation District gage rooftop at Modesto	373834121000601	Aug 14, 2002	10:00	Oct 4, 2002	11:15	<10	<30	<1
4	Modesto Irrigation District gage rooftop at Modesto	373834121000601	Oct 4, 2002	11:15	Jul 22, 2003	09:50	<10	<30	<1
4	Modesto Irrigation District gage rooftop at Modesto	373834121000601	Jul 22, 2003	10:15	Jul 8, 2004	10:00	<10	<30	<1
5	Modesto Irrigation District gage rooftop at Albers Road	373841120504801	Aug 14, 2002	11:00	Oct 4, 2002	12:00	<10	<30	<1
5	Modesto Irrigation District gage rooftop at Albers Road	373841120504801	Oct 4, 2002	12:00	Jul 22, 2003	10:15	<10	<30	<1
5	Modesto Irrigation District gage rooftop at Albers Road	373841120504801	Jul 22, 2003	11:00	July 8, 2004	10:30	<10	<30	<1
16	Gridley High School precipitation gage at Gridley	392205121410201	Feb 7, 2003	11:30	Jul 21, 2003	09:30	<10	<30	<1
16	Gridley High School precipitation gage at Gridley	392205121410201	Jul 21, 2003	09:30	Jun 3, 2004	10:15	<10	<30	<1
			Quality assurance surficial-soil control samples						
14	Soil Control Box at CSUS Placer Hall at Sacramento[2]	383343121252501	Aug 15, 2002	10:00	Oct 9, 2002	12:00	<10	<30	<1
14	Soil Control Box at CSUS Placer Hall at Sacramento[2]	383343121252501	Oct 9, 2002	12:00	Jul 31, 2003	09:50	<10	<30	<1
14	Soil Control Box at CSUS Placer Hall at Sacramento[2]	383343121252501	Jul 31, 2003	09:50	Jul 26, 2004	08:00	<10	<30	<1
			Quality assurance background concentration surficial-soil samples						
NA	Mix 1	373335121143001	Aug 1, 2002	12:00	NA	NA	<10	<30	<1
NA	Mix 2	371735121031201	Aug 1, 2002	12:15	NA	NA	<10	<30	<1
NA	Mix 3	373841120504801	Aug 1, 2002	12:30	NA	NA	<10	<30	<1
NA	Mix 4	392205121410201	Feb 7, 2003	11:30	NA	NA	<10	<30	<1

Appendix 9.   Analytical results presented in micrograms per kilogram (µg/Kg) for compounds with at least one detection from the soil box as surficial-soil samples collected at two sites in the San Joaquin Valley and one site in the Sacramento Valley, California, November 2002–March 2004.—Continued

[The background concentrations and control samples are considered quality assurance samples but are presented with the environmental samples for comparative purposes. All compounds were analyzed by gas chromatography/mass spectrometry (GC/MS) at the U.S. Geological Survey (USGS) National Water Quality Laboratory (NWQL). The Chemical Abstract Service Number is given below each compound name in brackets. The five digit alphanumeric number in parenthesis is a code used by the USGS to uniquely identify the given compound. Mix 1 collected near site 11 (Newman Raingage). Mix 3 collected near site 5 (Modesto Irrigation District at Albers Road). Mix 4 collected near site 16 (Gridley High School). Sample weight was wet weight in grams. Total area of soil box is 0.374 square meters. Abbreviations: CSUS, California State University, Sacramento; E, estimated reported laboratory value; NA, not applicable; [-], no Chemical Abstract Service Number available for given compound; <, less than laboratory reporting value]

Site number	Site name	Site identification number	3,4-Dichloroaniline (E) [95-76-1] (S0287)	4-Chloro-2-methylphenol (E) [1570-64-5] (S0288)	Carbaryl[3] (E) [63-25-2] (S0295)	Chlorpyrifos [2921-88-2] (S0296)	Chlorpyrifos oxon[1] (E) [5598-15-2] (S0297)	Dacthal (DCPA)[1] (E) [1861-32-1] (S0300)	Diazoxon[1] (E) [962-58-3] (S0303)
		Environmental surficial-soil samples—Continued							
4	Modesto Irrigation District gage rooftop at Modesto	373834121000601	E2	<10	E0.6	E0.8	<30	1	<5
4	Modesto Irrigation District gage rooftop at Modesto	373834121000601	<75	<10	E0.8	2	E3	1	<5
4	Modesto Irrigation District gage rooftop at Modesto	373834121000601	<75	<10	<4	2	<30	1	<5
5	Modesto Irrigation District gage rooftop at Albers Road	373841120504801	E3	<10	E0.3	1	<30	E0.9	<5
5	Modesto Irrigation District gage rooftop at Albers Road	373841120504801	<75	<10	E0.9	10	E5	E1	<5
5	Modesto Irrigation District gage rooftop at Albers Road	373841120504801	E2	<10	E1	7	<30	E0.9	<5
16	Gridley High School precipitation gage at Gridley	392205121410201	E1	<10	E2	1	<30	E0.8	<5
16	Gridley High School precipitation gage at Gridley	392205121410201	E1	<10	<2	E0.9	<30	E0.8	<5
		Quality-assurance surficial-soil control samples—Continued							
14	Soil Control Box at CSUS Placer Hall at Sacramento[2]	383343121252501	E4	<10	E2	<1	<30	E0.6	<5
14	Soil Control Box at CSUS Placer Hall at Sacramento[2]	383343121252501	<75	<10	E1	<1	<30	E0.7	<5
14	Soil Control Box at CSUS Placer Hall at Sacramento[2]	383343121252501	<75	<10	E2	E0.6	<30	E0.7	<5
		Quality-assurance background concentration surficial-soil samples—Continued							
NA	Mix 1	373335121143001	E5	<10	E1	<1	<30	E0.6	<5
NA	Mix 2	371735121031201	E6	<10	E2	E0.7	<30	E0.6	<5
NA	Mix 3	373841120504801	E4	<10	E2	E0.9	<30	E0.8	<5
NA	Mix 4	392205121410201	E0.9	<10	E2	<1	<30	<1	<5

Appendix 9. Analytical results presented in micrograms per kilogram (µg/Kg) for compounds with at least one detection from the soil box as surficial-soil samples collected at two sites in the San Joaquin Valley and one site in the Sacramento Valley, California, November 2002–March 2004.—Continued

[The background concentrations and control samples are considered quality assurance samples but are presented with the environmental samples for comparative purposes. All compounds were analyzed by gas chromatography/mass spectrometry (GC/MS) at the U.S. Geological Survey (USGS) National Water Quality Laboratory (NWQL). The Chemical Abstract Service Number is given below each compound name in brackets. The five digit alphanumeric number in parenthesis is a code used by the USGS to uniquely identify the given compound. Mix 1 collected near site 10 (Westly Raingage), Mix 2 collected near site 11 (Newman Raingage). Mix 3 collected near site 5 (Modesto Irrigation District at Albers Road). Mix 4 collected near site 16 (Gridley High School). Sample weight was wet weight in grams. Total area of soil box is 0.374 square meters. Abbreviations: CSUS, California State University, Sacramento; E, estimated reported laboratory value; NA, not applicable; [-], no Chemical Abstract Service Number available for given compound; <, less than laboratory reporting value]

Site number	Site name	Site identification number	Dichlorvos[1] (E) [62-73-7] (S0304)	Dicrotophos[1] (E) [141-66-2] (S0305)	Dieldrin [60-57-1] (S0306)	Dimethoate [60-51-5] (S0307)	Fenamiphos[1] (E) [22224-92-6] (S0310)	Fipronil sulfone [120068-36-2] (S0316)	Desulfinylfipronil [-] (S0317)
		Environmental surficial-soil samples—Continued							
4	Modesto Irrigation District gage rooftop at Modesto	37383412100060l	<30	<3	3	9	<30	1	E0.8
4	Modesto Irrigation District gage rooftop at Modesto	37383412100060l	<30	<3	2	<2	<30	1	E0.6
4	Modesto Irrigation District gage rooftop at Modesto	37383412100060l	<30	<3	2	<2	<30	1	E0.7
5	Modesto Irrigation District gage rooftop at Albers Road	37384112050480l	<30	<3	3	E6	<30	2	E0.9
5	Modesto Irrigation District gage rooftop at Albers Road	37384112050480l	<30	<3	2	<3	<30	1	E0.6
5	Modesto Irrigation District gage rooftop at Albers Road	37384112050480l	<30	<3	3	<4	<30	1	E0.7
16	Gridley High School precipitation gage at Gridley	39220512141020l	<30	<3	E0.6	<2	<30	<1	<1
16	Gridley High School precipitation gage at Gridley	39220512141020l	<30	<3	E0.5	<2	<30	<1	<1
		Quality-assurance surficial-soil control samples—Continued							
14	Soil Control Box at CSUS Placer Hall at Sacramento[2]	38334312125250l	<30	<3	3	10	<30	1	E0.6
14	Soil Control Box at CSUS Placer Hall at Sacramento[2]	38334312125250l	<30	<3	2	<2	<30	1	E0.6
14	Soil Control Box at CSUS Placer Hall at Sacramento[2]	38334312125250l	<30	<3	3	<2	<30	1	E0.6
		Quality-assurance background concentration surficial-soil samples—Continued							
NA	Mix 1	37333512114300l	<30	<3	3	E18	<30	1	E0.6
NA	Mix 2	37173512103120l	<30	<3	4	26	<30	1	E0.6
NA	Mix 3	37384112050480l	<30	<3	3	E23	<30	2	E0.8
NA	Mix 4	39220512141020l	<30	<3	E0.4	<2	<30	<1	<1

Appendix 9. Analytical results presented in micrograms per kilogram (µg/Kg) for compounds with at least one detection from the soil box as surficial-soil samples collected at two sites in the San Joaquin Valley and one site in the Sacramento Valley, California, November 2002–March 2004.—Continued

[The background concentrations and control samples are considered quality assurance samples but are presented with the environmental samples for comparative purposes. All compounds were analyzed by gas chromatography/mass spectrometry (GC/MS) at the U.S. Geological Survey (USGS) National Water Quality Laboratory (NWQL). The Chemical Abstract Service Number is given below each compound name in brackets. The five digit alphanumeric number in parenthesis is a code used by the USGS to uniquely identify the given compound. Mix 1 collected near site 10 (Westly Raingage). Mix 2 collected near site 11 (Newman Raingage). Mix 3 collected near site 5 (Modesto Irrigation District at Albers Road). Mix 4 collected near site 16 (Gridley High School). Sample weight was wet weight in grams. Total area of soil box is 0.374 square meters. Abbreviations: CSUS, California State University, Sacramento; E, estimated reported laboratory value; NA, not applicable; [-], no Chemical Abstract Service Number available for given compound; <, less than laboratory reporting value]

Site number	Site name	Site identification number	Iprodione[1] (E) [36734-19-7] (S0322)	Malathion [121-75-5] (S0325)	Metolachlor [51218-45-2] (S0328)	Metribuzin [21087-64-9] (S0329)	Myclobutanil (E) [88671-89-0] (S0330)	Parathion-methyl [298-00-0] (S0332)	Paraoxon-methyl (E) [950-35-6] (S0331)
		Environmental surficial-soil samples—Continued							
4	Modesto Irrigation District gage rooftop at Modesto	373834121000601	E6	10	55	E12	E10	<2	<5
4	Modesto Irrigation District gage rooftop at Modesto	373834121000601	E9	<2	9	E2	E9	<2	<5
4	Modesto Irrigation District gage rooftop at Modesto	373834121000601	E15	<2	5	<4	E10	<4	<5
5	Modesto Irrigation District gage rooftop at Albers Road	373841120504801	E6	12	66	E12	E17	<2	<5
5	Modesto Irrigation District gage rooftop at Albers Road	373841120504801	E28	<2	12	E5	E14	<3	E4
5	Modesto Irrigation District gage rooftop at Albers Road	373841120504801	E22	<2	7	E1	E17	<4	<5
16	Gridley High School precipitation gage at Gridley	392205121410201	E1	<2	<1	<4	<1	E6	E4
16	Gridley High School precipitation gage at Gridley	392205121410201	E0.9	<2	<1	<4	<1	<2	<5
		Quality-assurance surficial-soil control samples—Continued							
14	Soil Control Box at CSUS Placer Hall at Sacramento[2]	383343121252501	E6	6	76	E16	E16	<2	<5
14	Soil Control Box at CSUS Placer Hall at Sacramento[2]	383343121252501	E7	<2	37	E15	E13	<2	<5
14	Soil Control Box at CSUS Placer Hall at Sacramento[2]	383343121252501	E5	<2	31	E6	E13	<2	<5
		Quality-assurance background concentration surficial-soil samples—Continued							
NA	Mix 1	373335121143001	E5	18	58	E15	E18	<2	<5
NA	Mix 2	371735121031201	E9	25	128	E23	E14	<2	<5
NA	Mix 3	373841120504801	E8	27	108	E25	E12	<2	<5
NA	Mix 4	392205121410201	<10	<2	<1	<4	<1	<2	<5

Appendix 9. Analytical results presented in micrograms per kilogram (µg/kg) for compounds with at least one detection from the soil box as surficial-soil samples collected at two sites in the San Joaquin Valley and one site in the Sacramento Valley, California, November 2002–March 2004.—Continued

[The background concentrations and control samples are considered quality assurance samples but are presented with the environmental samples for comparative purposes. All compounds were analyzed by gas chromatography/mass spectrometry (GC/MS) at the U.S. Geological Survey (USGS) National Water Quality Laboratory (NWQL). The Chemical Abstract Service Number is given below each compound name in brackets. The five digit alphanumeric number in parenthesis is a code used by the USGS to uniquely identify the given compound. Mix 1 collected near site 10 (Westly Raingage). Mix 2 collected near site 11 (Newman Raingage). Mix 3 collected near site 5 (Modesto Irrigation District at Albers Road). Mix 4 collected near site 16 (Gridley High School). Sample weight was wet weight in grams. Total area of soil box is 0.374 square meters. Abbreviations: CSUS, California State University, Sacramento; E, estimated reported laboratory value; NA, not applicable; [-], no Chemical Abstract Service Number available for given compound; <, less than laboratory reporting value]

Site num- ber	Site name	Site identification number	Pendimethalin [40487-42-1] (S0333)	cis-Permethrin[3] [54774-45-7] (S0340)	trans-Perme- thrin[3] [51877-74-8] (S0341)	Phorate[1] (E) [298-02-2] (S0334)	Phorate oxon[1] [2600-69-3] (S0335)	Phosmet[1] [732-11-6] (S0336)
		Environmental surficial-soil samples—Continued						
4	Modesto Irrigation District gage rooftop at Modesto	3738341210006O1	E40	<5	<5	<5	<7	<30
4	Modesto Irrigation District gage rooftop at Modesto	3738341210006O1	E30	<5	<5	<5	<7	<30
4	Modesto Irrigation District gage rooftop at Modesto	3738341210006O1	E26	<5	<5	<5	<7	<30
5	Modesto Irrigation District gage rooftop at Albers Road	3738411205048O1	E62	<5	<5	<5	<7	<30
5	Modesto Irrigation District gage rooftop at Albers Road	3738411205048O1	E31	<5	<5	<5	<7	<30
5	Modesto Irrigation District gage rooftop at Albers Road	3738411205048O1	E26	<5	<5	<5	<7	<30
16	Gridley High School precipitation gage at Gridley	3922051214102O1	<1	<5	<5	<5	<7	<30
16	Gridley High School precipitation gage at Gridley	3922051214102O1	<1	<5	<5	<5	<7	<30
		Quality-assurance surficial-soil control samples—Continued						
14	Soil Control Box at CSUS Placer Hall at Sacramento[2]	3833431212525O1	E54	<5	<5	<5	<7	<30
14	Soil Control Box at CSUS Placer Hall at Sacramento[2]	3833431212525O1	E39	<5	<5	<5	<7	<30
14	Soil Control Box at CSUS Placer Hall at Sacramento[2]	3833431212525O1	E40	<5	<5	<5	<7	<30
		Quality-assurance background concentration surficial soil samples—Continued						
NA	Mix 1	3733351211430O1	58	E11	6	<5	<7	<30
NA	Mix 2	3717351210312O1	48	<5	<5	<5	<7	<30
NA	Mix 3	3738411205048O1	47	<5	<5	<5	<7	<30
NA	Mix 4	3922051214102O1	<1	<5	<5	<5	<7	<30

Appendix 9. Analytical results presented in micrograms per kilogram (µg/Kg) for compounds with at least one detection from the soil box as surficial-soil samples collected at two sites in the San Joaquin Valley and one site in the Sacramento Valley, California, November 2002–March 2004.—Continued

[The background concentrations and control samples are considered quality assurance samples but are presented with the environmental samples for comparative purposes. All compounds were analyzed by gas chromatography/mass spectrometry (GC/MS) at the U.S. Geological Survey (USGS) National Water Quality Laboratory (NWQL). The Chemical Abstract Service Number is given below each compound name in brackets. The five digit alphanumeric number in parenthesis is a code used by the USGS to uniquely identify the given compound. Mix 1 collected near site 5 (Modesto Irrigation District at Albers Road). Mix 2 collected near site 11 (Newman Raingage). Mix 3 collected near site 5 (Modesto Irrigation District at Albers Road). Mix 4 collected near site 16 (Gridley High School). Sample weight was wet weight in grams. Total area of soil box is 0.374 square meters. Abbreviations: CSUS, California State University, Sacramento; E, estimated reported laboratory value; NA, not applicable; [-], no Chemical Abstract Service Number available for given compound; <, less than laboratory reporting value]

Site number	Site name	Site identification number	Prometon (E) [1610-18-0] (S0338)	Simazine [122-34-9] (S0343)	Terbufos[1] [13071-79-9] (S0345)	Trifluralin [1582-09-8] (S0348)	Sample wet weight, grams (S0433)
			Environmental surficial-soil samples—Continued				
4	Modesto Irrigation District gage rooftop at Modesto	37383412100060l	<2	53	<1	17	32.2
4	Modesto Irrigation District gage rooftop at Modesto	37383412100060l	<2	9	<1	11	43.3
4	Modesto Irrigation District gage rooftop at Modesto	37383412100060l	<2	4	<1	8	37.8
5	Modesto Irrigation District gage rooftop at Albers Road	37384112050480l	<2	72	<1	17	32.6
5	Modesto Irrigation District gage rooftop at Albers Road	37384112050480l	<2	20	<1	8	41.9
5	Modesto Irrigation District gage rooftop at Albers Road	37384112050480l	<2	12	<1	7	41.9
16	Gridley High School precipitation gage at Gridley	39220512141020l	E2	11	<1	E0.6	33.7
16	Gridley High School precipitation gage at Gridley	39220512141020l	<2	2	<1	E0.3	33.1
			Quality-assurance surficial-soil control samples—Continued				
14	Soil Control Box at CSUS Placer Hall at Sacramento[2]	38333312125250l	<2	81	<1	21	42.2
14	Soil Control Box at CSUS Placer Hall at Sacramento[2]	38333312125250l	E2	54	<1	10	44.6
14	Soil Control Box at CSUS Placer Hall at Sacramento[2]	38333312125250l	<2	46	<1	11	45.7
			Quality-assurance background concentration surficial-soil samples—Continued				
NA	Mix 1	37333512114300l	E1	99	<1	17	40.9
NA	Mix 2	37173512103120l	E1	70	<1	31	42.2
NA	Mix 3	37384112050480l	E2	128	<1	25	34.6
NA	Mix 4	39220512141020l	E2	10	<1	E0.1	28.3

[1]Compound is biased low (B-L) because of low laboratory matrix and reagent spike recoveries, see *table 11*. Compounds with non-detections are presented in the appendix because concentrations may be present in low concentrations in the environment.

[2]Site 14 in the Sacramento Valley was considered a control and the results are presented with *appendix 8*, but discussed in the soil box field quality control section of this report.

[3]Compound is biased high (B-H) because laboratory matrix and reagent spike recoveries were greater than 150 percent, see *table 11*.